University Texts in the Mathematical Sciences

Editors-in-Chief

Raju K. George, Indian Institute of Space Science and Technology, Valiamala, India

S. Kesavan, Institute of Mathematical Sciences, Chennai, India

Sujatha Ramdorai, University of British Columbia, Vancouver, Canada

Shalabh, Indian Institute of Technology Kanpur, Kanpur, India

Associate Editors

Kapil Hari Paranjape, Indian Institute of Science Education and Research Mohali, Mohali, India

K. N. Raghavan, Institute of Mathematical Sciences, Chennai, India

V. Ravichandran, National Institute of Technology Tiruchirappalli, Tiruchirappalli, India

Riddhi Shah, Jawaharlal Nehru University, New Delhi, India

Kaneenika Sinha, Indian Institute of Science Education and Research, Pune, India

Kaushal Verma, Indian Institute of Science, Bengaluru, India

Enrique Zuazua, Friedrich-Alexander-Universität Erlangen-Nürnberg (FAU), Erlangen, Germany

Textbooks in this series cover a wide variety of courses in mathematics, statistics and computational methods. Ranging across undergraduate and graduate levels, books may focus on theoretical or applied aspects. All texts include frequent examples and exercises of varying complexity. Illustrations, projects, historical remarks, program code and real-world examples may offer additional opportunities for engagement. Texts may be used as a primary or supplemental resource for coursework and are often suitable for independent study.

S. P. Mukherjee · Asok K. Nanda

Probability Models in Reliability Analysis

 Springer

S. P. Mukherjee
Kolkata, West Bengal, India

Asok K. Nanda
Department of Mathematics and Statistics
Indian Institute of Science Education
and Research Kolkata
Kolkata, West Bengal, India

ISSN 2731-9318   ISSN 2731-9326 (electronic)
University Texts in the Mathematical Sciences
ISBN 978-981-96-3048-6   ISBN 978-981-96-3049-3 (eBook)
https://doi.org/10.1007/978-981-96-3049-3

Mathematics Subject Classification: 62E15, 62P10, 62P99, 62N05, 90B25, 62P30

© The Editor(s) (if applicable) and The Author(s), under exclusive license to Springer Nature Singapore Pte Ltd. 2025

This work is subject to copyright. All rights are solely and exclusively licensed by the Publisher, whether the whole or part of the material is concerned, specifically the rights of translation, reprinting, reuse of illustrations, recitation, broadcasting, reproduction on microfilms or in any other physical way, and transmission or information storage and retrieval, electronic adaptation, computer software, or by similar or dissimilar methodology now known or hereafter developed.
The use of general descriptive names, registered names, trademarks, service marks, etc. in this publication does not imply, even in the absence of a specific statement, that such names are exempt from the relevant protective laws and regulations and therefore free for general use.
The publisher, the authors and the editors are safe to assume that the advice and information in this book are believed to be true and accurate at the date of publication. Neither the publisher nor the authors or the editors give a warranty, expressed or implied, with respect to the material contained herein or for any errors or omissions that may have been made. The publisher remains neutral with regard to jurisdictional claims in published maps and institutional affiliations.

This Springer imprint is published by the registered company Springer Nature Singapore Pte Ltd.
The registered company address is: 152 Beach Road, #21-01/04 Gateway East, Singapore 189721, Singapore

If disposing of this product, please recycle the paper.

Preface

Reliability as reflecting quality of performance of products and services during use has become a major concern to all those connected with the design, production, testing, and maintenance of a wide variety of components and systems coming out of manufacturing and service enterprises. This has resulted in a very rapid growth of interest and engagement of both professionals and academics in the fields of engineering, applied probability and statistics, mathematics, computer science, operations research, and related disciplines in the widening ambit of reliability analysis (taken as a broad activity encompassing all its different facets).

A core area of reliability analysis is related to development and use of suitable probability models to represent the inherent uncertainty affecting various types of data that arise in the context of revealed performance of products during use. Some examples may be—time-to-failure or time between consecutive failures or time-to-repair or total time on a test for a sample of product units put on test, etc. A properly chosen model can help us in estimating reliability or making predictions about failure time or optimizing system reliability subject to constraints on cost, space, redundancy and the like. Over the years, a large number of probability models have been suggested by a whole host of research workers coming from different backgrounds, over and above the existing models available in the extant literature on probability and statistics.

Given the context of a virtual deluge of published materials on stochastic models for reliability analysis, the present volume claims several distinct features—by way of a brief and critical coverage of almost the entire published material on the subject that will enable an interested worker to use any of these models. The judicious choice of a particular model to fit a given set of data in a particular context depends on the relevant properties of any model and a chapter of this book is devoted to a discussion of some of the more important properties. This book takes off from well-known properties and results and goes on to cover recent and advanced findings about the models discussed. Relatively sophisticated and academically stimulating discussions on comparison (in terms of stochastic orders), classification and characterization of univariate life distributions have been included. It may be pointed out that there hardly exists any publication that puts together these three inter-related topics of great interest to advanced students and research workers.

Chapter 1 concretely spells out certain fundamental concepts and definitions to take students and research workers from different disciplines on a common platform in a somewhat unique manner. Types of data arising in reliability analysis and the relevance of probability models have been discussed. This is followed up in the next chapter that includes a detailed account of properties of lifetime models and their uses. Recent ideas of generalized failure rate, of local and global memory, of reversed hazard rate, convexity of distributions, and related concepts have been covered.

Chapters 3, 4, and 6 constitute the bulk of the book's content, devoted, respectively, to univariate continuous, discrete, and finite-range distributions. Generalizations and extensions of some well-known models like exponential, Weibull, gamma, log-logistic, Birnbaum-Saunders, and several others obtained through exponentiation, introduction of a new parameter, transmutation, and other techniques as well as several less-frequently-used distributions to suit special types of data like Topp-Leone distributions or Kumaraswamy distributions or beta-generated distributions have been reviewed, their important properties have been stated and methods for estimating parameters involved have been indicated. Sections on bathtub failure rate models as also models derived through compounding have been included. Generalized versions of common discrete distributions along with those obtained through the use of Mittag-Leffler functions and telescopic distributions have found a place. A few models with finite range and their distinctive properties are presented in Chap. 6. It may be pointed out that contents of Chaps. 4 and 6 are to be rarely found in books on the subject.

Chapter 5 is devoted to a detailed discussion on distributions which are derived from lifetime distributions and have been found to be useful in both diagnostic and prognostic analysis of reliability. Particularly interesting to advanced students and research scholars will be the sections dealing with failure rate transform, inactivity time equilibrium distribution, total-time-on-test, and generalized mean remaining life.

The next three chapters deal with stochastic orders among distributions, non-parametric classes of distributions and their properties, and characterization of life distributions. The treatment of stochastic orders has been quite rigorous and will surely appeal to those who would like to make use of such orders in establishing bounds on reliability and in related areas. Class properties, their inter-relations, and closure of the properties under various reliability operations have been studied by many research workers over a long period of time and the authors have attempted to present synoptic view of the more important aspects of this subject. Chapter 8 contains many interesting and informative results including properties of residual life, co-ordinate sub-tangent, and related properties.

Some special distributions of interest to research workers and inquisitive professionals are taken up for a brief discussion in Chap. 10. Distributions of quality-adjusted life, from damaged data on lifetime, from fuzzy data and for fatigue failure should attract the attention of workers who investigate reliability in such special situations.

The last chapter relates to the choice of a probability model among possible alternative candidates and the associated problem of testing the goodness of fit by the chosen model. Inherent difficulties in choosing among models with more or less similar behaviour have been mentioned and a few examples have been given to illustrate the issue.

With a rich collection of materials pertaining to almost all aspects of univariate stochastic models in reliability analysis presented in a cohesive manner avoiding redundancy and preserving rigour, this volume is expected to benefit a large number of professionals as well as graduate students and research workers in the domain of reliability analysis. It can be justifiably considered as a judicious blend of textbook and a reference book. Quite a few topics discussed by a host of research workers and scattered over a wide range of research articles—usually left out of the coverage by existing books on the subject—have been put together in the present volume.

Kolkata, India
November 2024

S. P. Mukherjee
Asok K. Nanda

Acknowledgements

The idea of writing this book was conceived by three of us nearly 5–6 years back in the context of numerous probability models to represent behaviours of failure times arising in diverse real-life situations. We lost one during the early phase of our work. Professor Aditya Chatterjee was snatched away by Covid-19 at a relatively young age. Two of us remember with gratitude his initial suggestions and contribution.

To the many associates who worked with one or both of us in the area of reliability, we express our thanks for silently motivating us to write this book.

We gratefully acknowledge the help we received from Aninda Kumar Nanda, research scholar at the Indian Statistical Institute, Delhi Center, in incorporating the figures in different chapters.

We will fail in our duty if we do not mention the guidance we received from Prof. Shalabh and his team from Springer in preparing the manuscript.

November 2024

S. P. Mukherjee
Asok K. Nanda

Contents

1	**Introduction**	1
	1.1 Some Preliminaries	1
	1.2 Definitions of Reliability	2
	1.3 System Reliability	5
	1.4 Data Bearing on Reliability	8
	1.5 Probability Models in Reliability Analysis	10
	1.6 Preview of the Book	11
2	**Properties of Life Distributions**	15
	2.1 Introduction	15
	2.2 Moments	16
	2.3 Measures of Shape	19
	2.4 Ordering and Ageing Properties	20
	2.4.1 Hazard Rate and Reversed Hazard Rate	21
	2.4.2 Mean Residual Life	25
	2.4.3 Generalized Failure Rate and Generalized Mean Residual Life	26
	2.4.4 Vitality Function	28
	2.4.5 Percentile Remaining Life Function	29
	2.4.6 Ageing Intensity	30
	2.4.7 Memory	32
	2.5 Entropy	34
	2.5.1 Derivation of Entropy	34
	2.5.2 Limitations of Entropy and its Generalizations	36
	2.5.3 Some Important Entropy-Expressions for Different Distributions	38
	2.6 Concavity and Log-Concavity	43
	2.7 A Reference to Life Tables	44

3 Univariate Continuous Distributions ... 47
- 3.1 Introduction ... 47
- 3.2 Development of Probability Models ... 49
- 3.3 Generating Classes of Distributions ... 52
- 3.4 Ubiquity of the Exponential Distribution ... 55
- 3.5 Models with Monotone Failure Rate ... 59
  - 3.5.1 Weibull Probability Model ... 59
  - 3.5.2 Gamma Distribution Model ... 64
  - 3.5.3 Other Probability Models ... 66
- 3.6 The Birnbaum-Saunders Distribution ... 73
- 3.7 Topp-Leone Distribution ... 75
- 3.8 Extensions and Generalizations ... 77
  - 3.8.1 Generalizations of the Exponential Model ... 79
  - 3.8.2 Generalizations of the Weibull Model ... 83
  - 3.8.3 Other Generalizations ... 85
  - 3.8.4 Transformed-Transformer Distribution ... 87
- 3.9 Compound Distributions ... 89
- 3.10 Models with Non-monotone Failure Rates ... 93
  - 3.10.1 Inverse Rayleigh and Inverse Weibull Models ... 93
  - 3.10.2 Log-Logistic Distribution ... 95
- 3.11 Distributions with Bathtub Failure Rate ... 96
- 3.12 Other Failure Rate Patterns ... 98

4 Some Discrete Distributions ... 101
- 4.1 Introduction ... 101
- 4.2 Common Discrete Distributions ... 102
  - 4.2.1 Binomial and Related Distributions ... 103
  - 4.2.2 Poisson and Related Distributions ... 105
  - 4.2.3 Distribution Using Mittag-Leffler Function ... 107
- 4.3 Discretization of Continuous Distributions ... 108
  - 4.3.1 Discretization-1 ... 109
  - 4.3.2 Discretization-2 ... 110
  - 4.3.3 Discretization-3 ... 110
  - 4.3.4 Discretization-4 ... 111
  - 4.3.5 Discretization-5 ... 111
- 4.4 Some Discretized Versions ... 112
  - 4.4.1 Discrete Lindley Distribution ... 113
- 4.5 Telescopic Distribution ... 114

5	**Distributions Derived from the Parent Distribution**		117
	5.1	Introduction	117
	5.2	Remaining (Residual) Life Distribution	118
	5.3	Equilibrium Distribution	122
	5.4	Distribution of Failure Rate Transform	124
	5.5	Inactivity Time or Past Residual Life	127
	5.6	Total Time on Test	129
	5.7	Order Statistics and Record Values	131
		5.7.1 Distribution of Order Statistics	131
		5.7.2 Record Values	133
6	**Finite-Range Life Distributions**		135
	6.1	Introduction	135
	6.2	Some Finite-Range Continuous Distributions	136
		6.2.1 Properties of Mukherjee-Islam Model	138
		6.2.2 Mukherjee-Roy Model and its Properties	143
		6.2.3 Some Characterization Results	145
	6.3	Stress-Strength Reliability	147
	6.4	Finite-Range Discrete Distribution	149
	6.5	Marshall-Olkin Extension	150
	6.6	Estimation of Parameters in Mukherjee-Islam Model	151
7	**Ordering Among Life Distributions**		153
	7.1	Introduction	153
	7.2	Usual Stochastic Order	154
	7.3	Hazard Rate Order	155
		7.3.1 Definitions and Preliminaries	156
		7.3.2 Component Redundancy Versus System Redundancy	158
	7.4	Likelihood Ratio Order	161
	7.5	Mean Residual Life Order	163
	7.6	Reversed Hazard Rate Order	165
		7.6.1 Definition and Preliminaries	165
		7.6.2 Component Redundancy Versus System Redundancy	168
	7.7	Mean Inactivity Time	169
	7.8	Generalized Orderings	172
		7.8.1 Closure Under Mixture of Distributions	190
		7.8.2 Characterizations in Terms of Residual Lives	197
		7.8.3 Characterizations in Terms of Equilibrium Distributions	202

		7.8.4	Characterizations in Terms of Laplace Transform	207
		7.8.5	Dispersion-Type Stochastic Orders	217
		7.8.6	Some Characterizations of Generalized Dispersion-Type Orders	220
	7.9	Generalized Ageing Properties		224
	7.10	Residual Life at Random Time		230
		7.10.1	Characterization in Terms of Laplace Transform	233
	7.11	Some Results in Terms of Excess Lifetime		239
		7.11.1	Characterizations in Terms of Residual Life Functions	240
	7.12	A Unified Study of Some Stochastic Orders		243
8	**Classes of Life Distributions**			**245**
	8.1	Relevance of Classification		245
	8.2	Classes Based on Ageing Properties		247
		8.2.1	Use of Failure Rate	247
		8.2.2	Classification of Discrete Life Distributions	249
	8.3	Classes Based on Residual Life and Inactivity Time		250
	8.4	Use of Coordinate Sub-tangent		253
	8.5	Classes Based on Stochastic Dominance		256
		8.5.1	The $\mathscr{L}$-Class	256
		8.5.2	The $\mathscr{M}$-Class and the $\mathscr{LM}$-Class	257
	8.6	Preservation of Class Properties		259
9	**Characterizations of Life Distributions**			**261**
	9.1	Introduction		261
	9.2	Characterizations of the Exponential Distribution		263
	9.3	Characterization Through Record Values		273
	9.4	Characterizations of the Weibull Distribution		275
	9.5	Characterizations of the Generalized Gamma Distribution		281
	9.6	Characterizations of Pearsonian Type XI Distribution		287
	9.7	Characterizations Through Ageing Intensity Function		289
	9.8	Characterizations Through Coordinate Sub-tangent		294
10	**Distributions of Special Interest**			**297**
	10.1	Introduction		297
	10.2	Distribution of Quality-Adjusted Life		298
	10.3	Life Distribution from Imprecise Data		298
	10.4	Reliability from Fuzzy Stress and Strength		299
	10.5	Reliability Under Exponential Stress and Exponential Strength		300
	10.6	Fatigue Failure Models		301
		10.6.1	Deterministic Models	302
		10.6.2	Stochastic Model	305

11	**Using an Appropriate Probability Model**	307
	11.1 Model Selection	309
	11.2 Testing Goodness of Fit	312
	11.3 Use of Characterization Results	313
	11.4 Some Examples	314

References ... 317

Index ... 343

Symbols

$f_X(\cdot)$	Probability density function (pdf) of a random variable X
$F_X(\cdot)$	Cumulative distribution function (cdf) of X
$\bar{F}_X(\cdot)$	Survival function of X
$E(X)$	Expectation of X
$V(X)$	Variance of X
$C(X)$	Coefficient of variation (CV) of X, i.e., $C(X) = \sqrt{V(X)}/E(X)$
$r_X(\cdot)$	Failure/hazard rate function of X
$\mu_X(\cdot)$	Reversed hazard rate function of X
X_t	$[X - t \mid X \geq t]$, Residual life random variable
$m_X(\cdot)$	Mean residual life, i.e., $m_X(t) = E(X_t)$
$M_X(\cdot)$	Moment generating function (MGF) of X
$V_X(\cdot)$	Variance residual life, i.e., $V_X(t) = V(X_t)$
$C_X(\cdot)$	CV of residual life, i.e., $C_X(t) = \sqrt{V_X(t)}/m_X(t)$
$v(\cdot)$	Vitality function, i.e., $v(t) = E(X \mid X \geq t) = m_X(t) + t$
${}_tX$	$(t - X \mid X < t)$, Inactivity time of the random variable X
$\alpha(t)$	$E({}_tX)$, Expected inactivity time
$m_L(t)$	$-m'_X(t)$, Local memory at t
M_G	$1 - C^2(X)$, Global memory
$v(t)$	Virtual age at time t
$M(t_1, t_2)$	(Interval) memory in the interval (t_1, t_2)
Me	Median
Mo	Mode
$\mathbb{R}$	$(-\infty, \infty)$
$\mathbb{R}^+$	$[0, \infty)$
$\mathbb{N}$	$\{1, 2, \ldots\}$
$\mathbb{Z}$	$\{0, \pm 1, \pm 2, \ldots\}$
$g^2(x)$	$(g(x))^2$ for any function g
$a \stackrel{sign}{=} b$	a and b have the same sign

$X \stackrel{a}{\sim} Y$ X and Y have asymptotically same distribution
$X \stackrel{D}{\to} Y$ X converges in distribution to Y
$X \stackrel{D}{=} Y$ X and Y have the same distribution

Chapter 1
Introduction

1.1 Some Preliminaries

The word 'reliability' is generally attributed to some concrete entity meant for 'use' or 'deployment'. The entity can be a manufactured product or a service. It can be argued that 'whatever is produced is a product' and that 'manufacture need not involve machines or engineering processes'. A product may imply a simple component or an assembly of several components or even a complex system of several assemblies. Similarly, a service may mean a simple installation or a minor repair, a thorough inspection or a major overhaul. However, to speak of reliability of any such entity, we leave out manufactured products which are used only as decorative items at home, in offices, or in public places which are not required to carry out any specified function(s). Such items are characterized by their esteem or prestige value, rather than any functional or use value.

To comprehend reliability as an important and, in some cases, crucial property of a manufactured product, it is desirable to first outline the 'life cycle' of such a product. In fact, this cycle involves the following four phases that may require amplifications appropriate in relation to a particular product. The last phase provides input for modifying the first and, thus, the phases constitute a cycle.

The first phase is **Mission**. In this phase, we bring in the concept of 'performance parameters' or 'functional characteristics' and impose restrictions on each such parameter to visualize its intended or satisfactory functioning or performance. For example, an automobile should have a minimum specified fuel efficiency (number of kilometres run per unit of fuel), a specified maximum speed attainable, besides some desired level of comfort and of safety. Among many performance parameters that describe the desired and/or the realized performance of a product, some are critical. We can then consider specifications for these parameters as critical events in the life of a particular product.

Denoting the critical parameters as $Y_1, Y_2, \ldots, Y_k$, we can write the mission plan as $Y_i \in S_i$, $i = 1, 2, \ldots, k$, where S_i is the specified or desired set of values for Y_i. Of course, we should also consider qualitative performance parameters.

The next phase that emanates from the Mission Plan or Profile is **Design**. Here we identify materials to be used, processes to be carried out, tests to be done, inspections to be performed on in-coming materials and products, in-process materials, and finished products in terms of their physical, chemical, biological, and other relevant properties, called 'quality parameters' and set desired tolerances for those. If we denote these 'quality parameters' as $X_1, X_2, \ldots, X_p$, then the design can be stated as the set of requirements $X_j \in T_j$, where T_j is the desired or designed set of values for the quality parameter X_j. Noting that performance parameters Y's depend on the quality parameters X's, it is possible to work out the sets T by examining the relation $Y = f(X)$, where X and Y are the vectors of quality and performance parameters, respectively.

Now follows the third phase, $viz.$ **Manufacture** to convert inputs into final products, taking care to conform to Design requirements in respect of different quality parameters. Given that the design does not specify a single value or level for any quality parameter but allows a range of variability and noting that a host of controllable (but uncontrolled) and uncontrollable factors operate on the manufacturing process, it is quite natural to expect variations in any quality parameter among units or copies of the product turned out by the same process following the same design. Such variations are detected during inspection carried out at various process locations.

The last phase of **Use** provides us with an assessment of the product's performance. A product unit is eventually put to use rightly or otherwise, under conditions as specified or otherwise. The unit ultimately fails to function satisfactorily after some usage time, but may continue to function somehow beyond this time. Time-to-failure or, in case of a repairable unit, time between consecutive failures, time taken for repair, and similar other features are observed during use or deployment.

We now recognize that random variations take place among units of production similar in respect of design and manufacture with respect to the specified quality parameters of inputs and that this will naturally lead to random variations in the performance of similar units during use, as reflected in realized values or levels of performance parameters. Thus, we recognize critical performance parameters as also the quality parameters controlled during manufacture as random variables.

1.2 Definitions of Reliability

Reliability can be broadly appreciated as the composite of Quality of Design and Quality of Conformance. Quality of Design as a conditional probability and Quality of Conformance as an unconditional probability are defined, respectively, as

$$P(Y_i \in S_i, \; i = 1, 2, \ldots, k | X_j \in T_j, \; j = 1, 2, \ldots, p)$$

1.2 Definitions of Reliability

and
$$P(X_j \in T_j, \ j = 1, 2, \ldots, p).$$

In a similar vein, we define Quality of Performance of a product (and not just a product unit) as the compound probability

$$P(Y_i \in S_i, \ i = 1, 2, \ldots, k; \ X_j \in T_j, \ j = 1, 2, \ldots, p),$$

and this is what is understood as Product Reliability in the most general way.

More often than not, performance has been understood narrowly as time up to which the product functions. It is presumed that if the product functions, it functions satisfactorily. Once any of the performance parameters deviates from its specified range or level, the product is taken to have failed. Thus, time up to which a product functions or survival time is the only parameter directly considered in evaluating the reliability of a product. In fact, to be reliable, performance parameters of the product have to satisfy the respective requirements at all times up to the 'mission time'.

Under the simplifying assumption that 'functioning' implies 'satisfactory functioning', reliability has been understood as Probability of Survival or functioning up to the specified 'mission time', say t_0. In fact, commonly reliability is defined as the probability that the product/equipment functions satisfactorily for a specified period of time, when used in the manner and for the purpose intended. Thus, time-to-failure, T say, or conversely survival time is a random variable. Hence, $\{T > t_0\}$ is a random event and its probability, $P[T > t_0]$, is taken as reliability. This definition relates to an infinite population of product units which are identically designed, manufactured, and tested. Reliability of a single product unit may be possibly worked out in a different way like fault-tree analysis (cf. Barlow and Proschan, 1981).

This definition of reliability starts with a directly observable time-to-failure, which is the same as length of life or simply life in the case of non-repairable (and particularly continuous-duty) products. And we have tacitly assumed that all the critical performance parameters lie within the specified intervals as long as the product functions. It is quite possible, however, that any such performance parameter Y changes with time (or age of the product) and usually changes randomly, the value or level realized at time t being $Y(t)$, which will generate a sample path. The first time that the path enters the absorbing state defined by a value or level outside the specified interval, we say that the product has failed. Thus, failure time should be defined as the minimum of first passage (to the absorbing state) times for the different processes $Y_i(t), \ i = 1, 2, 3, \ldots$

To define reliability more concretely, we need a classification of products based on the pattern of use and the possibility or otherwise of repair in case of partial or total failure.

Firstly, we note that certain products are meant to function only at one point of time, not necessarily immediately after production or delivery. The manufactured product may have to be stored for some time and required to function when actuated at some point of time. In such cases, survival up to a specified time is not a concern and probability that the product functions at the point when activated defines reliability.

Such items are called one-shot or instantaneous-duty products and are generally non-repairable. In case of a repairable one-shot product, reliability is a probability that relates to a population of occasions on which the product was activated.

Many products are meant to function during some interval of time, to remain switched off subsequently, and then again activated to function for some more time. This cycle of 'on' and 'off' situations continues. The time intervals during which it is required to function are not the same as the 'mission time', rather the mission time is stated in terms of the number of successful operation cycles and reliability is the probability that the number of successful operation cycles N equals or exceeds a specified number n_0. Thus, reliability becomes $P[N > n_0]$ and this probability may relate to a population of copies of the product. These are intermittent-duty products or equipment or repeated-shot ones. Such items are more often than not repairable. In case of a repairable product (also referred to as maintainable), quality of performance is judged in terms of 'availability'. The pair-wise availability of a product at time t is defined as the probability that it is in the functional state at time t, independently of whether the product was in the functional or failed state in the interval $[0, t)$. If we define the random variable $X(t)$ as

$$X(t) = \begin{cases} 1, & \text{if the product is in the functional state at time } t \\ 0, & \text{if it is either in the failed state and waiting for undergoing repair or} \\ & \text{is unde a planned maintenance,} \end{cases}$$

availability at time t is taken as $A(t) = P(X(t) = 1)$. Availability over an interval (t_1, t_2) is then

$$K(t_1, t_2) = \int_{t_1}^{t_2} A(t)\, dt.$$

In practice, availability is measured as a limiting concept and is computed as

$$\text{Availability} = \frac{\text{Mean uptime}}{\text{Mean uptime} + \text{Mean downtime}}.$$

If successive cycles of up and down times are considered and time-to-repair is taken to include waiting time for repair and time for a planned maintenance besides time-to-repair, then availability works out as

$$\text{Availabiity} = \frac{MTTF}{MTTF + MTTR},$$

where $MTTF$ represents the mean time-to-failure, whereas $MTTR$ is the mean time-to-repair.

A large majority of products or equipment are meant to function continuously after being activated or turned on. Examples are a power-generating system or a display board or monitoring system, etc. For such products, when these are non-repairable, we really have a 'mission time' specified and the product is required to function continuously till this time in order that the product is reliable. In case of repairable continuous-duty products or equipment, we have a different measure of reliability.

Thus, the commonly accepted definition of reliability is really applicable only to non-repairable continuous-duty products or equipment. In all other cases, we need to have different reliability measures.

In the broad category of repairable products, time-to-first failure is not the only variable of interest. Of equal importance in assessing the reliability of the product are times between successive failure and repair times. These, in turn, make for the up-time and down-time for the product during a finite interval or in the limit.

Another way to look at reliability is to recognize the fact that a product can function (satisfactorily) in terms of the strength built into it by the design and the production process. This strength has to exceed the stress to which the product is subjected during performance. While strength (could be breaking strength, compressive strength, tensile strength, or could be just some similar enabling property) is the outcome of design parameters as also of process parameters, stress is either environmental or incidental. Mostly, stress is imposed by the use environment, maybe by way of roughness, resistance, interference, excessive heat, etc. Some stress is developed incidentally as the product functions. Many electrical or electronic devices generate heat as these function (particularly for a long duration at a stretch) and this heat is detrimental to their further functioning.

This brings us to a definition of reliability as $R = P[Y > X]$, where X and Y stand, respectively, for stress and strength. Usually, X and Y will be treated as independently and identically distributed random variables. However, cases can arise where these become correlated, e.g., when units likely to be stronger are put to use in harsher or more stressful environments and the apparently weaker ones are used in less stressful situations.

1.3 System Reliability

Most of the products in use can be looked upon as systems involving several components (parts) configured according to the given designs. Quite obviously, the performance of such a system depends on the performance of the constituent components. A somewhat simplified though widely used reliability analysis of multi-component systems takes into account only two states of functioning or performance of a component or the entire system, $viz.$ a (fully) functional state and a failed state, at any point of time t.

Define, for $i = 1, 2, \ldots, n$,

$$X_i = \begin{cases} 1, & \text{if the } i\text{th component is working} \\ 0, & \text{otherwise.} \end{cases}$$

Consider a function $\phi : \{0, 1\}^n \to \{0, 1\}$ such that

$$\phi(x_1, x_2, \ldots, x_n) = \begin{cases} 1, & \text{if the system is working} \\ 0, & \text{otherwise,} \end{cases}$$

where x_i is the realization of X_i, for $i = 1, 2, \ldots n$. The function ϕ is known as the 'structure function' of the system formed out of the n components with ith component having lifetime X_i. The ith component of such a system is said to be irrelevant if

$$\phi(x_1, x_2, \ldots, x_{i-1}, 0, x_{i+1}, \ldots, x_n) = \phi(x_1, x_2, \ldots, x_{i-1}, 1, x_{i+1}, \ldots, x_n),$$

for all $(x_1, \ldots, x_{i-1}, x_{i+1}, \ldots, x_n) \in \{0, 1\}^{n-1}$. A component which is not irrelevant is called relevant. Now we define a coherent system as under.

Definition 1.3.1 A system is said to be coherent if each of the components is relevant, and its structure function is increasing in each argument keeping the others fixed.

Suppose we form a system where structure function is not increasing in at least one component. This means that with respect to that component the structure function will either be constant or decreasing. The structure function constant means that this component is irrelevant. In other words, this component has no effect on the performance or otherwise of the system. So, allowing such components to be included in the system will increase the complexity as well as the cost of the system without any benefit as far as performance of the system is concerned. This immediately tells us that a system should have only the relevant components. Further, if the structure function is decreasing with respect to a component, the system will work with a failed component (which is quite possible) but will fail once this particular component works, a case which is not expected. This clearly tells that the systems in practice should be coherent systems only.

Each of the components being relevant, it is guaranteed that each component has some importance whenever the functioning of the system is concerned. The feasibility of a system is guaranteed by considering the structure function non-decreasing. It is interesting to note that a coherent system drastically restricts the number of possible functions mapping $\{0, 1\}^n$ into $\{0, 1\}$. Of the 256 functions mapping from $\{0, 1\}^3 \to \{0, 1\}$, only 5 correspond to coherent systems. In spite of this drastic reduction, the coherent systems of order n grow rapidly with n. It can be noted that (cf. Samaniego, 2007) there are

(a) 2 coherent systems of order 2;
(b) 5 coherent systems of order 3;
(c) 20 coherent systems of order 4;
(d) 180 coherent systems of order 5;
(e) more than a billion coherent systems of order 30.

However, finding out an exact number of coherent systems of order n is an open problem.

1.3 System Reliability

A particular case of a coherent system is a k-out-of-n system which works as long as at least k components (out of n components forming the system) work with $n \geqslant k$. Such a system is also called a k-out-of-n : G system. It can be noted that, for a k-out-of-n : G system,

$$\phi(\mathbf{x}) = \begin{cases} 1, & \text{if } \sum_{i=1}^{n} x_i \geqslant k \\ 0 & \text{otherwise.} \end{cases}$$

A system which fails with the kth-component failure is known as a k-out-of-n : F system. Such a system works as long as the number of failures is less than k. This means that the number of working components is at least $n - k + 1$. Thus, we see that a k-out-of-n : F system is equivalent to an $(n - k + 1)$-out-of-n : G system. Thus, for a k-out-of-n : F system, we have

$$\phi(\mathbf{x}) = \begin{cases} 1, & \text{if } \sum_{i=1}^{n} x_i \geqslant n - k + 1 \\ 0 & \text{otherwise.} \end{cases}$$

It is to be noted that, whenever we say k-out-of-n system, we always mean k-out-of-n: G system. An n-out-of-n system (which fails with failure of any one component) is known as a series system. For such a series system, we have

$$\phi(\mathbf{x}) = \prod_{i=1}^{n} x_i.$$

A 1-out-of-n system (which works as long as at least one component works) is known as a parallel system. For such a system, we have

$$\phi(\mathbf{x}) = 1 - \prod_{i=1}^{n}(1 - x_i).$$

As against the parallel system where all the n components start functioning at the same point of time, we can have a configuration where only one component starts functioning, others remaining in a standby state, and as soon as the first component fails, a second component takes over its load instantaneously and as soon as the second component fails, another component becomes operative without any delay, and so on. At any point of time, only one component is functioning, others either have failed already or are waiting to take over the load of the functioning unit once the latter fails. For an n-component system where only one component functioning at any time is sufficient to ensure that the system is functioning, we have $(n - 1)$ redundant components. It may be assumed that the standby units remain in partly energized state so that on failure of the functioning component a standby component can be instantaneously switched on to take over the operating load without any warming-up delay. Given this definition of a standby system, one can recognize a parallel system as a hot standby system, while the classical standby systems are taken as cold standby systems. Warm standby systems are in between. Here once a component is functioning, other one is neither completely switched off nor in fully functioning state. This means that, if the original failure rate of the component is r, its working failure rate, r', is neither 0 nor r, rather it satisfies $0 < r' < r$. That is what we have mentioned above as partly energized state. This is used mostly in the places where no non-zero lead time (the time between failure of a component and starting of another component to work) is allowed, $viz.$ in case of shadowless lamp on an operation table.

1.4 Data Bearing on Reliability

Different types of data arise in different phases of Reliability Analysis to reflect directly or indirectly on some aspect of reliability as quality of performance. Phases include design and development (of prototypes), testing and evaluation prior to planned production, maintenance during use, etc. In fact, data types also vary from one purpose to another, e.g., assessment, analysis, control, and optimization of reliability. Some data may relate to inherent features of the product under consideration, some to the stress factors associated with its use, some to the time and nature of repair and replacement activities undertaken, and some may even relate to costs and other parameters appearing as constraints during reliability optimization exercises. Of course, data bearing on performance and usually reflected in times-to-failure or to counts of units failing by pre-specified points of time are the most widely used data in reliability analysis.

Some of the data features are to be kept controlled at specified levels, while most others are uncontrollable, being affected by chance factors over and above assignable ones, and hence behave as random variables. Observed values of such variables may be subject to errors, may be truncated by conditions imposed in the life-testing experiment, and may be even difficult to define unambiguously.

Data bearing on different facets of reliability analysis arise primarily from life-testing experiments, which presents a wide array and result in a correspondingly wide variety of data. This calls for the development and the use of a whole range of probability models for effective use of the observed (observable) data for appropriate purposes.

During the design and development phase, components and even systems are debugged or subjected to tests under simulated use conditions to identify components or systems which suffer from design or manufacturing deficiencies and are incompatible with the use environment so as to fail almost immediately when put on test. These early failures giving rise to a highly positively skewed distribution will provide some input for fixing the guaranteed life before which a product cannot fail. Only components and assemblies which survive the debugging period are subsequently assembled to manufacture products for release and use.

Thus, we come across a whole range of variables, some continuous, some discrete. Quite a few relate to time, while there are others which correspond to stress operating

1.4 Data Bearing on Reliability

on the product in terms of either the conditions of use and to strength built into the product to overcome operational and environmental stresses. Even time-to-failure as a random variable is likely to behave in a manner different from time between successive failures for a repairable product. Coming to a count of failures, the number arising from a life-test which is terminated at a pre-specified point of time (ahead of its natural end) will follow a discrete probability model. Otherwise also, if the stress applied to the units put on test is higher than the usual stress to hasten failures and complete the test earlier than otherwise, the observed pattern of variation in time observed in the accelerated life-test will have to be worked out for the normal stress situation. A different type of count data arises when we note the number of operation cycles during which a product functioned satisfactorily before it failed in the next cycle. Similarly, putting a batch of product units on test, we may observe the number of units surviving till the first unit failing. In dealing with fatigue failure or with failures caused by external shocks, we may involve both count data and time data along with data on magnitudes of stress or shock and their impacts. Data on environment characterizing features which can affect the functioning of the product and on some measure reflecting in-built strength may also be compiled and analysed to study reliability. Going beyond the two-state analysis, we can also observe the state (usually more than two) of functioning of a product at any given time, e.g., fully functioning, functioning with reduced efficiency (which again can be comprehended in terms of several levels) and failed. To the multiple states of functioning, we may associate different weights to work out some measure of performance like the disease-adjusted life-years for a human being.

Starting with a batch of, say, n units put on a life-test simultaneously, we may come across the following random variables which call for the use of appropriate probability models.

(*a*) Number of units failing by a specified test (inspection) time or, more generally, number of units failing during successive intervals (defined in terms of pre-specified inspection or monitoring times) or the number of units which functioned satisfactorily before the first unit failing.

(*b*) Times to failure for each of the items put on test or for each item failing by a specified time, others noted to have survived till the specified time or longer, or for each of a pre-specified number, say r, of failures.

(*c*) Number of shocks received by a product by a given time period.

(*d*) Points of time at which these shocks are received.

(*e*) Environmental and/or operational stress under which a product is functioning at any point of time t.

Data bearing on reliability do arise during operation and maintenance of products. In fact, for repairable items, time-to-repair once an item fails and time between two consecutive failures are typical random variables to be analysed in developing optimal maintenance policies (involving inspection of the system, repair of failing items, and replacement of failed items).

1.5 Probability Models in Reliability Analysis

Uncertainty affecting different attributes or variables specified during design or quality parameters during production or variables observed during use and maintenance of products as are reflected in random variations following, of course, some patterns invites applications of appropriate probability models. These models are required to bring out the specific pattern of variation representing each underlying phenomenon under experimentation and/or observation and to make relevant inductive inferences in the context of reliability analysis, the most important aspect of inferencing related to estimation of reliability. Such an estimate derived from the available data will be based on an assumed probability model that effectively represents the data-generating mechanism and the distinctive features of the observed data. Thus, the identification of an appropriate probability model with due consideration of the data features and the context (specifying the type of product, the nature of the experiment, the manner in which data have been recorded and the manner of its use) is a critical spell in the estimation exercise. A similar task is testing of relevant hypotheses concerning the model to be used. Quite an important exercise is system reliability optimization, subject to constraints on cost or volume or overall performance where again we need probability models to take care of given variables and given relations which are random and not exactly known.

Different models will have to be considered for Components versus system, systems with different configurations, count data versus continuous measurements, two-state versus multi-state analysis, recognizing simply a failed and a (fully) functional state or recognizing states of partially functioning in terms of some performance parameter(s).

Complete data allowing the life-test to continue till the last unit on test fails versus truncated or censored data from test which are terminated as soon as a specified number of units have failed or a specified time (irrespective of the number of failures) is elapsed. There could be data which are error-free versus those which are affected by errors of measurement or cannot be exactly noted, inviting fuzzy numbers.

There could be other distinctive features of data to be used in any reliability analysis. Thus, time-to-failure in an accelerated life-test subjecting the product units to a higher stress level than that required to be overcome during normal operation will call for a probability model different from the model for a usual life-test data. Data on early failures will justify a model different from that useful for time-to-chance failures or that due to wear-out failure.

Speaking of reliability of coherent systems, we note some general considerations linking component lifetimes with system lifetime.

Given the structure function, one can conveniently work out the form of system life as a function of component lives. Thus, for a series and a parallel system, system life can be expressed as

$$T_{series} = \min\{T_1, T_2, \ldots, T_n\} \quad \text{and} \quad T_{parallel} = \max\{T_1, T_2, \ldots, T_n\}.$$

The life of a (cold) standby system can be expressed as

$$T_{standby} = X_1 + X_2 + \cdots + X_n.$$

System lives can be similarly written out for other types of systems.

Component lifetimes within a system will have a joint probability distribution with the distribution function given by

$$F(x_1, x_2, \ldots, x_n) = P(T_1 \leqslant x_1, T_2 \leqslant x_2, \ldots, T_n \leqslant x_n).$$

From this joint distribution, we can conveniently work out distributions of the lifetime of any coherent system. In the relatively simple case of components functioning independently of one another and having independent life distributions, expressions for system life come out quite easily. To further simplify matters appropriate in some contexts, component lifetimes are assumed to be identically distributed. In such a situation, the life distribution of a standby system is simply the convolution of the common distribution assumed. Since the scope of the present volume is limited to univariate probability models, derivation of system life and its properties for various types of coherent systems has been kept outside the purview of our discussions here. However, in some simple cases of independently (and in most cases identically) distributed component lifetimes, expressions for system life have been just mentioned. The authors intend to discuss multivariate probability models in the second volume of this book.

The dominant position of the classical normal distribution in theory and applications of Statistics is not evident in the context of reliability and survival analysis, primarily because we deal with non-negative random variables in this context. Further, several features of the data of special interest in reliability analysis have led to the development and use of a virtual deluge of probability models. Considering the stress-strength approach, several authors have assumed stress X and strength Y as independently distributed to work out reliability (with no reference to any mission time) as $P(X < Y)$, some others assume a joint distribution for these two random variables. Further, some investigators even argue that stress operating on a single product unit during different operation cycles or working under different environments may vary randomly, though strength built into a given unit is expected to decrease with time of use in a deterministic manner. Strength could, of course, vary randomly across units coming out of the same design, production, and testing processes.

1.6 Preview of the Book

The present volume is intended to add some non-trivial material to the contents of a virtual deluge of books, monographs, and research as well as review articles on the subject of Reliability Analysis. The primary focus is on stochastic models used in the context of reliability analysis and to provide the reader with a more or less contemporary picture of developments in this restricted domain. While classical or later works cannot be dropped from any treatment of the topic, emphasis has been placed on the relatively recent contributions to different aspects of a comprehensive (though not complete) treatment of univariate probability models representing a wide spectrum of features revealed in the varied types of data arising in the context of reliability and survival analysis. Multi-component systems occupy a large segment of reliability analysis and multivariate probability models play a significant role in that context. Even for stress-strength reliability analyses, we have to deal with such models. However, these models do justify a comprehensive treatment, especially in view of the fact that quite a far additional concepts, methods, and results need incorporation into such a treatment.

The authors have no hesitation to admit that this volume is neither a textbook on the subject that is self-complete with all details including proofs of all results included nor an advanced-level monograph presenting a conspectus of some important aspects of the subject which can be a stand-alone support to research workers and/or scientists and engineers interested in the stochastic aspect of reliability analysis. At the same time, a modest claim can be made about inclusion of some topics which usually do not find a place in textbooks and of certain results worked out by the authors and their associates, besides those obtained by other distinguished research workers in the field.

Chapter 2 provides an intelligent and hopefully interesting summary of life distribution properties which are relevant in the context of reliability analysis. Certain less-known features which may attract the attention of potential research workers have been included. Topics like reversed hazard rate, ageing intensity, memory, probability-weighted moments, entropy, order statistics and record values, concavity and log-concavity, etc., have been briefly discussed.

The next chapter is devoted to a somewhat comprehensive account of continuous probability models which have been studied by a host of authors to describe random variations in lifetime, presented in a narrative manner, indicating the situations which motivated their origin and occasionally indicating methods for estimating the parameters involved (particularly methods other than the maximum likelihood method or the method of moments). Expressions for important reliability properties have been stated, but not derived. Different methods for generalization, modification, and extension of some of the well-known models using techniques like exponentiation or transmutation have been included. A brief account of mixture models and compound distributions along with some discussion on bathtub failure rate models also find a place in this chapter.

Discrete life distributions have been taken up in Chap. 4. Apart from the classical binomial, Poisson, geometric, negative binomial, and related distributions as also their generalized versions, the class of telescopic distributions has been considered. Different schemes for discretization of continuous probability distributions and discretized versions of some widely used models like the exponential and the Weibull find a place.

1.6 Preview of the Book

The content of Chap. 5 may not find a parallel in any publication and is largely based on research articles published by the authors and their associates. Distributions of remaining (residual) life, inactivity time, failure rate transform of the original lifetime variable, total-time-on-test transform, and equilibrium distribution have been discussed here. Possible uses of these distributions and their summary features in classifying and characterizing life distributions have been indicated. This area is open for further studies and the content of this chapter is expected to be useful to research workers.

Chapter 6 presents a somewhat detailed discussion of several finite-range life distributions, both continuous and discrete. Much of the material is derived from the published work of one of the authors. Various properties of these distributions of interest in reliability studies have been stated, though not derived, and some hints given for estimating the parameters involved. Expressions for stress-strength reliability with stress and strength following finite-range distributions have been provided.

Ordering among life distributions has been dealt with in Chap. 7. Here again, the reader will find a reflection of the research publications of the authors and their associates. Ordering based on hazard rate, reversed hazard rate, mean residual life, and mean inactivity time besides likelihood ratio ordering has been presented in working details. Stochastic ordering of the dispersion type as also some generalized ordering has been included. Incidentally, researches done by one of the authors constitute a part of the content.

Chapter 8 is devoted to classification of life distributions based on important reliability properties. Recent contributions in this regard leading to the $\mathcal{L}$-class, the $\mathcal{M}$-class, and the $\mathcal{LM}$-class, besides relatively less used classes like $NBUT$, have been incorporated. Preservation of class properties under reliability operations like coherent system formation, formation of mixtures, convolutions, etc., and inter-relations among different life distribution classes have been considered. Classification using coordinate sub-tangent is a novel feature of the content.

One step ahead of classification is characterization of individual life distributions and this is an area where volumes of published materials by a host of authors have come out over the years. Taking several routes to characterization, and confining to some oft-used life distributions, the authors have presented in Chap. 9 results primarily obtained by them with sketchy proofs in some cases and have presented without proofs several interesting results derived by other investigators. Obviously, this chapter is heavily loaded with contributions made by the authors.

Chapter 10 again is somewhat unusual in its content—being devoted to some special distributions arising in some typical situations besides some discussion on fatigue failure models, considering both static and dynamic failures. Distribution of quality-adjusted life (taking account of partly functional states in multi-state reliability analysis), lifetime models with data subject to errors of measurement and with fuzzy data have been briefly considered. Starting with some simple differential equations derived from fracture mechanics and introducing randomness in the parameters involved, appropriate lifetime distribution models have been presented.

In the fitness of things, one should address the problem of choosing an appropriate model out of the whole pack of models that broadly fit into the data scenario and this is what has been attempted in the last chapter. To be objective in our choice, we should take recourse to some criterion and having chosen a particular model we should apply some test for goodness-of-fit by the model chosen to the data at hand. Several criteria for model selection along with their limitations have been presented in terms of examples. Hints have been dropped in regard to limitations of the common goodness-of-fit tests.

The authors would like to point out the absence of detailed discussions on estimation of parameters involved in many models covered in this volume. One reason for this has been the fact that, with access to high-speed computers and efficient algorithms, maximum likelihood estimate of parameters can be obtained conveniently, even when the likelihood equations do not admit of analytic solutions. However, some particular methods not in common use have been briefly indicated in cases like the Weibull distribution. A distinct deficiency in the material content that can be easily identified is the lack of data sets—generated by the authors or borrowed from other authors—and illustrative calculations to fit some models to the selected data sets. This has been purposefully done—rightly or wrongly—keeping in mind two somewhat contradictory desiderata the authors took into consideration, *viz.*, a more or less complete account of models and relatively manageable size of the book. However, examples and counterexamples of statements and results have been added, wherever found desirable. It must be mentioned here that the words increasing (resp. decreasing) are interchangeably used as non-decreasing (resp. non-increasing).

Chapter 2
Properties of Life Distributions

2.1 Introduction

Life distributions are probability distributions of non-negative random variables with the positive half of the real line (or a part thereof) or the set of natural numbers (or a subset thereof) as the support. Various features of a probability distribution including distinctive features possessed by a single distribution or a particular group/class have been captured by several summary functions of the parameters involved like moments and quantiles, besides some properties of the density (mass) or the distribution functions and the tails of the corresponding curves. It may be incidentally pointed out that life distributions are generally asymmetric.

Length of life (or of duration of successful functioning) of a product exhibits features common to age of an individual in a human population subject to the prevailing pattern of mortality, as are revealed in a life table and, as in the latter case, many features of interest including rates of mortality are age-dependent. This way, ageing is an important phenomenon in reliability analysis. Rates of mortality find their analogue in rates of failure, and probabilities of survival up to specified ages are analogous to reliabilities. Different (deterministic) functions used to compute and predict mortality find their reflections in probability models to estimate and predict reliability.

Reliability estimation using a suitably chosen probability model has been a major task in reliability analysis. For this purpose, various properties of life distribution coupled with different methods of estimation—both traditional ones and those specially proposed in the reliability context—play the most important role. This volume, as indicated in the previous chapter, does not dwell upon relevant methods of estimation and properties of corresponding estimators. However, important properties of life distributions have been briefly taken up in this chapter. Those properties are quite interesting not just for estimating parameters but also in characterizing different univariate life distributions in grouping them in distinct classes and in establishing stochastic comparisons among the distributions. Classificatory properties also help in obtaining bounds on reliability in situations where we cannot identify the underlying life distribution uniquely.

2.2 Moments

Moments of different order r, defined for a continuous random variable X having the probability density function f, with support $S \subseteq \mathbb{R}^+ = [0, \infty)$, as

$$E(X^r) = \int_S x^r f(x)\, dx,$$

or for a discrete random variable with the probability mass function p given by $p(x) = P(X = x)$ as

$$E(X^r) = \sum_{x \in S} x^r p(x),$$

have been pretty widely used to reflect various properties of a life distribution and to estimate its parameters. Usually, r has been taken as a natural number. However, fractional moments allowing r to be any positive number (including a fraction) have been advocated for estimating parameters with higher efficiency compared to the usual moment estimators.

To estimate parameters in life distributions which do not admit of closed forms for inverses of the distribution functions, probability-weighted moments have been proposed (cf. Greenwood et al., 1979). In general, such a moment is defined and denoted as

$$M(p, r, s) = E\left[X^p F^r(X)(1 - F(X))^s\right],$$

where F is the distribution function of X, defined as $F(t) = P(X \leqslant t)$. Of usual interest are the moments

$$\alpha_s = M(1, 0, s) \quad \text{and} \quad \beta_r = M(1, r, 0).$$

Unbiased estimators of α_s and β_r can be conveniently obtained in terms of sample order statistics. Just to illustrate, $M(1, 0, k)$ for the Weibull distribution with parameters α (shape) and θ (scale) with survival function given by

$$\bar{F}(x; \alpha, \theta) = e^{-(x/\theta)^\alpha}, \quad x > 0, \, \theta > 0, \, \alpha > 0$$

comes out as (cf. Caiza and Ummenhofer, 2011)

$$M(1, 0, k) = \frac{\theta\, \Gamma\left(1 + \frac{1}{\alpha}\right)}{(1 + k)^{1 + \frac{1}{\alpha}}}.$$

2.2 Moments

Writing M_0 and M_1 for $M(1, 0, 0)$ and $M(1, 0, 1)$, respectively, we get closed-form expressions for the Weibull parameters in terms of M_0 and M_1.

Sometimes suitably chosen linear combinations of probability-weighted moments have been used to estimate parameters where maximum likelihood estimators cannot be worked out in closed forms. Such linear combinations have been referred to as L-moments. Moment ratios have also been used since long to reveal some distinctive features of univariate distributions. In fact, the simple coefficient of variation leads to interesting characterizing results. It is worth noting that the mean alone can determine the distribution uniquely in some cases, while more can be determined uniquely in terms of suitable moment ratios.

The following counterexample shows that even if all the moments exist, the distribution may not be uniquely determined (cf. Feller, 1971)).

Counterexample 2.2.1 *Let X_α be a random variable whose density function f_α is given by*

$$f_\alpha(x) = \frac{1}{24} e^{-x^{1/4}} - \frac{\alpha}{24} e^{-x^{1/4}} \sin\left(x^{1/4}\right), \ x > 0, \ 0 < \alpha < 1.$$

Then

$$\begin{aligned} E(X_\alpha^r) &= \frac{1}{24} \int_0^\infty x^r e^{-x^{1/4}} dx - \frac{\alpha}{24} \int_0^\infty x^r e^{-x^{1/4}} \sin\left(x^{1/4}\right) dx \\ &= \frac{1}{6} \int_0^\infty t^{4r+3} e^{-t} dt - \frac{\alpha}{6} \int_0^\infty t^{4r+3} e^{-t} \sin t \, dt \\ &= \frac{(4r+1)!}{6} < \infty, \end{aligned}$$

for any $r \in \mathbb{N}$, since the second integral is zero. Thus, for every $\alpha \in (0, 1)$, we get different distributions having the same set of moments. □

Below, we give a sufficient condition under which the moment sequence uniquely determines the distribution.

Theorem 2.2.1 *Let $\{m_k\}$ be the moment sequence of X. If*

$$\sum_{k=1}^\infty \frac{m_k}{k!} t^k \ \text{converges absolutely for some } t > 0,$$

then $\{m_k\}$ uniquely determines the distribution of X. □

Below are some more conditions under which the moments uniquely determine the distribution. Some of the conditions may be obtained in Stuart and Ord (1994).

(a) The moments uniquely determine the distribution if the range of the distribution is finite. This is because, if we take the origin at the start of the distribution (of range h, say), then $E(|X|^r) = \nu_r \leq h^r$ and hence

$$\sum_{k=1}^{\infty} \frac{\nu_k t^k}{k!} \leq \sum_{k=0}^{\infty} \frac{(ht)^k}{k!} = e^{ht} < \infty.$$

Hence, by Theorem 2.2.1, the result follows.
(b) The moments uniquely determine the distribution if

$$\limsup_{n \to \infty} \frac{\nu_n^{1/n}}{n} < \infty.$$

This is because a series, whose general term is $\nu_n t^n/n!$, converges if

$$\limsup_{n \to \infty} \left(\frac{\nu_n t^n}{n!} \right)^{1/n} < 1.$$

Replacing factorial by its Stirling's approximation, we have

$$\limsup_{n \to \infty} \left(\frac{\nu_n t^n}{\sqrt{2\pi} e^{-n} n^{n+1/2}} \right)^{1/n} < 1,$$

i.e., if, for some constant k, $\nu_n^{1/n} < nk/t$. Now, if the upper limit is finite, the inequality can be satisfied for some non-zero t.
(c) The moments uniquely determine the distribution if

$$\limsup_{n \to \infty} \left(\frac{\mu_{2n}^{\frac{1}{2n}}}{2n} \right) < \infty,$$

where $E(X^r) = \mu_r$. This is because the upper limits of $\mu_{2n}^{\frac{1}{2n}}/(2n)$ and $\nu_n^{1/n}/n$ are finite or infinite together.
(d) **Carleman's criterion:** A set of moments determines a distribution uniquely if

$$\sum_{j=0}^{\infty} \mu_{2j}^{-\frac{1}{2j}} = \infty \quad \text{if the support of } X \text{ is } \mathbb{R},$$

and

$$\sum_{j=0}^{\infty} \mu_j^{-\frac{1}{2j}} = \infty \quad \text{if the support of } X \text{ is } \mathbb{R}^+.$$

2.3 Measures of Shape

The measures of location give the position of a distribution on the real line, the measures of dispersion tell us how the values of a distribution are dispersed among themselves with respect to the measures of location. Now, in order to have some idea of the shape of a distribution, we must know whether the distribution is symmetric or not. In case it is not symmetric then we want to know the kind of asymmetry present there. After getting this idea, one may like to know how peaked a distribution is and how thick the tails are. These are two features we shall discuss now.

A random variable (or equivalently its distribution) is said to be symmetric if there exists a point a such that $P(X \leqslant a - h) = P(X \geqslant a + h)$ for all h. If the distribution has the density f, then it is said to be symmetric if $f(a + h) = f(a - h)$ for all h. If the distribution has the longer tail towards the right side it is said to be positively skewed and if it has longer tail towards the left side it is called negatively skewed.

Although it is observed in most of the cases that, for a unimodal distribution,

(i) $Mean = Median = Mode$ if the distribution is symmetric;
(ii) $Mean > Median > Mode$ if the distribution is positively skewed;
(iii) $Mean < Median < Mode$ if the distribution is negatively skewed,

it need not be so always. It is always possible to generate a distribution violating the above relationships.

Since a symmetric distribution has all odd order central moments zero, we may use the odd order central moments as measure of skewness. However, increasing the order of moment incurs more error in it. So, the least possible odd order moment (i.e., μ_3 because $\mu_1 = 0$ always) can be considered as a measure of skewness. But in order to make it free of unit of measurement, we consider a measure of skewness as $\gamma_1 = \mu_3/\mu_2^{3/2}$.

For a symmetric distribution $\gamma_1 = 0$; $\gamma_1 > 0$ for a positively skewed distribution, whereas $\gamma_1 < 0$ for a negatively skewed distribution. Based on the above discussion, another measure of skewness is defined as (Mean-Mode)/s.d. There are other measures of skewness available in the literature. However, we shall not discuss those here.

The main drawback of the moment measure of skewness is that $\mu_3 = 0$ does not always guarantee that the underlying distribution is symmetric. That's why there are different other measures of skewness available in the literature, $viz.$, measure based on quartiles.

It is known for a long time that the peakedness of a distribution is measured by the fourth-order central moment μ_4. However, to make it unit-free, the measure of kurtosis is sometimes defined as $\beta_2 = \mu_4/\mu_2^2$. It is well known that a normal distribution has $\beta_2 = 3$. To keep this distribution as a standard one, the measure of kurtosis is defined as $\gamma_2 = \beta_2 - 3$. The measure γ_2, being the excess of the value of moment measure of a normal distribution (which is 3), is sometimes called coefficient of excess. Thus, we have the following classification of distributions.

Table 2.1 Distributions with same mean, same variance, same measure of skewness and different measures of kurtosis

Probability density	γ_2	Max. height
$f_1(x) = \frac{1}{3\sqrt{\pi}} \left(\frac{9}{4} + x^4\right) e^{-x^2}$	−0.25	0.423
$f_2(x) = \frac{3}{2\sqrt{2\pi}} e^{-x^2/2} - \frac{1}{6\sqrt{\pi}} \left(\frac{9}{4} + x^4\right) e^{-x^2}$	0.125	0.387
$f_3(x) = \frac{1}{6\sqrt{\pi}} \left(e^{-x^2/4} + 4e^{-x^2}\right)$	1.5	0.470
$f_4(x) = \frac{3\sqrt{3}}{16\sqrt{\pi}} (2 + x^2) e^{-3x^2/4}$	−0.323	0.366
$f_5(x) = \frac{1}{\sqrt{2\pi}} e^{-x^2/2}$	0	0.399

Definition 2.3.1 A distribution is said to be

(a) mesokurtic if $\gamma_2 = 0$;
(b) platykurtic if $\gamma_2 < 0$;
(c) leptokurtic if $\gamma_2 > 0$. □

Note that, for a discrete distribution assuming values $x_1, x_2, \ldots$ with respective probabilities $p_1, p_2, \ldots$, the fourth-order central moment is defined as

$$\mu_4 = \sum_i (x_i - E(X))^4 p_i.$$

If more number of values x_i are far from $E(X)$ on either side, the value of μ_4 will be large. Thus, μ_4 (or equivalently γ_2) can be used to measure the tail heaviness of a distribution. It should be mentioned here that γ_2 alone cannot be considered as a measure of peakedness of a distribution. Table 2.1 justifies this by showing different distributions (borrowed from Kaplansky (1945)) which show that distribution having $\gamma_2 < 0$ has more height than that of a normal distribution whereas a distribution with $\gamma_2 > 0$ has less height than that of normal distribution. The following statement due to Kendall and Buckland (1971, p. 137) is interesting to note.

> Contrary to earlier belief, the ordinary measures of skewness and kurtosis based on moments are not necessarily good representations of shape.

2.4 Ordering and Ageing Properties

Ageing, an important phenomenon in reliability theory, is an inherent property of a unit that may be a living organism or a system of components. It largely describes how a unit ages with time. There are two kinds of ageing, *viz.* positive ageing and negative ageing. The concept of no ageing is also available in the literature. The no-ageing class mostly lies on the boundary of the positive ageing and the negative ageing classes. The no-ageing phenomenon is described by the exponential

2.4 Ordering and Ageing Properties

(for continuous lifetime) and the geometric (for discrete lifetime) distributions. By ageing we generally mean positive ageing. It describes a phenomenon whereby an older system has a shorter remaining lifetime, in some statistical sense, than a newer or younger one, i.e., it describes the situation where residual lifetime decreases, in some probabilistic sense, with increase in the age of a component. To be more specific, by positive ageing we mean a mathematical specification of degradation of life of a unit over time, whereas the negative ageing (also known as anti-ageing) is described by a mathematical specification of upgradation of life of a unit over time. The positive ageing is the adverse effect of age on the residual lifetime of the unit, whereas negative ageing refers to a beneficial effect of age. It may be considered to be axiomatic that any effect of age on a unit, which contributes to the reduction (resp. promotion) of its residual lifetime in some probabilistic sense is to be taken as an adverse (resp. beneficial) effect and the phenomenon is called positive (resp. negative) ageing. Different concepts of positive (and negative) ageing notions have been studied in the literature by different researchers.

2.4.1 Hazard Rate and Reversed Hazard Rate

For a lifetime random variable X having density function f, distribution function F and survival function $\bar{F}$ (defined as $\bar{F}(t) = P(X > t)$, although some authors define the survival function as $\bar{F}(t) = P(X \geq t)$ which does not make any difference in case of continuous random variables but may differ in case of discrete random variable), the failure rate of X at the point t, denoted by $r(t)$, is defined as $r(t) = f(t)/\bar{F}(t)$ which can be interpreted as under. Note that

$$\begin{aligned}
\lim_{\delta t \to 0+} P(t < X \leq t + \delta t | X > t) &= \lim_{\delta t \to 0+} \frac{P(t < X \leq t + \delta t)}{P(X > t)} \\
&= \lim_{\delta t \to 0+} \frac{F(t + \delta t) - F(t)}{\delta t} \cdot \frac{\delta t}{\bar{F}(t)} \\
&\approx \frac{f(t)}{\bar{F}(t)} . \delta t \\
&= r(t) . \delta t.
\end{aligned}$$

So, $r(t)\delta(t)$ gives us an approximate probability that a system fails during the time interval $(t, t + \delta t]$ when it is known that the system was in working state at time t, provided δt is a very small positive number. In other words, $r(t)$ gives the probability that the system fails instantaneously after time t. That's why $r(t)$ is called instantaneous failure rate or simply failure rate (also known as hazard rate).

Suppose a system is being checked periodically whether it is working or not. At a time point t it is observed that the system has failed. We might be interested to know what is the probability that the system actually has failed just before we observed it at time t. To get this, let us proceed as under.

$$\lim_{\delta t \to 0+} P(t - \delta t < X \leqslant t | X \leqslant t) = \lim_{\delta t \to 0+} \frac{P(t - \delta t < X \leqslant t)}{P(X \leqslant t)}$$

$$= \lim_{\delta t \to 0+} \frac{F(t) - F(t - \delta t)}{\delta t} \cdot \frac{\delta t}{F(t)}$$

$$\approx \frac{f(t)}{F(t)} . \delta t$$

$$= \mu(t) . \delta t,$$

where $\mu(t) = f(t)/F(t)$. So, $\mu(t)\delta t$ gives the approximate probability that the system fails within a small interval $(t - \delta t, t]$ when it is known that the system has failed at or before time t. This $\mu(t)$ is known as reversed hazard (or retro-hazard or backward hazard) rate.

Based on hazard rate and reversed hazard rate the following non-parametric classes have been defined in the literature. A life distribution is said to belong to the Increasing Failure Rate (IFR) class if the corresponding failure rate function is increasing. Obviously, the IFR class is a positive ageing class. A corresponding dual class is called DFR (Decreasing Failure Rate) if the corresponding failure rate is decreasing. The distributions having constant failure rate is called CFR (Constant Failure Rate) distributions. Note that, if $r(t) = \lambda$, a positive constant, then simple calculation shows that the corresponding continuous random variable X follows exponential distribution with mean time-to-failure ($MTTF$) $1/\lambda$ and conversely. However, one may think whether there exists any practical situation where life distribution follows exponential (since failure rate does not change with time). Below, we mention a few such situations.

- If the failure rates are determined by independent random events not related to the age of the system, a CFR model will result.
- Let a system be comprised of independent components and let failure of an individual component causes the system failure. In a renewal process where a component is replaced by a new one immediately after failure, the system will reach steady state after some time, i.e., after some time, a constant number of failures per unit time will be observed.

If a distribution has failure rate of the form $r(t) = a + bt$, for some $a \geqslant 0$ and $b > 0$, then the corresponding distribution is said to belong to the LFR (Linear Failure Rate) family. It is easy to see that the corresponding density function f is given by

$$f(t) = (a + bt)e^{-at - \frac{bt^2}{2}}, \ t \geqslant 0.$$

It is denoted by $X \sim LFR(a, b)$, where X is the underlying random variable. It is to be noted that the LFR family of distributions contains Rayleigh distributions ($a = 0$) whereas it is contained in the family of IFR distributions. The LFR distributions are skewed unimodal distributions over $[0, \infty)$. The family of LFR distributions is closed under the formation of series system. Further, if $X \sim LFR(a, b)$ the residual random variable $X_t = (X - t | X > t) \sim LFR(a + 2bt, b)$.

2.4 Ordering and Ageing Properties

A distribution is said to have bathtub-shaped (BT) failure rate, if there exist $0 \leqslant t_1 \leqslant t_2$ such that

(a) $\lambda(t)$ is (strictly) decreasing in $t \in [0, t_1]$;
(b) $\lambda(t)$ is constant for $t \in (t_1, t_2)$;
(c) $\lambda(t)$ is (strictly) increasing for $t \geqslant t_2$.

Thus, the failure rate

$$\lambda(t) = \begin{cases} \lambda_1(t), & 0 \leqslant t \leqslant t_1 \\ \lambda, & t_1 \leqslant t \leqslant t_2 \\ \lambda_2(t), & t \geqslant t_2, \end{cases}$$

where $\lambda_1(t)$ is decreasing in $[0, t_1]$ and $\lambda_2(t)$ is increasing in $[t_2, \infty)$, is BT. It is to be noted that if $t_1 = t_2 = 0$, the BT distribution becomes IFR; $t_1 = t_2 \to \infty$ gives DFR distribution whereas BT distribution becomes a member of CFR class if $t_1 = 0$ and $t_2 \to \infty$. It is known that the family of BT distributions is neither closed under convolution nor under mixture nor under the formation of coherent system. However, BT class is closed under increasing transformation, i.e., if X has BT failure rate then, for any increasing function ϕ, $\phi(X)$ is a member of BT family.

Note that exponential distribution lies at the border of IFR and DFR classes, as it can be considered a member of IFR class as well as of DFR class. It is well known that the IFR (resp. DFR) class is (resp. is not) closed under convolution whereas it is not closed (resp. is closed) under mixture of distributions. As far as the closure of the class under formation of coherent system is concerned, neither of IFR class and DFR class is closed under that. A natural question that arises is—what is the smallest class which is closed under the formation of coherent system? To answer this question, Barlow and Proschan (1975) defined a non-parametric class, called $IFRA$ (Increasing in Failure Rate Average). A distribution is said to belong to the $IFRA$ class if the average failure rate is increasing, i.e., $\frac{1}{t} \int_0^t r(u) du$ is increasing in $t > 0$. Its dual class, called $DFRA$ (Decreasing in Failure Rate Average), is defined as the collection of distributions for which $\frac{1}{t} \int_0^t r(u) du$ is decreasing in $t > 0$. Although $IFRA$ class is closed under the formation of coherent system, the $DFRA$ class is not. Barlow and Proschan (1981) have shown that the $IFRA$ (resp. $DFRA$) class is (resp. is not) closed under convolution. It is to be mentioned here that although the $IFRA$ class is not closed under mixture of distribution, the $DFRA$ class is.

Since the above non-parametric classes have been defined in term of monotonicity of the failure rate, one may like to know whether any such class can be defined based on the monotonicity of the reversed hazard rate. A class of distributions having the reversed hazard rate function decreasing is called Decreasing Reversed Hazard Rate ($DRHR$) class. Clearly, a distribution belongs to the $DRHR$ class if and only if the corresponding distribution function is log-concave. A distribution function F over the support $[0, \infty)$ is said to be log-concave if and only if, for any $\alpha \in [0, 1]$, $F(\alpha x + (1 - \alpha)y) \geqslant F^\alpha(x) F^{1-\alpha}(y)$, for all $x, y \in [0, \infty)$. One may be interested to know whether, similar to IFR class, there exists any distribution having reversed hazard rate function increasing over the support $[0, \infty)$, i.e., whether there exists any distribution function F which is log-convex. We answer this question below in negative (cf. Sengupta and Nanda, 1997).

Theorem 2.4.1 *There does not exist any non-degenerate life distribution which is log-convex over the entire domain* $[0, \infty)$.

Proof Suppose F is a distribution function such that $\ln F$ is convex over $S = [0, \infty)$. Let $t_1 \in S$ be such that $0 < F(t_1) < 1$. Then there exists a straight line, $\alpha + \beta t$, say, such that $\ln F(t) \geqslant \alpha + \beta t$ for all $t \in S$ with equality at t_1. Suppose $t_2 (> t_1) \in S$ where the line crosses the horizontal axis. Then we have $0 = \alpha + \beta t_2 \leqslant \ln F(t_2) \leqslant 0$ so that $F(t_2) = 1$. Take $t_3 = 2t_2 - t_1 (> t_2)$. So, we have $F(t_3) = 1$. Again, $\ln F$ being convex, we have

$$F(t_2) = F\left(\frac{t_1 + t_3}{2}\right) \leqslant \sqrt{F(t_1).F(t_3)}. \qquad (2.4.1)$$

Thus, we have $F(t_1) \geqslant 1$ giving $F(t_1) = 1$ which contradicts our assumption that $0 < F(t_1) < 1$. Hence, the proof follows by contradiction. $\square$

It is clear from the above theorem that if a reversed hazard rate is monotone over the entire (positive) real line, it must be decreasing. In other words, there does not exist any counterpart of 'negative ageing' as far as the reversed hazard rate is concerned. Thus, the $DRHR$ class cannot be considered as an ageing class.

Block et al. (1998) have shown that if a random variable has increasing reversed hazard rate ($IRHR$) over an infinite support, then the support must be of the form $(-\infty, a)$, where $a < \infty$. So, if a life distribution belongs to the $IRHR$ class, then its support will have to be finite. It is interesting to note that almost all useful distributions, $viz.$ exponential, Weibull, gamma, normal, log-normal, Pareto, and Gompertz distributions are members of $DRHR$ class. However, distribution having non-monotone reversed hazard rate is possible. Note that the $DRHR$ distributions are also known as log-concave distributions because 'reversed hazard rate decreasing' is equivalent to saying that 'the distribution function is log-concave'. It is not difficult to see that if a distribution has log-concave (resp. log-convex) density the corresponding distribution is IFR (resp. DFR). The class of distributions with log-concave density is called ILR (Increasing Likelihood Ratio) class, whereas the class of distributions with log-convex density is called DLR (Decreasing Likelihood Ratio) class. It is interesting to note that although $ILR \Rightarrow IFR \nRightarrow DRHR$ and the $DRHR$ class contains both ILR and DFR classes, the later contains DLR class. Sengupta and Nanda (1999) have noted that the $DRHR$ class is closed under the limits of distribution. To be specific, if, for $i = 1, 2, \ldots$, each F_i belongs to $DRHR$ class and if $F_n \xrightarrow{D} F$, as $n \to \infty$, then F also belongs to $DRHR$ class. It is also noted by them that this class is closed under the formation of a k-out-of-n system of iid components (actually, $DRHR$ class is closed under the formation of coherent system of iid components, see Nanda et al. (1998)). However, if the components forming the system are independent but not identically distributed, then this class is closed under the formation of parallel system but not under the formation of series system. Thus, the $DRHR$ class is not closed under the formation of coherent system.

2.4 Ordering and Ageing Properties

It is well known that the DFR class is not closed under convolution and DFR class is contained in $DRHR$ class. Then one immediate question that comes is whether the $DRHR$ class is closed under convolution. The answer is given in affirmative by Sengupta and Nanda (1999). It is also observed that the $DRHR$ class is not closed under mixture of distributions. They have also proved that the $DRHR$ class is the smallest class which is closed under the limits of distributions and under the formation of parallel systems, and contains all concave life distributions.

Let two systems have respective lives X and Y. Suppose X and Y have distribution functions F_X and F_Y, survival functions $\bar{F}_X$ and $\bar{F}_Y$, failure rate functions r_X and r_Y, and reversed hazard rate functions μ_X and μ_Y, respectively. Further, suppose that X and Y have the respective support S_X and S_Y, each being a subset of $\mathbb{R}^+ = [0, \infty)$. Since smaller the failure rate better is the system and larger the reversed failure rate better is the system, we define two partial orders between the random variables X and Y. The details are given in Chap. 7. For discrete distribution $F(k) = P(X \leqslant k)$ with support included in $\{0, 1, 2, \ldots\}$, the reversed hazard rate is defined as $\mu_F(k) = (F(k) - F(k-1))/F(k)$. In this case, F is said to be discrete decreasing reversed hazard (written as d-DRH) if the reversed hazard rate μ_F is a decreasing sequence.

It is noted in Nanda and Sengupta (2005) that many common discrete distributions, $viz.$, binomial, negative binomial, geometric, hypergeometric, Poisson, Hyper-Poisson, logarithmic series, zeta, and Yule distributions are d-DRH. For discrete distribution, we define the survival function as $\bar{F}(k) = 1 - F(k-1)$ and the hazard rate as $r_F(k) = (F(k) - F(k-1))/\bar{F}(k)$. A discrete distribution is said to be discrete increasing failure rate (written as d-IFR) if the hazard rate r_F is an increasing sequence.

2.4.2 Mean Residual Life

Whenever a system has survived for some t units of time, one might be interested to know how long this system will survive. This may be answered by studying the random variable $X_t = (X - t | X \geqslant t)$, the remaining life of the system surviving up to age t, where X denotes the life of the new system. Note that $E(X_t)$, known as mean remaining (or residual) life (MRL), gives us the average life left after the system has survived for t units of time, provided $E(X) < \infty$. We denote $E(X_t)$ by $m(t)$ and note that, for a finite-range distribution,

$$m(t) = \begin{cases} E(X_t), & \text{if } t < t_0 \\ 0, & \text{otherwise,} \end{cases}$$

where $t_0 = \sup\{t : \bar{F}(t) > 0\}$. It is to be noted that $m(t) < \infty$ for all $t < \infty$. However, $\lim_{t \to \infty} m(t) = \infty$ is possible. It is not difficult to see that, if the support of the underlying random variable is $\mathbb{R}^+$, then we have

$$m(t) = \frac{\int_t^\infty \bar{F}(u)\,du}{\bar{F}(t)}, \quad t > 0.$$

A random variable is said to belong to the $DMRL$ (resp. $IMRL$) class if the corresponding MRL function is decreasing (resp. increasing). It is known that if a life distribution belongs to the IFR (resp. DFR) class then it is $DMRL$ (resp. $IMRL$). It is to be mentioned here that if a random variable X is such that $V(X_t)$ is decreasing (resp. increasing) we say that X belongs to the $DVRL$ (resp. $IVRL$) class. Shaked and Shanthikumar (2007) have pointed out that, any non-negative continuous function $m(\cdot)$ can be considered as a mean residual life function of some random variable if it satisfies the following conditions.

(i) $m(0) > 0$.
(ii) $m(t) + t$ is increasing on $[0, \infty)$.
(iii) If $m(t) > 0$ for all $t > 0$, then $\int_0^\infty \frac{dt}{m(t)} = \infty$. However, if $m(t') = 0$ for some $t' \in (0, \infty)$, then $m(t) = 0$ for all $t \geq t'$.

It is to be mentioned here that neither $DMRL$ nor $IMRL$ class is closed under the formation of parallel system and hence is not closed under the formation of coherent system. It is to be noted here that if the components are iid then the $DMRL$ class is closed under the formation of parallel system but $IMRL$ class is not. Further, neither of the two classes is closed under the convolution of distributions. It is also shown that the $IMRL$ class is closed under the mixture of distributions whereas the $DMRL$ class is not.

Let X and Y be two non-negative random variables having survival functions $\bar{F}_X$ and $\bar{F}_Y$, and mean residual life functions m_X and m_Y, respectively. Then the random variable X is said to be smaller than Y in MRL order, written as $X \leq_{mrl} Y$, if $m_X(x) \leq m_Y(x)$ for all $x > 0$. It is not difficult to see that $X \leq_{mrl} Y$ if and only if

$$\frac{\int_x^\infty \bar{F}_Y(t)\,dt}{\int_x^\infty \bar{F}_X(t)\,dt} \text{ is increasing in } x \text{ over the set } \left\{ x : \int_x^\infty \bar{F}_X(t)\,dt > 0 \right\}.$$

2.4.3 Generalized Failure Rate and Generalized Mean Residual Life

The generalized failure rate (GFR) for a lifetime random variable X, as defined by Lariviere and Porteus (2001), given by $g(x) = x\,r(x)$ where $r(x)$ is the failure rate, has not found much use in reliability analysis, though the concept has some important applications in pricing and revenue management. The concept was introduced in economic applications by Singh and Maddala (1976) to model income distributions and subsequently by Belzunce et al. (1998) for ordering of truncated distributions through concentration curves.

2.4 Ordering and Ageing Properties

Coming to reliability analysis, an interesting result is that the GFR is increasing with age x for an exponential distribution. Thus, the exponential distribution with a constant failure rate is now an increasing GFR distribution. In fact, the only continuous distribution with constant GFR is the Pareto distribution with a distribution function given as $F(x) = 1 - (k/x)^\alpha$, $x \geq k$ and a failure rate $r(x) = \alpha/x$, the GFR being the constant α. For the Lomax distribution with survival function given by $\bar{F}(x) = (1 + x/\lambda)^{-\alpha}$, $x \geq 0$, the GFR is given by $g(x) = \alpha x/(\lambda + x)$ and this varies between 0 (for $x = 0$) and α (for $x \to \infty$). A detailed discussion on Pareto and Lomax distributions has been given in Sect. 3.5.3.

Just as life distributions with increasing and decreasing failure rates are well recognized in reliability analysis, we also have distributions with increasing GRF ($IGFR$) and decreasing GFR ($DGFR$). Clearly, every IFR distribution belongs to the $IGFR$ class but the converse is not necessarily true. Rather, there exist some DFR distributions which belong to the $IGFR$ class. Further all distributions may not be monotonic in terms of GFR functions. Actually we come across life distributions with non-monotonic GFR functions.

The following statements are equivalent. For a formal definition of TP_2 function, see Definition 7.8.7 and the order $\leq_{hr}$ is discussed in Sect. 7.3.

(a) X has increasing generalized failure rate.
(b) $\ln X$ has increasing failure rate.
(c) $X \leq_{hr} \lambda X, \lambda \geq 1$.

Like the GFR, the generalized mean residual life ($GMRL$) was initially studied in the context of market competition and supply chain coordination problems. In fact, $GMRL$ can be interpreted as the inverse of the price elasticity of expected demand. Lariviere (2006) notes that if X has increasing GFR but decreasing failure rate then there exists a δ such that $X + \delta$ does not have increasing GFR. This immediately tells that increasing GFR class is not closed under shift.

In the context of reliability and survival analysis, the $GMRL$ links the MRL at age t to the lifetime already survived up to age t and is formally defined as $N(t) = E(X_t)/t$. That way, the product $r(t)m(t)$ equals $g(t)N(t)$ and hence characterization results based on the product remain the same. Like the proportional hazards model which has been extended to the proportional generalized hazards model, we have the proportional MRL model linking the MRL with some covariates which can be stated as

$$m(t|Z) = m_0 t + \beta' Z,$$

starting with a baseline MRL function m_0 and incorporating the linear effects of a set of covariates Z. Similarly, we have the proportional $GMRL$ model also.

Life distributions have been classified as Decreasing Generalized Mean Residual Life ($DGMRL$) and Increasing Generalized Mean Residual Life ($IGMRL$), and relations connecting such classes have been examined. Thus, we note that X is $IGFR$ implies that X is $DGMRL$. However, $DGMRL$ does not necessarily imply $IGFR$ property. For example, the Birnbaum-Saunders distribution (see Sect. 3.6) often used in modelling lifetime data belongs to the $DGMRL$ class but is not $IGFR$ for some specific values of the parameters. In fact, X is $DGMRL$ and mean remaining life, $m(t)$, is log-convex together imply that X is $IGFR$. Further, $DMRL$ property neither implies $IGFR$ property nor is implied by the $IGFR$ property. It is worth noting that there exist many distributions with log-concave MRL functions which are also $IGFR$.

The GFR function at a random point of time, taken as $g(X) = X.r(X)$, can be looked upon as a transform of the original lifetime variable X and its distribution can be worked out. As indicated earlier, g will not have a degenerate distribution in case X has an exponential distribution with hazard rate λ. Instead, it will have a unit exponential distribution as $P[g(X) \geqslant x] = P[X \geqslant x/\lambda] = e^{-x}$ with the mean GFR being equal to unity.

Coming to the Weibull lifetime situation having scale parameter λ and shape parameter α, we get

$$P(g(X) \geqslant x) = P\left[X \geqslant \left(\frac{x}{\alpha\lambda}\right)^{1/\alpha}\right] = e^{-x/\alpha},$$

which corresponds to an exponential distribution with failure rate $1/\alpha$. It directly follows that $E(g(X)) = \alpha$.

This result was obtained by Dubey (1960) earlier without recognizing $Xr(X)$ as the generalized failure rate and using it to obtain an estimate of the Weibull shape parameter in terms of the observed (partly estimated) values of Z. One could make use of some non-parametric estimator of the failure rate. Given that values of Z can be computed for a (random) sample of X values, and assuming that X follows the Weibull distribution, one can estimate α in terms of the arithmetic mean of the computed z values.

Distributions of the GFR for other lifetime distributions can also be conveniently worked out. Thus, for the Lomax distribution with survival function given by $\bar{F}(x) = (1 + x/\lambda)^{-\alpha}$ and a failure rate $r(x) = \alpha/(\lambda + x)$, it follows that its expected value is $Eg(X) = \alpha/(\alpha + 1)$. Like in the case of the Weibull distribution, here also we can equate the mean of the observed-cum-estimated values of $g(x) = x\,r(x)$ to $\alpha/(\alpha + 1)$ to yield an estimate of the parameter α.

2.4.4 Vitality Function

The vitality function introduced by Kupka and Loo (1989) is a measure of ageing which complements the role of the failure rate, in some sense. The value of this function at age t is defined as the conditional mean life, i.e., $v(t) = E(X|X > t)$. While the failure rate $r(t)$ reflects the risk of a 'sudden' or 'instantaneous' failure or death, having survived up to age t, the vitality measure captures the reward in terms of an increase in mean life, having survived till age t. Mean remaining life, failure rate and vitality are simply inter-related as

2.4 Ordering and Ageing Properties

$$r(t) = \frac{v'(t)}{v(t) - t} = \frac{v'(t)}{m(t)}.$$

A geometric vitality function was considered by Nair and Rajesh (2000). In fact, the concept of vitality is linked to the fact that once units failing by some relatively small time are eliminated, the mean life for the remaining units will be much higher.

A characteristic feature of the exponential distribution is that the vitality measure at any age t is the mean life plus the age already survived.

2.4.5 Percentile Remaining Life Function

Several researchers have studied properties of quantiles (percentiles) of residual life distributions and their uses. The alpha-percentile residual life function $q_\alpha(t)$ for a life distribution with distribution function F is defined as $F^{-1}(\alpha + (1-\alpha)F(t)) - t$. In particular, they focused on median residual life in place of mean residual life as an important characteristic feature of a life distribution. Mention may be made of the contributions by Schmittlein and Morrison (1981), Arnold (1983), Joe and Proschan (1984), Joe (1985) and Lillo (2005) among others. Unlike the MRL, however, median remaining life cannot determine the life distribution uniquely. The tail behaviour of the residual life distribution does play a role. In fact, a life distribution cannot be determined uniquely by a single percentile of remaining life. Arnold and Brockett (1983) showed that two percentiles of orders α and β can determine the distribution uniquely provided α and β are incommensurate. Joe (1985) provided characterizations of the exponential, the beta, and the Pareto distributions in terms of two percentiles of their residual life distributions. Raja Rao et al. (2006) have identified families of distributions for which percentile residual life functions can be conveniently expressed.

The concept of percentile residual life function has been extended to that of a percentile residual life function up to a specified time t_0. Ordering and classification of life distributions based on these concepts have been discussed by a few authors. Two classes, $viz.$ decreasing percentile residual life ($DPRL$) and New better than used with respect to α-percentile residual life have been mentioned in relevant literature.

Just like the percentile residual life function, percentile inactivity time functions also have been studied by Mahdy (2013) and median inactivity time has been considered by Kandil et al. (2010).

While the loss-of-memory property is exhibited only by the exponential and the geometric distributions, 'putting the clock back to zero' property is shown by a few families of life distributions (cf. Raja Rao and Talwalker (1990)). This property requires that the conditional distribution of residual life, after the unit has survived till a certain time t, belongs to the same family as the original life distribution.

2.4.6 Ageing Intensity

Ageing is an important phenomenon in real life. Failure rate is one measure (among others) of ageing which plays an important role in analysing the failure pattern of a unit. A unimodal failure rate can be effectively viewed as either approximately decreasing or approximately increasing or approximately constant (cf. Jiang et al., 2003) which is very much qualitative in nature. To properly quantify this measure one may define the ratio of the failure rate $r(t)$ to a baseline failure rate which we take as average failure rate $\int_0^t r(u)\,du/t$. Let us call this ratio as ageing intensity $L(t)$ so that we get

$$L(t) = \frac{tr(t)}{\int_0^t r(u)\,du}, \quad t > 0,$$

which can be rewritten as

$$L(t) = -\frac{tf(t)}{\bar{F}(t) \ln \bar{F}(t)}, \quad t > 0. \qquad (2.4.2)$$

Unlike exponential distribution, which is characterized by constant failure rate, the constant ageing intensity does not characterize any specific distribution, rather it characterizes a family of distributions. It can be shown easily that, a lifetime random variable having $L(t) = c$, a constant, characterizes a family of Weibull distributions with shape parameter c (irrespective of its scale parameter). Rather, two distributions having the failure rates proportional to one another will have the same ageing intensity.

It has been observed in Nanda et al. (2007) that the monotonicity of failure rate is not transmitted to that of ageing intensity function. They have shown that an IFR distribution can have increasing, decreasing or non-monotone ageing intensity function. They have noted that Erlang distribution having density $f(t) = 4te^{-2t}$, $t \geq 0$, has increasing failure rate whereas its ageing intensity function is decreasing. It is not difficult to see that both failure rate function and ageing intensity function of uniform distribution are increasing. Further, if $X \sim N(4, 1)$ then the failure rate of $Y = (X|X \geq 4)$ is increasing but its ageing intensity function is non-monotone. To see this, note that, for any $t > 0$, the failure rate of Y is $r(t) = \phi(t)/(1 - \Phi(t))$, where ϕ and Φ are, respectively, the pdf and the cdf of the standard normal distribution. Then, a simple algebra shows that

$$\frac{dr(t)}{dt} \stackrel{sign}{=} (1 - \Phi(t))\phi'(t) + \phi^2(t)$$
$$= \phi^2(t) - t\phi(t)(1 - \Phi(t))$$
$$\stackrel{sign}{=} 1 - \frac{t(1 - \Phi(t))}{\phi(t)}.$$

It is not difficult to show that, for all $t > 0$,

2.4 Ordering and Ageing Properties

$$\left(\frac{1}{t} - \frac{1}{t^3}\right) \leqslant \frac{1 - \Phi(t)}{\phi(t)} \leqslant \frac{1}{t}.$$

This immediately gives that $r'(t) \geqslant 0$. So, the failure rate of Y is increasing. However, the ageing intensity function is obtained as

$$L(t) = \frac{-(t+4)\phi(t)}{(1 - \Phi(t))\ln[2(1 - \Phi(t))]}, \quad t > 0,$$

which is not monotone because it can be seen that $L(1) \cong 6.64330$, $L(37.75) \cong 2.19747$ and $L(38.186) \cong 2.23845$. They have noted that a similar property is observed in the class having decreasing failure rate. Let us consider a linear decreasing failure rate function which starts from 0.8 at $t = 0$ and decreases linearly to 0.62 at the point $t = 0.6$, and then becomes constant for all $t \geqslant 0.6$. Some algebraic calculation gives the corresponding ageing intensity function as

$$L(t) = \begin{cases} \frac{8-30t}{8-15t}, & \text{if } 0 \leqslant t \leqslant 0.06 \\ \frac{3100t}{3100t+27}, & \text{if } t \geqslant 0.06, \end{cases}$$

which is not monotone as Fig. 2.1 shows. They have also shown that the increasing property of ageing intensity function is preserved neither by mixture, nor by convolution nor by formation of a k-out-of-n system. Further, it is shown that deceasing property of ageing intensity function may not be transmitted through mixture of distributions and formation of k-out-of-n system.

Since smaller the ageing intensity better is the system, Nanda et al. (2004) defined an ordering, called ageing intensity ordering, as under.

Definition 2.4.1 A random variable X having ageing intensity function L_X is said to be smaller than another random variable Y having ageing intensity function L_Y in the ageing intensity order (denoted by $X \leqslant_{AI} Y$) if $L_X(t) \geqslant L_Y(t)$ for all $t > 0$. □

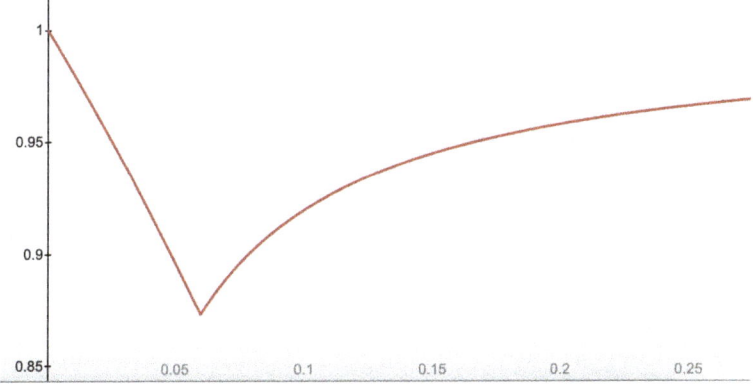

Fig. 2.1 DFR distribution having non-monotone ageing intensity function

They have noted that $X \leqslant_{AI} Y$ if and only if $\int_0^t r_X(u)\,du / \int_0^t r_Y(u)\,du$ is increasing in $t > 0$, where r_X and r_Y are, respectively, the hazard rate functions of X and Y. They have also characterized the ageing intensity ordering as under.

Theorem 2.4.2 *Let X and Y be two continuous random variables. Then $X \leqslant_{AI} Y$ if and only if $\phi(X) \leqslant_{AI} \phi(Y)$ for any strictly increasing continuous function $\phi : \mathbb{R}^+ \to \mathbb{R}^+$ with $\phi(0) = 0$.* □

Bhattacharjee et al. (2013a) have shown that a linear ageing intensity $L(t) = b + \lambda t$ characterizes a 3-parameter Weibull model having survival function given by

$$\bar{F}(t) = \exp\left\{-at^b e^{\lambda t}\right\}, \ a > 0, \ b > 0, \ \lambda \geqslant 0, \ t \geqslant 0,$$

given in Lai et al. (2003). They have also characterized a distribution in terms of ageing intensity function, which is called a 5-parameter Weibull by Pham and Lai (2007) and its survival function is given by

$$\bar{F}(t) = \exp\left\{-\lambda \left(\frac{t-a}{b-t}\right)^\beta\right\}, \ 0 \leqslant a < t < b, \ \lambda > 0, \ \beta > 0.$$

One may be interested to look into the paper by Giri et al. (2023) where quite a few different kinds of Weibull distributions, *viz.* Additive Weibull, Modified Weibull, Generalized Weibull, Generalized Modified Weibull, Exponentiated Weibull, Inverse Weibull, Exponentiated Inverse Weibull, Beta Inverse Weibull, Log-Weibull, Odd Weibull, Flexible Weibull, Extended Weibull etc. have been characterized in terms of ageing intensity functions.

2.4.7 Memory

Although the memoryless or loss-of-memory property of the exponential (and the geometric) distribution has been widely discussed, a formal definition of 'memory' of a probability distribution (particularly of a lifetime distribution), its extension to the multivariate setup and its possible uses in characterizations of distributions merit some consideration. One of the earliest references in this connection is Muth (1977) who provided a formal definition of memory and presented some related results. Subsequently, Ebrahimi and Zahedi (1992) considered memory-ordering of survival functions, Arnold and Zahedi (1988) dealt with multivariate mean residual life functions, without explicitly coming out with a multivariate extension of the memory function, and Mukherjee and Roy, in some of their papers on characterisations, referred to the memory property. Ebrahimi and Zahedi (1992) showed that failure rate ordering and memory ordering are not the same. One possible reason for the somewhat slim literature on memory is the fact that memory is derived from MRL function which has been comprehensively discussed by a host of authors for

2.4 Ordering and Ageing Properties

various purposes and that new results by using memory instead of this MRL function are rather unlikely to come up.

Muth defined virtual age at time t as $\nu(t) = m_X(0) - m_X(t)$ where $m_X(t) = E(X|X \geq t) - t$ is the mean residual life beyond age t so that $m_X(0) = E(X)$. He defines the following three types of memory:

- **Interval memory:** Considering the time interval (t_1, t_2), $t_1 < t_2$, interval memory in the interval is
$$M(t_1, t_2) = \frac{\nu(t_2) - \nu(t_1)}{t_2 - t_1}.$$

- **Local memory:** Allowing the above interval to have length tending to zero, local memory at age t is
$$m_L(t) = \frac{d\nu(t)}{dt} = -m'_X(t).$$

- **Global memory:** It is defined for a distribution, over its entire support, as a parameter
$$M_G = \int_0^\infty m_L(t) w(t)\, dt,$$

where $w(t)$ is a suitable weight function.

Muth takes the weight as $w(t) = \frac{2}{\mu^2} m_X(t) \bar{F}(t)$. It thus follows that

$$M_G = 2 - \frac{E(X^2)}{\mu^2} = 1 - C^2(X),$$

where $C(X)$ is the coefficient of variation of X. When $\nu(t) = t$, then $M(0, t) = 1$ and the distribution fully remembers the time which has elapsed and this is called perfect memory. When $\nu(t) = 0$ then $M(0, t) = 0$ and there is no memory in $(0, t)$. This way, we have perfect memory if $M(0, t) = 1$, positive memory if $M(0, t) > 0$, negative memory if $M(0, t) < 0$ and no memory if $M(0, t) = 0$.

Nair (1983) pointed out that Muth's definition of memory does not apply to discrete distributions, since $M(0, t) = 0$ does not imply geometric law. A given value of M_G does not enable identifying the underlying distribution. Even, the choice of the weight function by Muth was criticized as being arbitrary. Nair et al. (2022) defined memory from a renewal-theoretic approach which, of course, eventually yield the same results as the definition by Muth.

For the Finite-range life distribution suggested by Mukherjee and Islam (1983) with distribution function given by $F(x) = (x/a)^p$, $0 \leq x \leq a$, it can be shown that $C^2(X) = 1/(p(p+2))$ and this equals 1 if and only if $p \approx 0.4142$. Thus, the property of 'zero global memory' holds only for a specified member of this family of distributions. It may be remembered that the memoryless property is true for the exponential distributions, irrespective of the parameter value. Further, even for the shape parameter value 0.4142, loss-of-memory is observed only at one point. In fact,

Finite-range distributions do not conform to some of the relations which hold for infinite-range life distributions.

2.5 Entropy

Physicists originally developed, in the context of equilibrium thermodynamics, the notion of entropy. This was later extended through the development of statistical mechanics and information theory which was made mathematically rigorous in 1940s. Information theory deals with all theoretical problems connected to the transmission of information over communication channels and includes information processing, information storage, information retrieval and decision making. It studies the uncertainty (information) measures and various practical and economical methods of coding information for transmission. In 1865, Rudolf Julius Emanuel Clausius, one of the founders of thermodynamics, coined the term *entropy* derived from the Greek word *en-trepein* meaning energy turned to waste, although he introduced it in the context of classical thermodynamics in the year 1850. A statistical basis to entropy was given later by Ludwig Boltzmann, Willard Gibbs and James Clerk Maxwell (cf. Nanda, 2006; Nanda and Maiti, 2006). An analog to thermodynamic entropy is information entropy. Shannon (1948), an electrical engineer in Bell Telephone Laboratory and the Father of Information Age, developed a general concept of information entropy to mathematically quantify the statistical nature of lost information in phone-line signals.

At the same time, Wiener (1948) also considered the communication situation and came up, independently, with results similar to those of Shannon. Both Shannon and Wiener considered the communication situation as one in which a signal, chosen from a specified class, is to be transmitted through a channel. The basic problem of communication is to reconstruct, as clearly as possible, the input signal after observing the received signal at the output. However, the approach used by Shannon differs from that of Wiener in the nature of the transmitted signal and in the type of decision made at the receiver. The Noisy Channel Coding Theorem, the basic theorem of Information Theory, states that any message of sufficiently large length can be transmitted through a channel having capacity C at the rate $R(< C)$ and can be decoded with any desired degree of accuracy. Most of the results of this section may be available at Nanda and Chowdhury (2021).

2.5.1 Derivation of Entropy

Let us explain the concept of entropy with an example given in Nanda and Chowdhury (2019). If the probability of getting a job by a candidate is very high, the amount of unpredictability in getting the job is very less. On the other hand, very low probability of getting the job results in high level of unpredictability. Therefore, one can conclude

2.5 Entropy

that the more is the chance of getting a job, the less is the entropy. This tells that the entropy, $h(p)$, is a deceasing function of p, the probability of occurrence of an event. Further, since any small amount of additional information on the occurrence of the event should reduce the amount of uncertainty prevailing before getting the additional information, $h(p)$ must be continuous in p, with obvious condition $h(1) = 0$.

As far as the occurrence of an event E is concerned, the information to be received is either $h(p)$ or $h(1-p)$, and this remains unknown to us until the occurrence of E or E^c. Hence expected information received concerning the event E, known as entropy corresponding to E, is

$$ph(p) + (1-p)h(1-p), \ 0 < p < 1.$$

Further, for two independent events E_1 and E_2 with $P(E_i) = p_i, i = 1, 2$, the information received by the occurrence of the two events together is same as the sum of the information received when they occur separately, i.e.,

$$h(p_1 p_2) = h(p_1) + h(p_2).$$

The above properties of h together with the fact that the sets of rational numbers and the irrational numbers are dense in the set of real numbers, we see that $h(p) = -c \log_a p$, where a and c are some positive constants. Without any loss of generality, we take $c = 1$ and $a = 2$ so that $h(p) = -\log_2 p$. In case of n events with probability vector $\mathbf{p} = \{p_1, p_2, \ldots, p_n\}$, the Shannon's Entropy is obtained as

$$H(\mathbf{p}) = -\sum_{i=1}^{n} p_i \log_2 p_i,$$

with $p_i \geq 0$, $\sum_{i=1}^{n} p_i = 1$.

The original postulates proposed by Shannon are as given below.

(a) $H(p_1, p_2, \ldots, p_n)$ should be continuous in $p_i, i = 1, 2, \ldots, n$.
(b) If $p_i = \frac{1}{n}$ for all i, then H should be a monotonic increasing function of n.
(c) $H(tp_1, (1-t)p_1, p_2, \ldots, p_n) = H(p_1, p_2, \ldots, p_n) + p_1 H(t, 1-t)$ for all probability vectors $\mathbf{p} = \{p_1, p_2, \ldots, p_n\}$ and all $t \in [0, 1]$.

Alfréd Rényi (1961) has observed that different sets of postulates may characterize the Shannon's entropy. The following set of postulates given by Feinstein (1958) also generate the Shannon's entropy.

(a) $H(\mathbf{p})$ is symmetric in its arguments.
(b) $H(p, 1-p)$ is continuous in $p \in [0, 1]$.
(c) $H\left(\frac{1}{2}, \frac{1}{2}\right) = 1$.
(d) $H(tp_1, (1-t)p_1, p_2, \ldots, p_n) = H(p_1, p_2, \ldots, p_n) + p_1 H(t, 1-t)$, for all probability vectors $\mathbf{p}$ and all $t \in [0, 1]$.

2.5.2 Limitations of Entropy and its Generalizations

Shannon's entropy has been used by different people in different contexts. However, this is also not free from drawbacks which are reported by Awad (1987). By considering the probability distribution $\mathbf{p} = \{p_1, p_2, p_3\} = \{0.25, 0.25, 0.5\}$, Awad has noted that although $p_1 \neq p_3$, their contribution in developing entropy is the same. This means that different probability may contribute to the same entropy, which is not a good property one measure should possess. He has also noted by taking

$$\mathbf{p} = \{0.5, 0.125.0.125, 0.125, 0.125\} \quad \text{and} \quad \mathbf{q} = \{0.25, 0.25, 0.25, 0.25\}$$

that $H(\mathbf{p}) = H(\mathbf{q})$ although $\mathbf{p} \neq \mathbf{q}$. This indicates some identifiability problem in the definition of Shannon's entropy. Further, a measure should not change its value when a characteristic is measured in different units. In case of a continuous random variable X, by writing $Y = a + bX$, with a and b non-zero constants, we get that $H(Y) \neq H(X)$ (see below the definition of entropy for continuous random variables). Sometimes Shannon's entropy can also be negative (which does not make sense as far as its purpose is concerned). The above mentioned limitations are removed by Awad by suggesting a different entropy, known as Sup-entropy, defined by

$$A_n(\theta) = -\sum_{i=1}^{n} E\left[\log\left(\frac{f(X_i; \theta)}{\delta}\right)\right],$$

where $\delta = \sup_{x_i} f(x_i; \theta)$.

For continuous random variables, different entropies are defined analogously by replacing sum by integral. Shannon's and Rényi's entropies for continuous random variable X are defined as

$$H(X) = -\int_{-\infty}^{\infty} f(x) \log f(x) dx$$

and

$$H_\alpha(X) = \frac{1}{1-\alpha} \log \int_{-\infty}^{\infty} f^\alpha(x) dx, \quad \alpha(\neq 1) > 0,$$

respectively. Since changing log with base 2 to log with any other base is equivalent (in the sense that one is a constant multiple of the other), people generally take natural logarithm in defining entropy.

Khinchin (1957) has considered an entropy as

$$H_g(f) = \int_{-\infty}^{\infty} f(x) g(f(x)) dx,$$

2.5 Entropy

for any convex function g with $g(1) = 0$, which generalizes Shannon's entropy because $g(x) = -\log x$ gives Shannon's entropy. A more general entropy, known as (h, ϕ)-entropy, has been defined by Pardo et al. (1995). They define

$$H_\phi^h(X) = h\left(\int_{-\infty}^{\infty} \phi(f(x))dx\right),$$

where $\phi : \mathbb{R}^+ \to \mathbb{R}$ is concave and $h : \mathbb{R} \to \mathbb{R}$ is increasing and concave or, ϕ is convex and h is decreasing and concave. Shannon's entropy ($h(x) = x$, $\phi(x) = -x \log x$), Rényi entropy $\left(h(x) = \frac{1}{1-\alpha} \log x, \phi(x) = x^\alpha\right)$ and some other entropies available in the literature can be obtained as particular cases of the above (h, ϕ)-entropy.

By writing $F(x) = p$, we get

$$H(X) = \frac{1}{1-0} \int_0^1 \log\left(\frac{dF^{-1}(p)}{dp}\right) dp,$$

which is estimated by

$$H_{mn} = \frac{1}{n} \sum_{i=1}^{n} \log\left(\frac{n}{2m}\left(x_{(i+m)} - x_{(i-m)}\right)\right)$$

by Vasicek (1976) and used for testing normality, which has been reused by different researchers in other kind of testing problems. Whenever we have additional information on the life of any system, Shannon's entropy may not be a good measure. In this case, one needs to take the remaining life of the system $X_t = [X - t | X > t]$, in place of X, to construct a suitable measure. This has been reported by Ebrahimi and Pellerey (1995) who have defined a modified version, suitable for a used system, as

$$H(X; t) = -\int_t^\infty \frac{f(x)}{\bar{F}(t)} \log\left(\frac{f(x)}{\bar{F}(t)}\right) dx,$$

which can equivalently be written as

$$H(X; t) = 1 - \frac{1}{\bar{F}(t)} \int_t^\infty f(x) \log r_F(x) dx,$$

where $r_F(\cdot)$ is the corresponding failure rate function. This is known as Residual Entropy. Corresponding discrete version is denoted by $H^d(X; k)$, and is defined as

$$H^d(X; k) = -\sum_{i=k}^{\infty} \frac{p_i}{\bar{P}(k)} \ln\left(\frac{p_i}{\bar{P}(k)}\right),$$

where $p_i = P(X = i)$ and $\bar{P}(k)$ is the corresponding survival function. Under what condition, an absolutely continuous (resp. a discrete) distribution is uniquely determined by $H(X; t)$ (resp. $H^d(X; k)$) is given below, which is due to Belzunce et al. (2004). This result corrects mistakes in Ebrahimi (1996) and Rajesh and Nair (1998).

Theorem 2.5.1 *If X has an absolutely continuous (resp. a discrete) distribution with an increasing residual entropy $H(X; t)$ (resp. $H^d(X; k)$), then the underlying distribution is uniquely determined.* □

Below is one counterexample which shows that the condition '$H^d(X; k)$ is increasing' in the above theorem cannot be dropped.

Counterexample 2.5.1 *Let $X \sim B(p)$, Bernoulli distribution with success probability p. Then*

$$H^d(X; k) = \begin{cases} -q \log q - p \log p, & \text{if } k = 0 \\ 0, & \text{if } k = 1, \end{cases}$$

where $q = 1 - p$. Here $H^d(X; k)$ is decreasing in k, and $H^d(X; k)$ gives that $X \sim B(p)$ or $B(q)$.

2.5.3 Some Important Entropy-Expressions for Different Distributions

The Rényi entropy for a non-negative random variable having density f is defined as

$$H(X, \beta) = \frac{1}{1 - \beta} \ln \int_0^\infty f^\beta(x)\,dx, \ \beta \neq 1, \ \beta > 0,$$

whereas the residual Rényi entropy is defined as

$$H(X, \beta; t) = \frac{1}{1 - \beta} \ln \int_t^\infty \left(\frac{f(x)}{\bar{F}(t)}\right)^\beta dx, \ \beta \neq 1, \ \beta > 0,$$

where $\bar{F}$ is the corresponding survival function. Different properties of $H(X, \beta; t)$ for proportional hazards model have been discussed in Gupta and Nanda (2002), whereas characterizations of different univariate distributions through residual Rényi entropy have been considered by Belzunce et al. (2004).

By writing $\dot{H}(X, 1) = \lim_{\beta \to 1} \frac{d}{d\beta} H(X, \beta)$, Song (2001) has noted that $\dot{H}(X, 1)$ is the negative half of the variance of the loglikelihood. Since $\mathcal{S}_f = -2\dot{H}(X, 1)$ is independent of location and scale, we can use $\mathcal{S}_f$ as a measure of intrinsic shape of the distribution (cf. Bickel and Lehmann, 1975). We generally use the fourth-order central moment of a distribution to measure tail-heaviness. However, this cannot be used for a distribution for which fourth-order moment does not exist. In such cases, $\mathcal{S}_f$ can be used to measure tail-heaviness of such distributions. For distribution of a

2.5 Entropy

unit which has survived for some t units of time, one can define a measure analogous to $\mathcal{S}_f$ as

$$\mathcal{S}_f(t) = -2\dot{H}(X, 1; t) = -2 \lim_{\beta \to 1} \frac{d}{d\beta} H(X, \beta; t).$$

Below, we give certain expressions for $H(X, \beta; t)$ and hence of $\mathcal{S}_f(t)$ which may be obtained in Nanda and Maiti (2007).

For a uniform random variable with support (a, b), one can easily get $H(X, \beta; t) = \ln(b - t)$ with $\mathcal{S}_f(t) = 0$ for all t.

For Folded-normal distribution having density

$$f(x) = \frac{2}{\sqrt{2\pi}\sigma} \exp\left(-\frac{(x-\mu)^2}{\sigma^2}\right), \quad x > \mu \geq 0,$$

the residual Rényi entropy is given by

$$H(X, \beta; t) = \frac{1}{1-\beta} \left[(1-\beta)\frac{1}{2}\ln\left(\frac{\pi\sigma^2}{2}\right) - \beta\ln\left(2\left\{1 - \Phi\left(\frac{t-\mu}{\sigma}\right)\right\}\right) + \ln\left(2\left\{1 - \Phi\left(\frac{\sqrt{\beta}(t-\mu)}{\sigma}\right)\right\}\right) - \frac{1}{2}\ln\beta\right],$$

where $\Phi(\cdot)$ denotes the cdf of $N(0, 1)$. This immediately gives us

$$\mathcal{S}_f(t) = \frac{1}{2} + \frac{A(1)A''(1) - (A'(1))^2}{A^2(1)},$$

where

$$A(1) = 2\left\{1 - \Phi\left(\frac{t-\mu}{\sigma}\right)\right\},$$

$$A'(1) = -\frac{t-\mu}{\sqrt{2\pi}\sigma} \exp\left(-\frac{(t-\mu)^2}{2\sigma^2}\right),$$

$$A''(1) = \frac{t-\mu}{2\sigma\sqrt{2\pi}} \exp\left(-\frac{(t-\mu)^2}{2\sigma^2}\right) \left\{1 + \frac{(t-\mu)^2}{\sigma^2}\right\}.$$

The residual Shannon entropy is obtained as

$$H(X; t) = \ln \bar{F}(t) + \frac{1}{2} \ln\left(\frac{\pi\sigma^2}{2}\right) + \frac{\Gamma\left(\frac{3}{2}; \frac{1}{2}\left(\frac{t-\mu}{\sigma}\right)^2\right)}{\sqrt{\pi}.\bar{F}(t)},$$

where

$$\bar{F}(t) = 2\left[1 - \Phi\left(\frac{t-\mu}{\sigma}\right)\right], \quad t > \mu.$$

For the generalized exponential distribution with pdf given by

$$f(x) = \frac{\alpha}{\theta}\left(1 - e^{-(x-\mu)/\theta}\right)^{\alpha-1} e^{-(x-\mu)/\theta}, \quad x > \mu > 0, \; \alpha, \theta > 0,$$

the residual Rényi entropy is given by

$$H(X, \beta; t) = \frac{1}{1-\beta}\left[-\beta \ln \bar{F}(t) + \beta \ln(\alpha/\theta) + \ln B\left(\beta(\alpha-1)+1, \beta; 1 - e^{-t/\theta}\right)\right],$$

where

$$\bar{F}(t) = 1 - \left(1 - e^{-t/\theta}\right)^{\alpha} \quad \text{and} \quad B(a, b, c) = \frac{\Gamma(a+b)}{\Gamma(a)\Gamma(b)} \int_c^{\infty} x^{a-1}(1-x)^{b-1} dx.$$

The corresponding shape measure for a used item is calculated as

$$\mathcal{S}_f(t) = \dot{\psi}\left(1; 1 - e^{-\frac{t}{\theta}}\right) + (\alpha - 1)^2 \dot{\psi}\left(\alpha; 1 - e^{-\frac{t}{\theta}}\right) - \alpha^2 \dot{\psi}\left(\alpha + 1; 1 - e^{-\frac{t}{\theta}}\right).$$

The corresponding residual Shannon entropy is obtained as

$$H(X; t) = \ln \bar{F}(t) - \ln(\alpha/\theta) - \psi\left(1; 1 - e^{-t/\theta}\right) - (\alpha - 1)\psi\left(\alpha; 1 - e^{-t/\theta}\right) + \alpha\psi\left(\alpha + 1; 1 - e^{-t/\theta}\right),$$

where $\psi(\cdot; \cdot)$ is the di-gamma function given by $\psi(p; x) = (\partial/\partial p) \ln \Gamma(p; x)$ with $\Gamma(p; x) = \int_x^{\infty} w^{p-1} e^{-w} dw$, and $\dot{\psi}(p; x) = (\partial/\partial p)\psi(p; x)$.

For Folded-Cauchy distribution with pdf given by

$$f(x) = \frac{2}{\pi\sigma\left\{1 + \left(\frac{x-\mu}{\sigma}\right)^2\right\}}, \quad x > \mu \geq 0, \; \sigma > 0,$$

the residual Rényi entropy is given by

$$H(X, \beta; t) = \frac{1}{1-\beta}\left[-\beta \ln \bar{F}(t) + (1-\beta) \ln\left(\frac{\sigma}{2}\right) - \beta \ln \pi + \ln B\left(\frac{1}{2}, \beta - \frac{1}{2}; t_1\right)\right],$$

where

$$\bar{F}(t) = 1 - \frac{2}{\pi} \tan^{-1}\left(\frac{t-\mu}{\sigma}\right) \quad \text{and} \quad t_1 = \left(\frac{t-\mu}{\sigma}\right)^2 \left(1 + \left(\frac{t-\mu}{\sigma}\right)^2\right)^{-1}.$$

The corresponding shape measure for a used item is obtained as

$$\mathcal{S}_f(t) = \dot{\psi}\left(\frac{1}{2}; t_1\right) - \dot{\psi}(1; t_1),$$

while the expression for residual Shannon entropy is

2.5 Entropy

$$H(X;t) = \ln \bar{F}(t) + \ln\left(\frac{\pi\sigma}{2}\right) - \psi\left(\frac{1}{2};t_1\right) + \psi(1;t_1).$$

For Lévy distribution having density

$$f(x) = \left(\frac{\sigma}{2\pi}\right)^{1/2} \frac{1}{(x-\mu)^{3/2}} e^{-\frac{\sigma}{2(x-\mu)}}, \quad x > \mu \geq 0,$$

the residual Rényi entropy is given by

$$H(X,\beta;t) = \ln\left(\frac{\sigma}{2}\right) + \frac{1}{1-\beta}\left[-\beta\ln\bar{F}(t) - \frac{\beta}{2}\ln\pi - \left(\frac{3\beta}{2}-1\right)\ln\beta + \ln\left\{\Gamma\left(\frac{3\beta}{2}-1\right) - \Gamma\left(\frac{3\beta}{2}-1;\frac{\beta\sigma}{2(t-\mu)}\right)\right\}\right],$$

where

$$\bar{F}(t) = 2\left[\Phi\left(\sqrt{\frac{\sigma}{2(t-\mu)}}\right) - \frac{1}{2}\right].$$

The corresponding shape measure for a used item is given by

$$S_f(t) = -\frac{5}{2} + \frac{a(t)a_2(t) - (a_1(t))^2}{a^2(t)},$$

where

$$a(t) = \int_0^{\frac{\sigma}{2(t-\mu)}} e^{-w} w^{-1/2} dw,$$

$$a_1(t) = \frac{3}{2}\int_0^{\frac{\sigma}{2(t-\mu)}} e^{-w} w^{-1/2} \ln w \, dw + \sqrt{\frac{\sigma}{2(t-\mu)}} \cdot e^{-\frac{\sigma}{2(t-\mu)}},$$

$$a_2(t) = \frac{9}{4}\int_0^{\frac{\sigma}{2(t-\mu)}} e^{-w} w^{-1/2} \ln^2 w \, dw + \sqrt{\frac{\sigma}{2(t-\mu)}} \cdot e^{-\frac{\sigma}{2(t-\mu)}}\left\{-\frac{\sigma}{2(t-\mu)} + 3\ln\frac{\sigma}{2(t-\mu)} - \frac{1}{2}\right\}$$

and the residual Shannon entropy is obtained as

$$H(X;t) = \ln\bar{F}(t) + \frac{1}{2}\ln\frac{\pi e\sigma^2}{4} - \frac{3}{2}\left[\frac{\psi\left(\frac{1}{2}\right)\Gamma\left(\frac{1}{2}\right) - \Gamma\left(\frac{1}{2};\frac{\sigma}{2(t-\mu)}\right)\psi\left(\frac{1}{2};\frac{\sigma}{2(t-\mu)}\right) - \sqrt{\frac{\sigma}{2(t-\mu)}}e^{-\frac{\sigma}{2(x-\mu)}}}{\Gamma\left(\frac{1}{2}\right) - \Gamma\left(\frac{1}{2};\frac{\sigma}{2(t-\mu)}\right)}\right].$$

For beta distribution with parameters (p,q), we have the residual Rényi entropy

$$H(X,\beta;t) = \frac{1}{1-\beta}[-\beta\ln B(p,q;t) + \ln\Gamma(\beta(p-1)+1;t) + \ln\Gamma(\beta(q-1)+1;t) - \ln\Gamma(\beta(p+q-2)+2;t)],$$

which gives the shape measure for a used item as

$$S_f(t) = (p-1)^2\dot{\psi}(p;t) + (q-1)^2\dot{\psi}(q;t) - (p+q-2)^2\dot{\psi}(p+q;t).$$

The residual Shannon entropy as obtained as

$$H(X;t) = \ln B(p,q;t) - (p-1)\psi(p;t) - (q-1)\psi(q;t) + (p+q-2)\psi(p+q;t).$$

For a Finite-range distribution with density

$$f(x) = \frac{\alpha}{\gamma}\left(1 - \frac{x}{\gamma}\right)^{\alpha-1}, \quad \alpha > 1, \ 0 \le x \le \gamma < \infty,$$

the residual Rényi entropy is given by

$$H(X,\beta;t) = \ln\left(1 - \frac{t}{\gamma}\right) + \frac{1}{1-\beta}\left[\beta\ln\frac{\alpha}{\gamma} + \ln\gamma - \ln(\beta(\alpha-1)+1)\right],$$

which gives the shape measure of a used item as

$$\mathcal{S}_f(t) = \left(1 - \frac{1}{\alpha}\right)^2.$$

This is same as the shape measure for the life distribution of a new item $\mathcal{S}_f$. The residual Shannon entropy is obtained as

$$H(X;t) = \ln\left(1 - \frac{t}{\gamma}\right) + \left(1 - \frac{1}{\alpha}\right) - \ln\frac{\alpha}{\gamma}.$$

For Generalized Gamma distribution having density

$$f(x) = \frac{c\sigma^a}{\Gamma(a)} x^{ca-1} e^{-\sigma x^c}, \quad x > 0, a, c, \sigma > 0,$$

the residual Rényi entropy is given by

$$H(X,\beta;t) = \frac{1}{1-\beta}\left[-\beta\ln\Gamma(a;\sigma t^c) + \beta(\ln c + a\ln\sigma) - \ln\beta c\sigma - \frac{1}{c}\{\beta(ca-1) - c + 1\}\{\ln\beta + \ln\sigma\}\right]$$
$$+ \frac{1}{1-\beta}\left[\ln\Gamma\left(\frac{\beta(ca-1)+1}{c}; \beta\sigma t^c\right)\right].$$

which gives the shape measure of a used item as

$$\mathcal{S}_f = \frac{2}{c} - a + \left(a - \frac{1}{c}\right)^2 \dot\psi(a).$$

This gives the residual Shannon entropy as

$$H(X;t) = \ln\Gamma(a;\sigma t^c) - \ln c - \frac{1}{c}\ln\sigma + a - \left(a - \frac{1}{c}\right)\psi(a;\sigma t^c) + \frac{(\sigma t^c)^a e^{-\sigma t^c}}{\Gamma(a;\sigma t^c)}.$$

2.6 Concavity and Log-Concavity

One can easily obtain the expressions corresponding to different measures for Gamma distribution by putting $c = 1$, whereas those for Weibull distribution can be obtained by putting $a = 1$.

The expression for the residual Rényi entropy for Inverse Rayleigh distribution having density

$$f(x) = 2\theta^2 x^{-3} e^{-\theta^2/x^2}, \quad x > 0, \theta > 0$$

is obtained as

$$H(X, \beta; t) = \frac{1}{1-\beta}\left[-\beta \ln \bar{F}(t) - (1-\beta)\left(\ln\left(\frac{2}{\theta\beta^{3/2}}\right)\right) - \ln \beta + \ln\left\{\Gamma\left(\frac{3\beta-1}{2}\right) - \Gamma\left(\frac{3\beta-1}{2}; \frac{\beta\theta^2}{t^2}\right)\right\}\right],$$

where $\bar{F}(t) = 1 - e^{-\theta^2/t^2}$. The residual Shannon entropy is obtained as

$$H(X; t) = 1 + \ln \bar{F}(t) + \ln(\theta/2) - \frac{3}{2\bar{F}(t)}\left[\psi(1) - (1 - \bar{F}(t))\left\{\psi\left(1; \theta^2/t^2\right) + 2\theta^2/(3t^2)\right\}\right].$$

The corresponding shape measure for a used item is obtained as

$$S_f(t) = -2 + \frac{\bar{F}(t)h_1(t) - h^2(t)}{\bar{F}^2(t)},$$

where

$$h(t) = \frac{3}{2}\int_0^{\frac{\theta^2}{t^2}} e^{-w} \ln w \, dw + \frac{\theta^2}{t^2} e^{-\frac{\theta^2}{t^2}}$$

$$h_1(t) = \frac{9}{4}\int_0^{\frac{\theta^2}{t^2}} e^{-w} \ln^2 w \, dw + \frac{\theta^2}{t^2} e^{-\frac{\theta^2}{t^2}}\left(3 \ln \frac{\theta^2}{t^2} - \frac{\theta^2}{t^2}\right).$$

2.6 Concavity and Log-Concavity

A distribution function F with support over $\mathbb{R}^+$ is said to be concave if, for $\alpha \in (0, 1)$ and $0 \leqslant x, y < \infty$,

$$F(\alpha x + (1-\alpha)y) \geqslant \alpha F(x) + (1-\alpha)F(y).$$

If the pdf exists, then the distribution function is concave if and only if the corresponding pdf is non-increasing. For example, the exponential and the half-normal distributions are concave. A distribution function F with support over $\mathbb{R}^+$ is said to be log-concave if

$$F(\alpha x + (1-\alpha)y) \geqslant F^\alpha(x) F^{1-\alpha}(y),$$

for $\alpha \in (0, 1)$ and $0 \leqslant x, y < \infty$. Sengupta and Nanda (1999) have shown that the above definition is equivalent to the concavity of $\log F$ over $\mathbb{R}^+$, if $\log 0$ is defined as $-\infty$. It is easy to see that a concave life distribution is also log-concave. Further, if the pdf exists, then log-concavity of the density (also known as the PF_2, Polya frequency function of order 2, property) implies log-concavity of the distribution. A wide range of distributions happen to have log-concave distribution functions (cf. Finner and Roters, 1993). These include two-parameter Weibull, Gamma, and Makeham families of distributions (for all positive values of the respective shape parameters), Pareto, Log-normal, and Linear Failure Rate distributions. Log-concavity of the density function has been a subject of considerable interest (cf. Das Gupta and Sarkar, 1984). It is well known that distributions with log-concave density happen to be IFR. Sengupta and Nanda (1999) have shown that such distributions are also log-concave. They have also shown that if X is a non-negative random variable, having distribution function F, then (a) F is log-concave if and only if the distribution of $_tX = (t - X | X \leqslant t)$ is stochastically increasing in t over the support, (b) If the reversed hazard rate of X exists, then F is log-concave if and only if the reversed hazard rate is decreasing (non-increasing) in x.

In contrast to the log-concavity of the distribution functions, log-concavity of the survival function $\bar{F}$ corresponds to the IFR class. Although an ageing interpretation of the class of log-concave distributions would be rather far-fetched, the log-concavity property allows one to derive inequalities analogous to those for the IFR class. It is easy to see that the concave life distributions are closed under mixture and also closed under the formation of series system which can be seen by noting the fact that product of non-increasing convex functions is non-increasing convex. Further, if the distribution F of a non-negative random variable X is concave so is the distribution of its remaining life X_t at any time point t.

2.7 A Reference to Life Tables

Much earlier than the formal development of reliability analysis in respect of manufactured products was introduced, demographers had been constructing life tables for living beings. Following up a cohort of l_0 births subjected to a mortality pattern in terms of prevailing death rates, a life table is constructed with columns which correspond to different properties of a lifetime distribution. While a set of age-specific mortality rates provides the basis for developing a life table, reliability analysis starts from the probability distribution assumed for time-to-failure (which is the same as length of life for a non-repairable continuous-duty equipment). This distribution is specified by the cdf F.

The ratio $l_x / l_0 = {}_xp_0$ in a life table, where l_x stands for the number of persons surviving up to exact age x, is the analogue of $\bar{F}(x) = 1 - F(x)$. The failure rate at age x is thus same as $r_x = -dl_x/dx$, called the instantaneous rate of mortality or the force of mortality at age x by demographers. In this context, we find that the curve of human mortality in terms of the plot of r_x against x resembles a bathtub, the

2.7 A Reference to Life Tables

force of mortality initially decreasing till about age 10, thereafter remaining more or less constant till middle life or age 45 (or 50) and finally increasing rather rapidly at higher ages.

The complete expectation of life at age x denoted as e_x^0 is just the (conditional) mean remaining life at age x. This is what is called MRL at age x in reliability analysis. Thus, the mean age at death of people who have survived up to a certain age is higher than the average age at death in the general population (including those who die before this age). In the context of a life table, this is because $x + e_x^0$ is larger than e_0^0. In fact, the former measures what demographers have been referring to as vitality at age x. This measure is also used in reliability analysis, though rather infrequently, as the conditional mean remaining life plus the age survived. Complete expectation of life at birth (age 0) defines $E(X)$ where X stands for lifetime.

As pointed out earlier, life table measures have recently been extended to take account of morbidity (sickness due to diseases) besides mortality, reliability analysis also has come to recognize states of less-than-satisfactory performance, going beyond the classical two-state analysis.

Chapter 3
Univariate Continuous Distributions

3.1 Introduction

Univariate probability models are involved in the analysis of failure times (or survival times) in two distinct domains of application. We may be dealing with a single component or with a multi-component assembly or system. And, in the second case, we usually start with a multivariate probability distribution to model the joint behaviour of component failure times and then work out the univariate distribution of the life of some system made up of these components according to some configuration. The configuration derives system life as a function of component lives. Thus, a large part of reliability and survival analysis is based on univariate probability models with relevant properties.

As pointed out earlier, random variables involved in reliability analysis could be either discrete or continuous, depending on the nature of the underlying experiment. Quite often, we are interested in time-to-failure or time-between-consecutive failures or time-to-repair (or replace) and all these illustrate continuous random variables. However, starting with a specified number n of items (units) put on test at the same time, we may be interested in noting the number of items failing by a specified (test) time or the number of items which survive the test time before the first item failing by the test time or even the number of items (not pre-specified) required to be put on test in order to have a specified number r of failures. And whenever a count is the outcome of the experiment, we are dealing with a discrete variable. Basic to the generation of count data is an experiment where each item put on test is just dichotomized in terms of an indicator variable X which takes the value 1 if the item fails by (survives till) the test time and the value 0 otherwise.

In this chapter, we briefly consider some univariate probability models for continuous random variables arising from reliability experiments and mention their reliability properties only, without any mathematical derivation. Most of these models are commonly used in probability and statistics, while a few were developed in the reliability analysis context and have been primarily used to represent random variations in data arising from life-testing experiments or in the context of survival analysis. Some classical probability models have been generalized to widen the ambit of their applications, through incorporation of some additional parameter(s) or through exponentiation of a component of the density or the distribution function or through some modification of the original form of the density or the distribution function. In fact, recent times have witnessed a plethora of such extensions and generalizations and we have simply indicated the more important ones, and that too in a sketchy manner. We focus on the need to accommodate different types of data, and specially different behaviours of the failure rate function, rather than on mathematical sophistication.

To fit any probability model, we have to estimate the parameters in the probability model from observed data sets. In some cases, these estimators have been modified to make these unbiased. In most cases, the likelihood equations do not admit of explicit solutions and have to be solved numerically. With the current availability of efficient softwares, numerical solutions have become more convenient. In case of some probability models, the maximum likelihood estimate of some parameter can be explicitly obtained in terms of the estimate(s) of some other parameter(s). Coming to the method of moments, the use of probability-weighted moments and linear combinations of order statistics have yielded more efficient estimates compared to the classical moment method. To increase efficiency, moments of suitably chosen fractional orders have also been advocated. Besides these two widely used methods, estimation based on quantiles has also been studied in respect of certain distributions yielding explicit forms of the quantile of order p, with proper choice(s) of the order(s) of the quantile(s) to be used. The order(s) can be found out to maximize the relative efficiency of the quantile-based estimates compared to maximum likelihood estimates. And this requires the computation of the asymptotic variance-covariance matrix of the maximum likelihood estimates. Non-parametric estimates of parameters involved in a few life distributions can be obtained by using the expressions for generalized failure rate. We have avoided details of parameter estimation for the models considered in this book, except when some non-traditional methods come into play.

We begin with developments in generating distributions or families of distributions starting from some deterministic mechanism like some differential equation to be satisfied by the outcome variable of a life-testing or reliability experiment or some differential equation to be satisfied by the probability density function or the distribution function of a continuous random variable. Mixtures of probability distributions as well as compound distributions have also been briefly dealt with. While non-monotone failure distributions have been considered in this chapter, distributions with bathtub failure rate have been taken up in a separate chapter. Similarly, a separate chapter has been devoted to a discussion on finite-range life distributions. Distributions considered in the present chapter have support on the entire positive half of the real line.

3.2 Development of Probability Models

In some cases, the behaviour of a phenomenon may be, at least to a reasonable degree of approximation, represented by a simple deterministic model which can be subsequently modified to accommodate random variations exhibited by the phenomenon. It may be incidentally mentioned here that solutions of some differential equations to be satisfied by the probability density function $y = f(\cdot)$ of a random variable X with a unimodal distribution has yielded some well-known families of distributions. Thus, the normal distribution $N(\mu, \sigma^2)$ can be derived as the solution of

$$\frac{dy}{dx} = -\frac{y(x-a)}{b_0},$$

with $\mu = a$ and $\sigma = b_0$. Generalizing, we may set up the equation as

$$\frac{dy}{dx} = -\frac{y(x-a)}{b_0 + b_1 x + b_2 x^2},$$

which has resulted in the Pearsonian system of unimodal distributions with high degree of contact at both the ends. This means that all the differential coefficients vanish at the point of contact. This family of continuous probability distributions was first published by Karl Pearson in 1895 and subsequently extended by him in 1901 and 1916 in a series of articles on bio-statistics. A comprehensive account was given by Elderton (1906) in the context of graduating observed mortality data. Although the normal distribution corresponds to the choice $b_1 = b_2 = 0$, it is not generally included in the Pearsonian family. Characterized in terms of moments, different types of distributions within this family (symmetrical, skewed, infinite-range as well as finite-range) have been identified in terms of their (β_1, β_2) values. Commonly used exponential and gamma distributions are members of this family. A choice of the appropriate type is based on the sample estimates of β_1 and β_2 and subsequently parameters in the model are estimated in terms of sample moments.

A second flexible system of distributions was proposed by Johnson (1949) based on three families of transformation that translate an observed non-normal variable to the standard normal variate. The exponential, the logistic and the hyperbolic sine transformations are used to generate log-normal (S_L), Unbounded (S_U) and Bounded (S_B) distributions. Unbounded, S_U, is the set of distributions which go to infinity on both the upper tail and the lower tail. Bounded, S_B, is the set with fixed boundaries on either the upper or the lower tail or both. The transformations used for S_U and S_B are

$$z = \lambda + \delta \sinh^{-1}\left(\frac{x-\alpha}{\beta}\right) \quad \text{and} \quad z = \lambda + \delta \ln\left(\frac{x-\alpha}{\delta + \lambda - x}\right),$$

respectively, where z stands for the standard normal variate. It may be pointed out that except for the log-normal, distributions in the S_U and the S_B classes are rarely used in the context of reliability. A Johnson distribution involves two shape parameters, besides a location and a scale parameter. The family comprising these three distributions, viz. S_U, S_B and S_L along with the normal distribution covers the whole skewness-kurtosis region. Wheeler (1980) derived an alternative method of fitting Johnson distributions to data based on quantiles instead of moments.

The Burr family of 12 positively skewed, heavy-tailed distributions to represent various types of data arising in different contexts including survival analysis was developed by Burr (1942). It includes, overlaps or as a limiting case, several univariate distributions which are bell-shaped and J-shaped (but not U-shaped) like gamma, log-normal, log-logistic etc. Of great interest is Type XII Burr distribution (sometimes simply referred to as the Burr distribution) which is fairly well used in reliability analysis. The distribution functions $y = F(\cdot)$ for the distributions in this family were derived from the differential equation

$$\frac{dy}{dx} = y(1-y)g(x, y)$$

by considering different forms of the function $g(x, y)$ which is positive for y in the interval $(0, 1)$ and x in the support of $F(x)$. For example, Burr Type II distribution corresponds to the choice

$$g(x, y) = g_1(x) = ke^{-x}\frac{(1+e^{-x})^{k-1}}{(1+e^{-x})^k - 1}, \quad -\infty < x < \infty,$$

with $k > 0$. Incidentally, Burr Type I is just the Uniform distribution, Type II is the Generalized Logistic, Type III is the Dagum distribution (see Sect. 3.12), and Type X is the Pareto (or the generalized Rayleigh) distribution. Quite a few extensions and modifications of Burr distributions have been proposed over the years.

The two-parameter Burr XII distribution is similar to the log-normal distribution with a non-monotone failure rate function (see Fig. 3.1 for similarity). Its advantages are that the density and the reliability functions have closed forms and the likelihood function for censored data can also be conveniently written out. The density function is given by

$$f(x; \alpha, \beta) = \alpha\beta x^{\beta-1}\left(1+x^\beta\right)^{-(\alpha+1)}, \quad x \geqslant 0,$$

with $\alpha > 0$ and $\beta > 0$ as the two shape parameters, β playing the more important role.

The Burr Type XII distribution is unimodal with the mode at $(\beta-1)/(\alpha\beta+1)^{1/\beta}$ if $\beta > 1$ and the distribution is inverted J-shaped if $\beta \leqslant 1$. The raw moment of order r works out as

$$E(X^r) = \beta\Gamma\left(\beta - \frac{r}{\alpha}\right)\cdot\frac{\Gamma\left(\frac{r}{\alpha}+1\right)}{\Gamma(\beta+1)},$$

if $\alpha, \beta > 1$.

3.2 Development of Probability Models

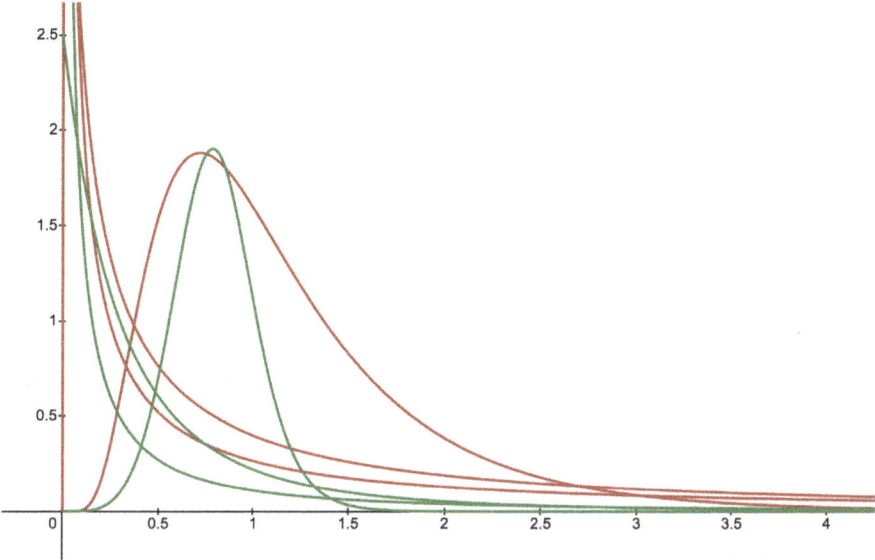

Fig. 3.1 Different shapes of Burr Type XII distribution with $\alpha = 2$ (plotted in green) and for $\beta = 0.5, 1, 5$, and those of log-normal distribution (plotted in red) with $\mu = 0$ and $\sigma = 0.25, 1, 1.5$.

The log-logistic distribution corresponds to the special case $\alpha = 1$ (see Sect. 3.10.2). The reliability function has the form

$$\bar{F}(x) = 1 - F(x) = \left(1 + x^\beta\right)^{-\alpha}.$$

The failure rate function is either decreasing or upside down bathtub-shaped. In fact, this distribution is better suited for representing situations where failure times occur with lesser frequencies compared to models with exponential tails.

Isaic and Voda (2008) derived the one-parameter Burr-Hutke exponential distribution from the generalized Burr-Hutke differential equation, given by the distribution function

$$F(x) = 1 - \frac{e^{-\alpha x}}{1 + x}, \quad x \geqslant 0,$$

having a decreasing failure rate

$$r(x) = \alpha + \frac{1}{1 + x}.$$

This was extended to a two-parameter distribution by Yousof et al. (2018) with a survival function given by

$$\bar{F}(x) = \frac{e^{-(\alpha x)^\beta}}{x+1}, \quad x \geq 0.$$

Another two-parameter version was proposed by El-Morshedy (2021) with the distribution function given by

$$F(x) = \left(1 - \frac{e^{-\alpha x}}{1+x}\right)^\beta$$

with a somewhat complex form of the failure rate function. A versatile inverse Power Burr-Hatke distribution (cf. Abdelaziz et al., 2024) with the density function given by

$$f(x) = \frac{\eta e^{-\alpha x^{-\eta}}[\alpha + (\alpha+1)x^\eta]}{x(x^\eta + 1)^2}, \quad x > 0, \ \alpha, \eta > 0,$$

has a failure rate function which can be increasing or unimodal or decreasing and the distribution curve which can be positively or negatively skewed or inverted J-shaped. The distribution was found to fit a data set on strength of glass fibres quite well.

3.3 Generating Classes of Distributions

Let X be a random variable with distribution function F and survival function $\bar{F}$. Marshall and Olkin (1997) added a parameter α, called tilt parameter, to generate a new family of distributions with survival function given by

$$\bar{G}(x; \alpha) = \frac{\alpha \bar{F}(x)}{1 - (1-\alpha)\bar{F}(x)}, \quad \alpha > 0.$$

This immediately gives the distribution function as

$$G(x; \alpha) = \frac{F(x)}{1 - (1-\alpha)\bar{F}(x)}$$

with the corresponding density function as

$$g(x; \alpha) = \frac{\alpha f(x)}{(1 - (1-\alpha)\bar{F}(x))^2}.$$

The failure rate function has the form

$$r(x; \alpha) = \frac{r(x)}{1 - (1-\alpha)\bar{F}(x)},$$

3.3 Generating Classes of Distributions

where $r(x)$ is the failure rate of the parent distribution. The Marshall and Olkin family (henceforth written as M-O family) includes the parent distribution as the special case of $\alpha = 1$. With one more shape parameter $\tau > 0$, the M-O family has been further generalized to the Harris-Extended family (henceforth written as H-E family) with the survival function given as

$$\bar{G}(x; \alpha, \tau) = \left[\frac{\alpha \bar{F}^\tau(x)}{1 - (1-\alpha)\bar{F}^\tau(x)} \right]^{1/\tau}.$$

The failure rate function works out as

$$r(x; \alpha, \tau) = \frac{r(x)}{1 - (1-\alpha)\bar{F}^\tau(x)}.$$

When $\tau = 1$, we get the M-O family. A host of authors has come up with the M-O distributions with parent distributions like gamma, Pareto, Weibull, Rayleigh, Lomax, etc. Similarly, several authors studied the H-E family of distributions with various life distributions.

Eugene et al. (2002) used the beta distribution with shape parameters α and β to develop the beta-generated distributions. The distribution function of a beta-generated random variable X is defined as

$$G(x) = \int_0^{F(x)} b(t)\,dt,$$

where $b(\cdot)$ is the density function of the beta distribution and F is the distribution function of any other distribution. Some more discussion on this is given in Sect. 3.8.4. Quite a few distributions developed on this line have since been studied. Jones (2008) and Cordeiro and de Castro (2011) extended the beta-generated class by replacing the beta density with the density function for the Kumaraswamy distribution given by

$$b(t) = \alpha \beta x^{\alpha-1}(1-x^\alpha)^{\beta-1},\ 0 \leqslant x \leqslant 1, \alpha, \beta > 0.$$

Kumaraswamy-Weibull and Kumaraswamy-generalized gamma distributions have been used to fit some reliability data. It is to be noted that the shape of the density given by $b(t)$ takes different shapes for different values of the parameters α and β. To be specific, it becomes constant for $\alpha = \beta = 1$, increasing straight line for $\alpha = 2, \beta = 1$, decreasing straight line for $\alpha = 1, \beta = 2$. It becomes increasing concave for $1 < \alpha < 2, \beta = 1$, while decreasing concave density is obtained for $\alpha = 1, 1 < \beta < 2$. Other shapes, $viz.$, J-shaped and inverted J-shaped, bell-shaped and U-shaped, are shown in Fig. 3.2.

The transformations $G(x, \eta) = [F(x; \theta)]^a$, $a > 0$, and $\bar{G}(x; \eta) = [\bar{F}(x; \theta)]^b$, $b > 0$, define the proportional reversed hazards model (Lehmann Type I distribution) and proportional hazards model (Lehmann Type II distribution), respectively. Marshall and Olkin (2007) refer to a and b as the resilience and the frailty parame-

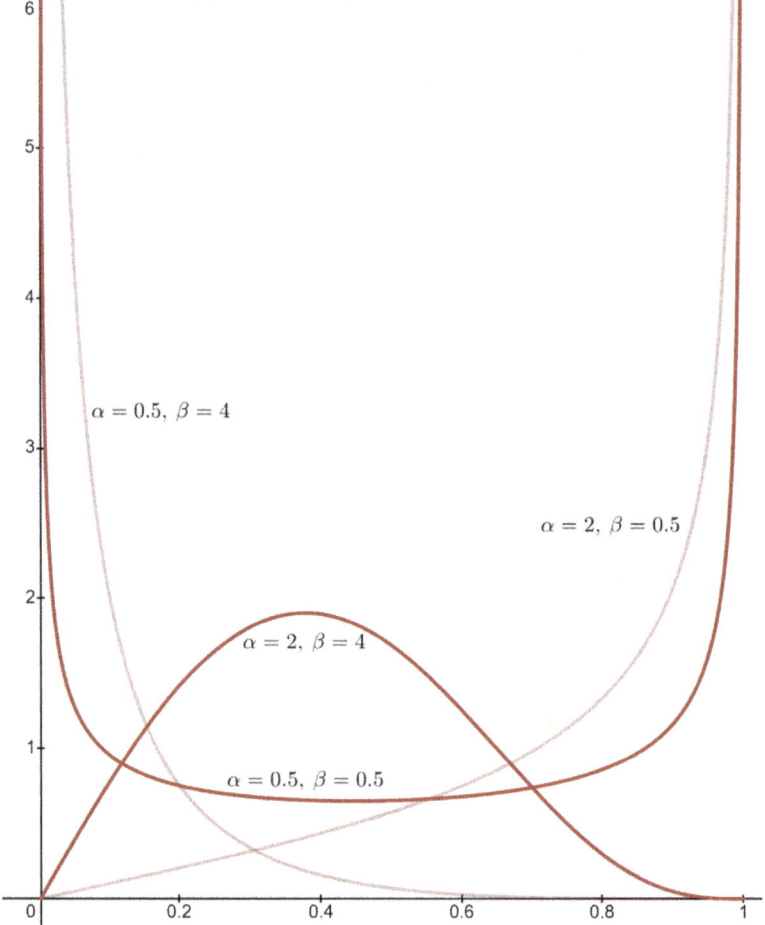

Fig. 3.2 Different shapes of Kumaraswamy distribution for different values of the parameters.

ters, respectively. A combination gives the beta-generated methods. A complete list of beta-generated distributions and an interesting review of this method for generating families of distributions can be found in Lai (2013b). Alexander et al. (2012) used the generalized beta of first type, introduced by Mcdonald (1984), to increase the flexibility of the generator. The emerging generated distribution has the density function given by

$$f(u; a, b, c) = \frac{c}{B(a, b)} u^{ac-1} (1 - u^c)^{b-1},$$

for $u \in (0, 1)$ and $a, b, c > 0$. We eventually have

$$f(x; \eta) = \frac{c}{B(a, b)} (1 - F(x; \theta))^{ac-1} \left(1 - \bar{F}^c(x; \theta)\right)^{b-1} f(x; \theta).$$

This class of distributions is called beta-exponential-X family. With $c = 1$, we get the beta-generalized distribution. The case $a = 1$ can be thought of as Reflected Kumaraswamy-generated distributions[1].

3.4 Ubiquity of the Exponential Distribution

Just as the normal distribution occupies a central place in the broad arena of Statistical Methods, particularly for statistical inference, the exponential distribution plays a pivotal role in modelling data arising in the context of reliability, safety and survival analysis. Both these distributions have been found to provide reasonably good fits to real-life data sets in diverse problems. The exponential distribution is possibly having the largest number of results characterizing it. Of course, there are quite a few interesting characterization results for the normal distribution. As will be generally accepted, characterization results which can be conveniently verified in terms of observed data or even on the basis of prior knowledge regarding the underlying phenomenon on which observations are being generated are more useful than results which are mathematically appealing but not amenable to ready verification. That way, several simple and conveniently verifiable characterization results are available for the exponential distribution.

The normal distribution can be derived from several standpoints, including a differential equation to be satisfied by the probability density function and the behaviour of points actually hit in a bull's eye experiment. The latter is intimately connected with the Theory of Errors. While the symmetric normal distribution has not been much used to model random failure time, the positively skewed log-normal distribution has found many applications in dealing with failure times of high-technology products, semi-conductors and other electronic devices, specially in modelling failures due to stress. In fact, in estimating reliability from a stress-strength analysis, stress as a random variable has often been assumed to follow a log-normal distribution. The density function for this distribution (also called Galton distribution) is given by

$$f(x; \mu, \sigma) = \frac{1}{x\sigma\sqrt{2\pi}} e^{-\frac{(\ln x - \mu)^2}{2\sigma^2}}, \ \mu \in \mathbb{R}, \ \sigma > 0, \ x > 0$$

---

[1] $Y = -X$ is said to have a reflected distribution, say Weibull, if X has a Weibull distribution.

with mean $= m = e^{\mu+\sigma^2/2}$, median $= e^\mu$ and mode $= e^{\mu-\sigma^2}$. Evidently, μ is the location parameter and σ the shape parameter. The variance works out as

$$V(X) = \left(e^{\sigma^2} - 1\right) e^{2\mu+\sigma^2} = \left(e^{\sigma^2} - 1\right) m^2.$$

Parameters of the distribution can be conveniently estimated by using probability plotting technique. The distribution has an upside down bathtub-shaped failure rate curve.

It may be of some interest to note that the Generalized Rayleigh distribution (also called Burr Type X distribution) proposed by Surles and Padgett (2001) having the distribution function

$$F(x; \alpha, \lambda) = \left(1 - e^{-\lambda x^2}\right)^\alpha, \ x > 0, \ \lambda > 0, \ \alpha > 0$$

somehow resembles the log-normal distribution (for some specific values of the parameters). The plot of log-normal and Burr Type X are given in Fig. 3.3a to see that they resemble (up to certain extent) for some values of the parameters. The failure rate function for the Burr Type X distribution is increasing if $\alpha > 1/2$ and has the shape of a bathtub if $\alpha \leqslant 1/2$ (see Fig. 3.3b). The exponential distribution was

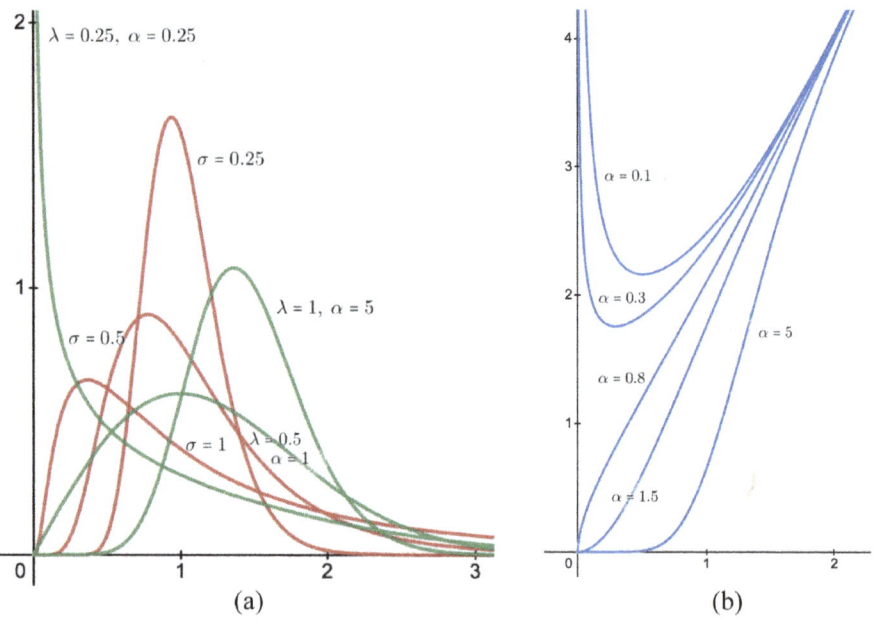

Fig. 3.3 **a** shows the shapes of the density function of log-normal distribution (in red) taking $\mu = 0$ and different values of σ, and the shapes of the density of Burr Type X distribution (in green) with different values of the shape and the scale parameters, whereas **b** shows different shapes of the failure rate function of Burr Type X distribution with $\lambda = 1$ and different values of α.

3.4 Ubiquity of the Exponential Distribution

initially being used to represent random variations in time to the next event and was simply and directly derived from the Poisson distribution or from a homogeneous Poisson process. The exponential distribution is taken as a reference distribution against which other life distributions are compared in respect of different properties and classified accordingly.

All this and more account for the fact that volumes have been devoted to comprehensive studies of different aspects of the exponential distribution. And new results are coming up every now and then. The one-parameter exponential distribution with the density and the survival functions as

$$f(x; \lambda) = \lambda e^{-\lambda x} \quad \text{and} \quad \bar{F}(x) = e^{-\lambda x},$$

respectively, with $x \geq 0, \lambda > 0$, has been very widely used in the broad domain of reliability and survival analysis, besides its use in many other contexts, e.g., in queueing models. Among its many appealing properties that explain its wide usage, the following are worth mentioning.

(a) One can conveniently introduce a location parameter μ in the density or the distribution function, to take care of situations where chance failures cannot take place prior to time μ, maybe because of debugging or burning-in for such a period to weed out 'early' or 'instantaneous' failures.

(b) It has a constant failure rate λ, independent of age x.

(c) Going further, the ageing intensity function which represents the ratio between the instantaneous failure rate and the average failure rate up to the present instant has the value unity. Both these results are unique for the exponential distribution (in the class of continuous distributions).

(d) It is characterized by the loss-of-memory property indicated by

$$\bar{F}(x+t) = \bar{F}(x)\bar{F}(t),$$

for all $x, t \geq 0$. This implies $\bar{F}^n(t) = \bar{F}(nt)$, for $n \in \mathbb{N}$.

(e) Residual (remaining) life at age x, i.e, $Y = [X - x | X > x]$ has the same exponential distribution, independent of age x survived. Hence, mean remaining life at age x, $E(X - x | X > x)$ is the same as the original mean life $E(X) = 1/\lambda$ for all x.

Median (shelf) life, moments, entropy and similar other measures have very simple expressions. The moment of order r is given by $E(X^r) = \Gamma(r+1)/\lambda^r$ and one easily gets the coefficient of variation as unity. The median works out as $Me = \ln 2/\lambda$. In a similar vein, distributions of order statistics and, that way, system life distributions can be easily worked out. In fact, the minimum of several independently and exponentially distributed random variables X_i with mean $1/\lambda_i$ is also distributed exponentially, implying that series system life with exponentially distributed independent component lives has an exponential distribution. This follows directly from the fact that

$$P(Y \geq t) = P(X_1 \geq t, X_2 \geq t, \ldots, X_n \geq t)$$
$$= \prod_{i=1}^{n} P(X_i \geq t)$$
$$= e^{-(\sum_{i=1}^{n} \lambda_i)t}.$$

This clearly tells that the failure rate of a series system is the sum of the component failure rates. It is worth mentioning here that the reversed hazard rate of a parallel system is the sum of the component reversed hazard rates. That the converse of both the above results is true, i.e., if the hazard rate (resp. reversed hazard rate) of a system is the sum of the component hazard rates (resp. reversed hazard rates) then the system is a series (resp. parallel) system, has been shown in Nanda et al. (1998). Thus, a series (resp. parallel) system is characterized in terms of the sum of the component hazard (resp. reversed hazard) rates.

Distributions of parallel and standby systems also can be easily worked out. The exponential distribution is characterized by maximum Shannon entropy (viz., 1-ln λ) among all continuous distributions with a given mean.

The exponential distribution is not the same as the Exponential Family which is a set of parametric distribution families including binomial, normal, exponential, gamma, Poisson, and several other distributions which do admit of a sufficient statistic in the context of parameter estimation. While the exponential distribution is the only continuous distribution with the loss (lack)-of-memory property, distributions with periodic failure rates exhibit what has been called in the literature as the almost-lack-of-memory (ALM) property. A distribution is said to have a periodic failure rate property with period c if $r(x+c) = r(x)$ for all $x \geq 0$ or equivalently if $r(x+nc) = r(x)$ for all $n = 0, 1, 2, \ldots$. This means that, for a period c, a distribution will satisfy $P(X \geq x+c) = P(X \geq x).P(X \geq c)$ for all x (but infinite values of c, although not for all c). This property is defined as ALM property by Chukova and Dimitrov (1992) who give necessary and sufficient conditions for a distribution to have this property. It can be seen that the waiting time associated with a counting process in the context of minimal repair policy will exhibit the periodic failure rate property. Application in monitoring environmental evolution with periodic behaviour has been reported by Dimitrov and Khalil (1990).

Fitting the exponential distribution to observed data in a complete or even a censored life-test is rather simple. In fact, for the one-parameter exponential, one can use a probability plotting paper to estimate the only parameter λ from the relation $\ln \bar{F}(t) = -\lambda t$, yielding $\bar{F}(1/\lambda) = 1/e \cong 0.367879$. The maximum likelihood estimate of λ comes out as $\hat{\lambda} = n/\sum_{i=1}^{n} x_i$ and the uniformly minimum variance unbiased estimate of λ is given by $\tilde{\lambda} = (n-1)/\sum_{i=1}^{n} x_i$. In case of a right-censored data set where, say, r failure times less than a pre-specified test-time t_0 (out of n items put on test) have been observed, the maximum likelihood estimate of the only parameter λ comes out as $n/[\sum_{j=1}^{r} t_j + (n-r)t_0]$, the denominator being the total time on test. However, estimation of a location parameter in the two-parameter case is a slightly ticklish issue which has engaged the attention of several researchers including Ghosh and Razmpour (1982) and Samaniego (1985) besides a few later workers. Some of them assumed the coefficient of variation to be known.

It may be added that many recent contributions to the field have pointed out deficiencies in fitting the exponential distribution to data which really should have been better modelled by some other distributions, not so simple as the exponential though.

3.5 Models with Monotone Failure Rate

Early attempts to look beyond the exponential distribution for representing random variations in lifetimes was motivated by the observed (estimated) behaviour of the failure rate function in many real-life situations. In relatively simple cases, this function was monotonic increasing or decreasing. And there were cases where the function was non-monotonic and, in particular, bathtub-shaped or even inverted bathtub-shaped. On the one hand, distributions with some such behaviour of the failure rate was developed, while many other developments came up with general families of distributions which could accommodate different patterns of failure rate behaviour.

While analytical tractability is a major advantage in the choice of the exponential model, its constant failure rate or the memoryless property circumscribes its application in many situations. This fact has led to many modifications, extensions and generalizations of the exponential distribution and recent decades have witnessed a deluge of published materials which deal with general families of distributions where the exponential distribution plays the role of a pivot. While representation of situations with age-dependent failure rate or memory and similar features of observed failure time data on systems with different configurations provided a motivation to go beyond the exponential distribution and seek a more general distribution that could include the exponential model as a special case, purely theoretical interest in developing a very general distribution family that could cover many other well-known distributions as special cases led some investigators to come up with families of distributions with quite a few parameters and somewhat intractable forms for some important properties of such models. Adding a parameter, exponentiating the distribution or the survival function, generating a new survival function from the existing one, transmutation, etc. are some of the mechanisms adopted to provide general families or even classes of life distributions.

3.5.1 Weibull Probability Model

Waloddi Weibull, a Swedish physicist, noted that while the exponential model was pretty good to represent random variations in failure times of many varieties of electrical and electronic component, the model could not provide a good fit to observed failure times of systems of such components. It may be incidentally noted that failure times of series, parallel and standby systems of n independent components are really given by the minimum, the maximum and the sum of component failure times, respectively. Weibull found that a simple generalization of the exponential distribution with an additional shape parameter α (λ is the scale parameter) and having the probability density function f given by

$$f(t; \alpha, \lambda) = \alpha \lambda t^{\alpha-1} e^{-\lambda t^\alpha}, \quad \alpha, \lambda, t > 0,$$

provides a satisfactory fit to observations on system lives which had age-dependent failure rates. In Fig. 3.4, we plot the above pdf for different values of α taking $\lambda = 1$. The failure rate function works out as $r(t) = \alpha \lambda t^{\alpha-1}$ which is monotonic increasing (decreasing) in t as $\alpha > (<) 1$. Thus, the Weibull distribution can be used to represent both degradation or wear-out failures as well as early failures. The ageing intensity function has a constant value α.

This generalization due to Weibull includes the exponential distribution as the special case $\alpha = 1$. In both the exponential and the Weibull distributions, one can introduce a location parameter δ (linked to the warranty time before which failures cannot take place) on replacing t by $t - \delta$.

We have seen that the Weibull distribution takes various shapes depending on the choice of the parameter values. This is unlike the monotonically diminishing expo-

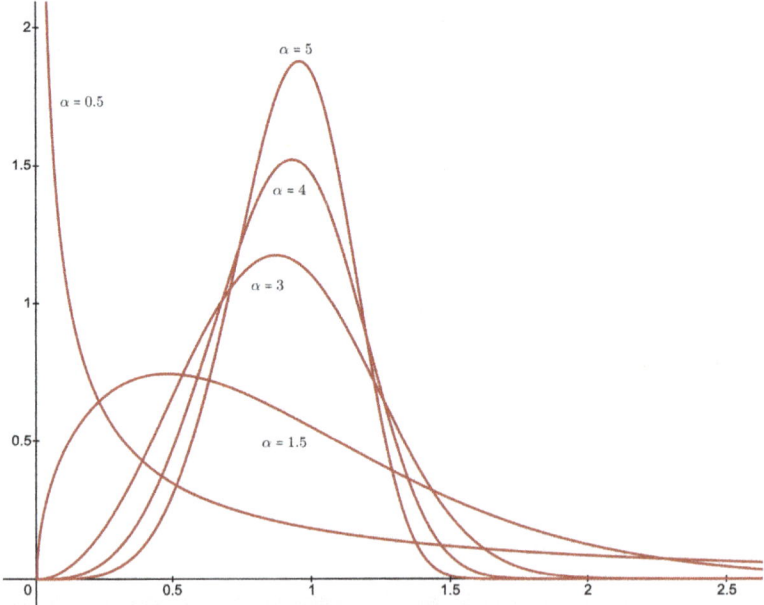

Fig. 3.4 Different shapes of the probability density functions of Weibull distribution with scale parameter $\lambda = 1$ and different values of the shape parameter α

3.5 Models with Monotone Failure Rate

nential distribution. As mentioned earlier, $\alpha = 1$ corresponds to a constant failure rate λ. The raw moment of order r works out as

$$E(X^r) = \frac{\Gamma\left(\frac{r}{\alpha}+1\right)}{\lambda^{r/\alpha}},$$

yielding the mean life as $E(X) = \Gamma\left(\frac{\alpha+1}{\alpha}\right)/\lambda^{1/\alpha}$. The median simplifies to $-\left(\frac{\ln 2}{\lambda}\right)^{1/\alpha}$. Moment measures of skewness and kurtosis are functions of the shape parameter α only.

The likelihood equations

$$\frac{n}{\alpha} + \sum_{i=1}^{n} \ln t_i - \sum_{i=1}^{n} t_i^{\alpha} \ln t_i = 0$$

$$\frac{n}{\lambda} - \sum_{i=1}^{n} t_i^{\alpha} = 0$$

do not admit of closed form solutions. In case α is known, the maximum likelihood estimator (MLE) of λ comes out simply as $\widehat{\lambda} = n/\sum_{i=1}^{n} t_i^{\alpha}$.

The asymptotic variance-covariance matrix of the MLEs is given by

$$\begin{pmatrix} \frac{6\alpha^2}{n\pi^2} & -\frac{6\alpha\lambda\Delta}{n\pi^2} \\ -\frac{6\alpha\lambda\Delta}{n\pi^2} & \frac{\lambda^2}{n}\left(1+\frac{6\Delta^2}{\pi^2}\right) \end{pmatrix}$$

where $\Delta \cong 0.42 - \ln \lambda$.

The shape parameter α is more important and $V(\widehat{\lambda}) \cong 0.6079\alpha^2/n$. Various alternative methods have been used to estimate α and λ, based on sample quantiles, sample moments, order statistics, etc.

For Weibull case, method of moments uses the fact that

$$C^2 = \frac{\Gamma\left(\frac{2}{\alpha}+1\right)}{\Gamma^2\left(\frac{1}{\alpha}+1\right)-1} = \Psi(\alpha), \text{ say,}$$

which has been graphed against α by Cohen and Whitten (1988). Equating the square of sample coefficient of variation to $\Psi(\alpha)$ we get an estimate of α.

We note that $Y = \lambda X^{\alpha}$ has the standard exponential distribution and $V(\ln Y) = \pi^2/6$. Further,

$$\alpha = \left(\frac{V(\ln X)}{V(\ln Y)}\right)^{1/2} = \left(\frac{6}{\pi^2} V(\ln X)\right)^{1/2}.$$

Based on this fact, Menon (1963) proposes the estimator

$$\widehat{\alpha} = \left[\frac{6}{\pi^2(n-1)} \left\{ \sum_{i=1}^{n} \ln^2(x_i) - \frac{1}{n} \left(\sum_{i=1}^{n} \ln x_i \right)^2 \right\} \right]^{1/2}$$

and shows that

$$\widehat{\alpha} \stackrel{a}{\sim} N\left(\alpha, \frac{11\alpha^2}{10n}\right)$$

with an asymptotic relative efficiency (ARE) of 55%. In case λ is known, Menon's estimator is

$$\widehat{\alpha} = \frac{1}{\Phi(1)n} \sum_{i=1}^{n} \ln x_i \stackrel{a}{\sim} N\left(\alpha, \frac{0.66\alpha^2}{n}\right)$$

with an ARE of 84%.

Mukherjee and Sasmal (1984) used fractional moments of orders r_1 and r_2 for estimating α and λ. To use sample moments $S_1 = \sum_{i=1}^{n} x_i^{r_1}/n$ and $S_2 = \sum_{i=1}^{n} x_i^{r_2}/n$, $0 < r_1 < r_2$, they determine the optimal choice of r_1 and r_2 such that the joint ARE of the estimators (α^*, λ^*) compared to the MLEs is maximized. The joint ARE is found to depend only on the shape parameter α as

$$\text{Joint } ARE = \frac{|Cov(\widehat{\alpha}, \widehat{\lambda})|}{|Cov(\alpha^*, \lambda^*)|} = \Phi(\alpha, r_1, r_2), \text{ say.}$$

In case $\alpha \leqslant 1$, use of fractional moments with $r_1 < 1$ and $r_2 > 1$ always results in higher efficiencies than the classical choice of $r_1 = 1$ and $r_2 = 2$.

In the IFR case, $r_1 = 0.5$ and $r_2 = 1.5$ yield the prepared results with joint ARE of more than 95% over a large range of (α, λ) values. One can consider the average

$$\int_{k_1}^{k_2} \Phi(\alpha, r_1, r_2) g(\alpha) \, d\alpha = h(r_1, r_2)$$

for maximization with respect to r_1 and r_2, where $g(\alpha)$ is the prior density of α and k_1 and k_2 are appropriate limits within which α lies.

Abe (1993) extends this method to estimation of parameters in any distribution, minimizing the mean square errors of the estimators.

The initial (ordinary) moment estimates or approximations to maximum likelihood estimates are used to indicate the optimal orders, as also to eliminate the first-order biases of the estimators. Abe also proves the uniqueness of the estimator of the shape parameter derived this way.

3.5 Models with Monotone Failure Rate

Mukherjee and Sinha (1978) suggested two different estimators based on the log-linear regression

$$\ln r(t) = (\ln \alpha + \ln \lambda) + (\alpha - 1)\ln t = A + Bt$$

and using estimates of $\ln r(t)$ on the lines of Hotelling's and Fisher's estimates of $\frac{dp(t)}{dt}/p(t)$ in connection with the fitting of logistic curve to population data. However, one has to note the problem of heteroscedasticity in $\ln r(t)$.

Dubey (1967) considered percentile estimators and showed that the estimator of α based on the 17th and the 97th percentiles yields an ARE of 66%. When both α and λ are to be estimated, 24th and 93rd percentiles correspond to the best choice with a lower ARE of 41%. Seki and Yokoyama (1996) arbitrarily took the 63rd and the 31st percentiles to estimate λ and α, respectively.

Murthy (1968) points out that $E(g(t)) = \alpha$ where $g(t) = tr(t)$. He therefore proposes

$$\widehat{\alpha} = \frac{1}{n}\sum_{i=1}^{n} t_i r_n(t_i)$$

with the density estimators $r_n(t_i) = f_n(t_i)/\bar{F}_n(t_i)$. He proves that $\sqrt{n}(\widehat{\alpha} - \alpha)$ is asymptotically normal.

Keller and Kamath (1982) used the inverse Weibull model to describe the degradation mechanism of mechanical components such as crank shafts and pistons of diesel engines as also to describe breakdown of insulating fluids. This has a density function given by

$$f(x; \alpha, \beta) = \alpha\beta x^{-(\beta-1)} e^{-\alpha x^{-\beta}}, \quad \alpha, \beta > 0, \ x > 0.$$

The plot of the density with scale parameter unity is shown in Fig. 3.5. The one-parameter Rayleigh distribution, which can be derived in the same way as a chi-square distribution with 2 degrees of freedom or can be looked upon as a Weibull distribution with shape parameter $\alpha = 2$, with the pdf given by

$$f(x; \lambda) = 2\lambda x e^{-\lambda x^2}, \quad \lambda > 0, \ x > 0,$$

has been used to represent failure times of resistors, capacitors and transformers in aircraft radar sets.

Going from the constant failure rate exponential model to the versatile, monotone failure rate Weibull model, we pass through an interesting intermediate, $viz.$, the linear failure rate model (introduced by Bain, 1974), also known as the linear exponential distribution, that admits of a situation where the failure rate at time $t = 0$ is non-zero (unlike in a Weibull model) and may be lower than the constant failure rate, typical of the exponential model. The linear failure rate model having the survival function (see also Sect. 2.4.1)

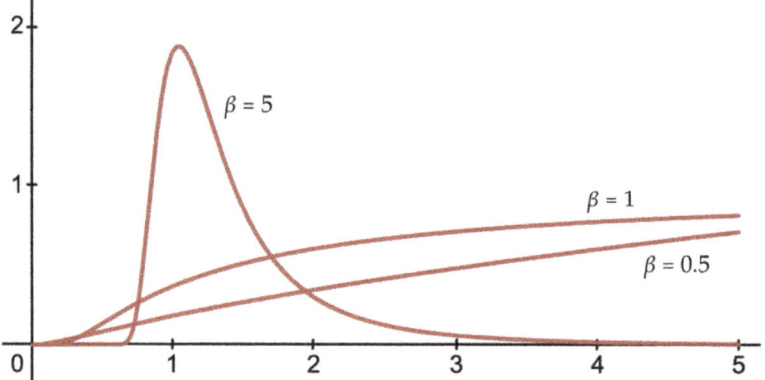

Fig. 3.5 Shapes of the inverse Weibull distribution for different values of the shape parameter β with scale parameter $\alpha = 1$

$$\bar{F}(t; \alpha, \beta) = \exp\left\{-\alpha x - \frac{1}{2}\beta x^2\right\}, \ \alpha, \beta > 0, \ x > 0,$$

includes the Rayleigh distribution as $\alpha = 0$, and the exponential distribution as $\beta = 0$. Obviously, the failure rate function has the form $r(x) = \alpha + \beta x$. Of course, none of these simple and widely used models can be applied in situations where the failure rate is non-monotonic. For some properties of linear failure rate model, one may refer to Sen (2005).

3.5.2 Gamma Distribution Model

The two-parameter gamma distribution has often been used to represent failure times, specially when a failure of the item/unit on test occurs only after several (a specified number k, say) partial failures have taken place to result in the item/unit to finally fail. This is the case when we have k redundant components in the item or unit under test. Partial failure times are taken to be exponentially distributed and the failure time being analysed is their sum.

Let X_i, $i = 1, 2, \ldots, n$, be iid random variables each following exponential distribution with failure rate λ. Then, from the MGF of the exponential distribution, we see that $X = \sum_{i=1}^{n} X_i$ has the MGF

$$M_X(t) = \prod_{i=1}^{n} M_{X_i}(t) = \left(\frac{\lambda}{\lambda - 1}\right)^n.$$

One can easily verify, from the uniqueness property of MGF, that X has the pdf given by

3.5 Models with Monotone Failure Rate

$$f_X(x) = \frac{\lambda^n}{\Gamma(n)} e^{-\lambda x} x^{n-1}, \ \lambda > 0, \ x \geqslant 0.$$

In the above pdf, $n \in \mathbb{N}$. However, if we replace n by α we get the pdf as

$$f_X(x) = \frac{\lambda^\alpha}{\Gamma(\alpha)} e^{-\lambda x} x^{\alpha-1}, \ \lambda > 0, \ x \geqslant 0. \tag{3.5.1}$$

One can easily verify that the integral

$$\int_0^\infty e^{-\lambda x} x^{\alpha-1} dx$$

converges if $\alpha > 0$. In this case, we write $X \sim Gamma(\lambda, \alpha)$ if X has the pdf given in (3.5.1) and we say that X follows gamma distribution with parameters λ (scale) and α (shape). In this case, we have $E(X) = \alpha/\lambda$, $V(X) = \alpha/\lambda^2$ and $C(X) = \sqrt{\alpha}$. The shape of the density function varies with shape parameter α, which makes it decreasing convex to bell-shaped. It is a positively skewed distribution. For $\alpha \leqslant 1$, the distribution is decreasing convex, whereas $\alpha > 1$ makes it bell-shaped unimodal distribution which is positively skewed (see Fig. 3.6).

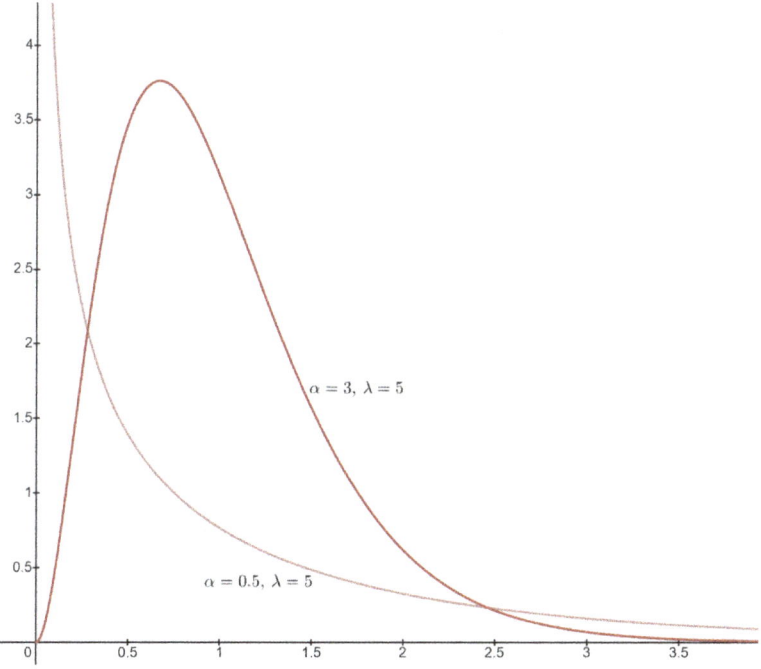

Fig. 3.6 Different shapes of gamma distribution for different values of the parameters

3.5.3 Other Probability Models

Gui (2013) has used the two-parameter Half-exponential Power distribution with density function given by

$$f(x; \lambda, \alpha) = \frac{\alpha^{1-1/\alpha}}{\lambda \Gamma(1/\alpha)} \exp\left\{-\frac{x^\alpha}{\alpha \lambda^\alpha}\right\}, \ x \geq 0, \ \alpha > 0, \ \lambda > 0,$$

for representing lifetime data. With the shape parameter $\alpha = 1$, we get the exponential distribution and with $\alpha = 2$ we get the Half-normal distribution. With $\alpha \geq 1$ the distribution has an increasing failure rate while it is decreasing for $0 < \alpha < 1$.

The moment of order k is given by

$$E(X^k) = \frac{\alpha^{k/\alpha} \lambda^k}{\Gamma(1/\alpha)} \Gamma\left(\frac{k+1}{\alpha}\right).$$

The one-parameter Lindley distribution (Lindley, 1958) which has a fairly established usage in reliability analysis besides its applications in many other aspects of statistical inference has the density function given by

$$f(x; \alpha) = \frac{\alpha^2}{1+\alpha}(1+x)e^{-\alpha x}, \ \alpha > 0, \ x \geq 0$$

and the survival function given by

$$\bar{F}(x) = \frac{\alpha + 1 + \alpha x}{1 + \alpha} e^{-\alpha x}$$

so that the failure rate function comes out as

$$r(x) = \frac{\alpha^2(1+x)}{\alpha + 1 + \alpha x},$$

which is monotonic increasing. The distribution has the coefficient of variation less than unity. Figure 3.7 shows different shapes of the Lindley distribution for different values of the parameter. The distribution can be derived as a mixture of the exponential distribution having hazard rate α and the $Gamma(2, \alpha)$ distribution with respective weights $\alpha/(1+\alpha)$ and $1/(1+\alpha)$.

Shanker (2015) proposed the Akash distribution by way of a mixture of the exponential distribution with failure rate θ and a gamma distribution with scale parameter θ and shape parameter 3, using proportions $\theta^2/(\theta^2 + 2)$ and $2/(\theta^2 + 2)$, respectively, to yield the density function

$$f(x; \theta) = \frac{\theta^3}{\theta^2 + 2}(1 + x^2)e^{-\theta x}, \ \theta, x > 0.$$

3.5 Models with Monotone Failure Rate

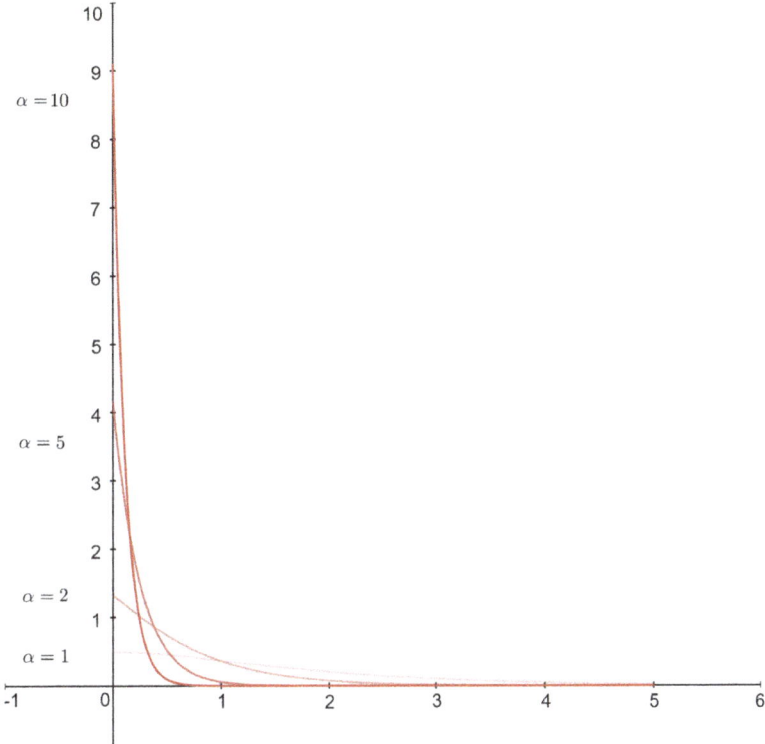

Fig. 3.7 Different shapes of the Lindley distribution for different values of the parameter α

This is more flexible than the exponential as well as the Lindley distribution. Different shapes of Akash distribution (taking $\theta = 0.5, 1, 3$) are shown in Fig. 3.8. Like the Lindley distribution, the Akash distribution also has an increasing failure rate. Shanker and Fesshaye (2016) considered several data sets, in some of which the Lindley distribution gave a better fit compared to the exponential distribution and in certain others the Akash distribution provided a better fit. They present the comparison among the three distributions given in Table 3.1. While $r(t)m(t) = 1$ for exponential distribution, it is less than 1 for the Lindley distribution.

The Pareto distribution (sometimes referred to as the Pareto Type I distribution) proposed originally in the context of income distributions has been used in reliability analysis as a distribution with slowly declining tail. With the survival function given by $\bar{F}(x) = (k/x)^\alpha$, it has a decreasing failure rate, $viz.$, $r(x) = \alpha/x$, $x \geqslant k$. The distribution has the mean $E(X) = k\alpha/(\alpha - 1)$ for $\alpha > 1$ and is not finite otherwise.

In fact, $E(X^r) = k\alpha^r/(\alpha - r)$ if $\alpha > r$. The tells that the MGF does not exist. The mode of this distribution is obviously the parameter k and it can be easily shown that the median is at $k(1/2)^{-1/\alpha}$. An interesting characteristic property of the Pareto distribution is the fact that the generalized failure rate $g(x) = x r(x)$ is a constant

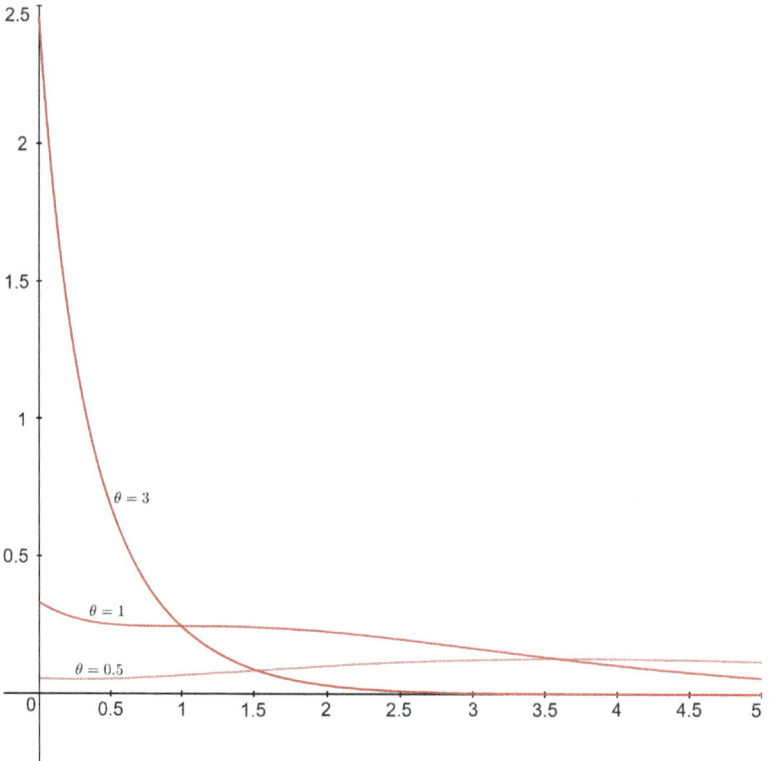

Fig. 3.8 Different shapes of Akash distribution for different values of the parameter

Table 3.1 Comparison among Exponential, Lindley and Akash distributions

Function	Exponential	Lindley	Akash
$\bar{F}(t)$	$e^{-\theta t}$	$e^{-\theta t}\left(1+\frac{\theta t}{1+\theta}\right)$	$e^{-\theta t}\left(1+\frac{\theta t(\theta t+2)}{\theta^2+2}\right)$
$r(t)$	θ	$\frac{\theta^2(1+t)}{1+\theta+\theta t}$	$\frac{\theta^3(1+t^2)}{\theta t(\theta t+2)+(\theta^2+2)}$
$m(t)$	$1/\theta$	$\frac{2+\theta+\theta t}{\theta(1+\theta+\theta t)}$	$\frac{\theta^2 t^2+4\theta t+(\theta^2+6)}{\theta[\theta t(\theta t+2)+(\theta^2+2)]}$
$E(T)$	$1/\theta$	$\frac{\theta+2}{\theta(\theta+1)}$	$\frac{\theta^2+6}{\theta(\theta^2+2)}$
$V(T)$	$1/\theta^2$	$\frac{\theta^2+4\theta+2}{\theta^2(\theta+1)^2}$	$\frac{\theta^4+16\theta^2+12}{\theta^2(\theta^2+2)^2}$

α, as mentioned in Sect. 2.4.3. The failure rate has a probability distribution with a distribution function

$$P(r(X) < t) = \left(\frac{kt}{\alpha}\right)^\alpha$$

and the MRL function has the form

3.5 Models with Monotone Failure Rate

$$m(t) = E(X|X > t) - t = \frac{t}{\alpha + 1}.$$

The parameters k and α can be estimated in several different ways. We can take the smallest sample observation or the most frequently observed value as an estimate of k and then use it to get an estimate of α by equating the sample mean to its population counterpart. A non-traditional way to estimate the shape parameter would take it as the mean of the generalized failure rate $Z = t\,r(t)$ computed at several values of t (which can allow a non-parametric estimate of $r(t)$ to be obtained).

In view of its established usefulness in representing random variables in completed length of series (in the context of labour mobility studies) and because of the analogy between the rate of wastage and the rate of failure, Pearsonian Type XII distribution (also known as Pareto Type II or Lomax distribution) has found some applications in reliability studies. This is a two-parameter heavy-tailed distribution with survival function given by

$$\bar{F}(x) = (1 + \beta x)^{-\alpha}, \quad \alpha, \beta > 0, \ x \geq 0.$$

The failure rate function works out as

$$r(x) = \frac{\alpha \beta}{1 + \beta x},$$

which is monotonic decreasing in x. As shown by Silcock (1954), this distribution can be obtained as the unconditional distribution of X following an exponential distribution with the parameter having a gamma distribution with scale parameter β and shape parameter α.

The failure rate decreases monotonically from $\alpha\beta$ at $x = 0$ to 0 as $x \to \infty$. The raw moment of order r is given by

$$E(X^r) = \frac{\Gamma(\alpha - r)\Gamma(1 + r)}{\beta^r \Gamma(\alpha)}, \quad \text{if } \alpha > r.$$

Thus, the mean comes out as $E(X) = 1/[\beta(\alpha - 1)]$ if $\alpha > 1$, undefined otherwise. The median is located at $Me = (1/\beta)(2^{1/\alpha} - 1)$. The generalized failure rate works out as $g(x) = \alpha\beta x/(1 + \beta x)$ to be estimated by using computed values of the random variable $Z = X.\hat{r}(x)$, with some non-parametric estimate of $r(x)$.

Further,

$$P(g(X) \geq x) = P\left(X \geq \frac{x}{\beta(\alpha - x)}\right)$$

with the expected value $\alpha/(\alpha - 1)$.

Parameters of the distribution may be estimated by using the sample mean and the sample median. The shape parameter α may also be estimated by equating the expected value of the generalized failure rate to the mean of Z computed at several observed failure times.

Assuming strength X and stress Y to be independently distributed as Lomax with parameters (α_1, β) and (α_2, β), respectively, it can be easily shown that

$$R = P(X > Y) = \frac{\alpha_1}{\alpha_1 + \alpha_2}.$$

For a discussion on this distribution, reference may be made to Dey et al. (2017).

Several different modifications and extensions of the Lomax distribution have been offered by research workers over the years, including exponentiated Lomax (cf. Abdul-Moniem, 2012), gamma Lomax (cf. Cordeiro et al., 2013), generalized beta Lomax (Singh and Zhang, 2022), inverse Lomax (cf. Kleiber and Kotz, 2003), Kumaraswamy-generalized Lomax (Shams, 2013), Kumaraswamy-generalized inverse Lomax (cf. Ogunde et al., 2023), Kumaraswamy-generalized power Lomax (Nagarjuna et al., 2021), Marshall-Olkin extended Lomax (cf. Gupta et al., 2007), Maxwell-Lomax (cf. Abiodun and Ishaq, 2022), McDonald-Lomax (cf. Lemonte and Cordeiro, 2013), modified Lomax (cf. Alnssyan, 2023), power Lomax (cf. Rady et al., 2016), slashed Lomax (cf. Li and Tian, 2020), Weibull-Lomax (cf. Tahir et al., 2015), and weighted Lomax (cf. Kilany, 2016) distributions.

One generalization that can accommodate increasing, decreasing and constant failure rate behaviour for different parameter values is the Gompertz-Lomax (sometimes referred to as GoLomax distribution) proposed by Oguntunde et al. (2017). This has a distribution function given by

$$F(x) = 1 - e^{(\delta/\lambda)\left(1-(1+\beta x)^{\alpha\lambda}\right)}, \; x \geqslant 0, \alpha > 0, \delta > 0, \lambda > 0.$$

The distribution has a failure rate function involving four parameters in the form

$$r(x) = \delta\alpha\beta(1+\beta x)^{\alpha\lambda - 1}.$$

The reversed hazard rate has, of course, a somewhat complicated expression. It can be easily seen that the failure rate is constant in case $\alpha\lambda = 1$, is increasing or decreasing depending on whether $\alpha\lambda$ is more than or less than 1. It may be incidentally noted that median life for this distribution has a relatively simpler form as

$$Me = \frac{1}{\beta}\left[\left(1 + \frac{\lambda}{\delta}\ln 2\right)^{\frac{1}{\alpha\lambda}} - 1\right].$$

Another compound distribution involving the Lomax distribution is the exponential Lomax distribution proposed by El-Bassiouny et al. (2015a) with the survival function given by

$$\bar{F}(x) = \exp\left[-\lambda\left(\frac{\beta}{x+\beta}\right)^{-\alpha}\right]$$

and a failure rate function as

3.5 Models with Monotone Failure Rate

$$r(x) = \frac{\alpha\lambda}{\beta}\left(\frac{\beta}{\beta+x}\right)^{-\alpha+1}.$$

The authors considered failure times of 84 aircraft windshields and found that the exponential Lomax gave a better fit to the data set compared to several other distributions, judged by criteria like AIC, BIC etc. (discussed in Chap. 9).

The log-logistic distribution, also known as the Fisk distribution used in Economics to represent wealth or income distribution, is a unimodal, positively skewed and heavy-tailed distribution of a non-negative random variable X such that $\ln X$ has the Logistic distribution commonly used in Demography. If X follows a log-logistic distribution with parameters (α, β), then $Y = \ln X$ will have a Logistic distribution with a location parameter $\ln \alpha$ and a scale parameter $1/\beta$. This can be written in terms of the accelerated failure time model by allowing the scale parameters to differ between groups. Several results characterizing the log-logistic distribution have been provided by Ahsanullah and Alzaatreh (2018) among others.

It has been pretty widely used in survival analysis. With a shape parameter β and a scale (rate) parameter α, it has the density function given by

$$f(x;\alpha,\beta) = \frac{\beta\alpha^\beta x^{\beta-1}}{\left(\alpha^\beta + x^\beta\right)^2}, \quad x \geq 0, \alpha, \beta > 0$$

and the survival function as

$$\bar{F}(x) = \frac{\alpha^\beta}{\alpha^\beta + x^\beta}$$

so that the failure rate function comes out as

$$r(x) = \frac{\beta x^{\beta-1}}{\alpha^\beta + x^\beta},$$

which is unimodal if $\beta > 1$ and is monotonic decreasing if $\beta \leq 1$. This distribution can be and has been reparametrized conveniently. Different shapes of the density function and the failure rate function of log-logistic distribution are shown in Fig. 3.9.

The parameter α is interestingly the median of the distribution. Also to be noted are the results: First Quartile $(Q_1) = 3^{-1/\beta}\alpha$ and the $3rd$ Quartile $(Q_3) = 3^{1/\beta}\alpha$. In fact, the quantile of order p works out as $x_p = \alpha p^{1/\beta}(1-p)^{-1/\beta}$. The mode of the distribution is located at

$$\text{Mode} = \alpha\left(\frac{\beta-1}{\beta+1}\right)^{1/\beta}$$

if $\beta > 1$ and zero otherwise. The mean and the variance are given by

$$E(X) = \alpha B\left(\frac{\beta+1}{\beta}, \frac{\beta-1}{\beta}\right)$$

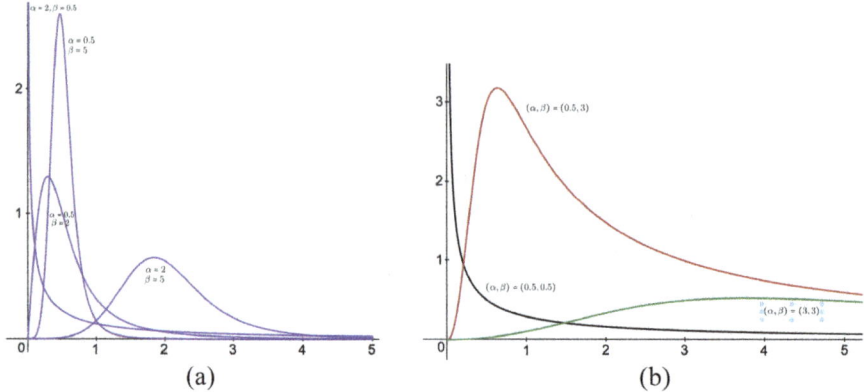

Fig. 3.9 **a** shows different shapes of the density function of the Fisk distribution, whereas **b** shows different shapes of the failure rate function of the same distribution.

and

$$V(X) = \alpha^2 B\left(\frac{\beta+2}{\beta}, \frac{\beta-2}{\beta}\right) - E^2(X),$$

where B stands for the beta function. Entropy for this distribution is given by $\ln(\alpha/\beta) + 2$.

We can estimate the parameters by using median and quartiles. It is also possible to explore optimal orders of the quantiles based on which the two parameters can be estimated. The parameter α can be estimated by the sample median or by the square root of the product of first and third sample quartiles. In fact, the square root of the product of two symmetrically placed quantiles of orders p and $1 - p$ can also serve the purpose. Thereafter, the shape parameter can be estimated from any sample quantile and the estimated median. Of course, the estimates can be obtained numerically through the method of maximum likelihood.

Several extensions or generalizations of this distribution have been suggested and a review has been attempted by Muse et al. (2021). Notable among these are the Marshall-Olkin extended log-logistic distribution (Gui, 2013), the two-parameter exponentiated log-logistic distribution (Choudhary, 2019) and the Zografos-Balakrishnan log-logistic distribution (Zografos and Balakrishnan, 2009). The respective authors have discussed applications of the models proposed by them in real-life data. The pdf of the extension suggested by Gui (2013), involving an additional parameter λ, is given by

$$f(x; \alpha, \beta, \lambda) = \alpha^\beta \beta \lambda \frac{x^{\beta-1}}{\left(x^\beta + \lambda\alpha^\beta\right)^2}, \quad x > 0,$$

which has more or less the same properties as the log-logistic distribution. The two-parameter exponentiated log-logistic has the distribution function

$$F(x; \alpha, \beta) = \left(\frac{x^\beta}{1+x^\beta}\right)^\lambda$$

while Zografos-Balakrishnan model has an additional shape parameter δ in the density function given by

$$f(x; \alpha, \beta, \delta) = \left(\frac{\beta}{\alpha^\beta \Gamma(\delta)}\right) x^{\beta-1} \left[1 + \left(\frac{x}{\alpha}\right)^\beta\right]^{-2} \cdot \left[\ln\left(1 + \left(\frac{x}{\alpha}\right)^\beta\right)\right]^{\delta-1}, \quad x > 0, \; \alpha, \beta, \delta > 0.$$

The failure rate function of this distribution can be increasing, decreasing, unimodal and bathtub-shaped depending on different values of the parameters. For a detailed discussion one may refer to Ramos et al. (2013). The log-logistic distribution is obtained by taking $\delta = 1$. Some more properties of log-logistic distribution are discussed in Sect. 3.10.2.

3.6 The Birnbaum-Saunders Distribution

The Birnbaum-Saunders (BS) distribution, introduced by Birnbaum and Saunders (1969) and discussed comprehensively in the book by Leiva (2015), has been very widely used to represent fatigue failure with particular reference to failure due to growth of cracks developing in structures subjected to cyclic stress (see also Sect. 10.6 where a detailed discussion on fatigue failure model may be obtained). A comprehensive review of this distribution, its statistical properties and applications has been provided in Balakrishnan and Kundu (2018). The density function of the BS distribution with a scale parameter β and a shape parameter α, denoted by $BS(\alpha, \beta)$, is given by

$$f(x; \alpha, \beta) = \frac{1}{2\sqrt{2\pi}\alpha\beta} \left[\left(\frac{\beta}{x}\right)^{1/2} + \left(\frac{\beta}{x}\right)^{3/2}\right] \exp\left[-\frac{1}{2\alpha^2}\left(\frac{x}{\beta} + \frac{\beta}{x} - 2\right)\right],$$

which is shown in Fig. 3.10 for $\beta = 1$. This is a unimodal distribution, with mode x_0 for $\beta = 1$ given as the solution of

$$x_0^3 + x_0^2(\alpha^2 + 1) + x_0(3\alpha^2 - 1) = 1.$$

The mean and the variance come out as

$$E(X) = \beta\left(1 + \frac{\alpha^2}{2}\right) \quad \text{and} \quad V(X) = (\alpha\beta)^2\left(1 + \frac{5\alpha^2}{4}\right),$$

respectively. The failure rate function is unimodal, initially increasing and thereafter decreasing (or remaining more or less flat).

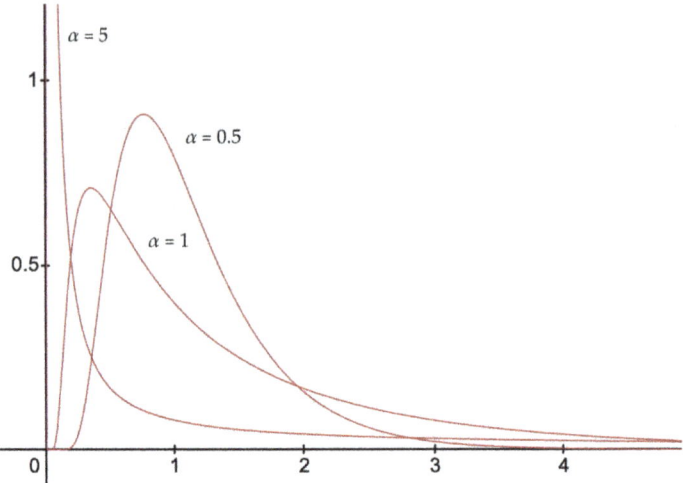

Fig. 3.10 Different shapes of Birnbaum-Saunders distribution

The parameters can be conveniently estimated in terms of the sample arithmetic mean $\bar{X}$ and harmonic mean h as

$$\widehat{\alpha} = \left(2\left[\left(\frac{\bar{X}}{h}\right)^{1/2} - 1\right]\right)^{1/2} \quad \text{and} \quad \widehat{\beta} = (\bar{X} \cdot h)^{1/2}.$$

If $T \sim BS(\alpha, \beta)$ then

(a) $Z = \frac{1}{\alpha}\left[\left(\frac{T}{\beta}\right)^{1/2} - \left(\frac{T}{\beta}\right)^{-1/2}\right] \sim N(0, 1);$

(b) $1/T \sim BS(\alpha, 1/\beta).$

Assuming that the stress in the jth cycle leads to an increase X_j in crack length and noting that the structure fails as soon as the crack length exceeds a critical figure of L, we can study the failure time distribution in terms of the minimum number of cycles n such that $T = \sum_{j=1}^{n} X_j > L$. Assuming X_j to be independently distributed as $N(\mu, \sigma^2)$, the failure time distribution can be worked out as

$$P(T \geqslant t) = 1 - \Phi\left(\frac{t - n\mu}{\sigma\sqrt{n}}\right),$$

where t corresponds to L.

3.7 Topp-Leone Distribution

Topp and Leone (1955) proposed a J-shaped distribution to deal with empirical data with J-shaped histograms over a finite-range, e.g., failure times of powered hand tools and calculating

$$f(x; b, \alpha) = 2\left(\frac{\alpha}{b}\right)\left(\frac{x}{b}\right)^{\alpha-1}\left(1 - \frac{x}{b}\right)\left(2 - \frac{x}{b}\right)^{\alpha-1}, \quad 0 \leqslant x \leqslant b < \infty, \ 0 < \alpha < 1.$$

This gives the distribution function as

$$F(x) = \left(\frac{x}{b}\right)^{\alpha}\left(2 - \frac{x}{b}\right)^{\alpha}.$$

Putting the scale parameter b as unity, we get, for this distribution,

$$F(x; \alpha) = x^{\alpha}(2 - x)^{\alpha}$$

and density

$$f(x; \alpha) = 2\alpha x^{\alpha-1}(1 - x)(2 - x)^{\alpha-1}.$$

This gives the failure rate as

$$r(t; \alpha) = \frac{2\alpha t^{\alpha-1}(1 - t)(2 - t)^{\alpha-1}}{1 - t^{\alpha}(2 - t)^{\alpha}},$$

which has a bathtub shape, whereas the density function is inverted J-shaped as Fig. 3.11 confirms, where we have taken $b = 1$. The median can be found numerically by solving the equation

$$\ln[x(2 - x)] = -\frac{1}{\alpha}\ln 2.$$

The mean has the expression

$$E(X) = 1 - 4^{\alpha}\frac{\Gamma^2(1 + \alpha)}{\Gamma(2 + 2\alpha)}.$$

Thus, we get an explicit expression for the moment estimate of the shape parameter α from the equation

$$\Gamma(2 + 2\alpha)(m - 1) + 4^{\alpha}\Gamma^2(1 + \alpha) = 0,$$

where m is the sample mean ($\bar{x}$). The maximum likelihood estimate of α (with $b = 1$) can be obtained as

$$\widehat{\alpha} = -\frac{n}{\sum_{i=1}^{n} \ln[x_i(2 - x_i)]}.$$

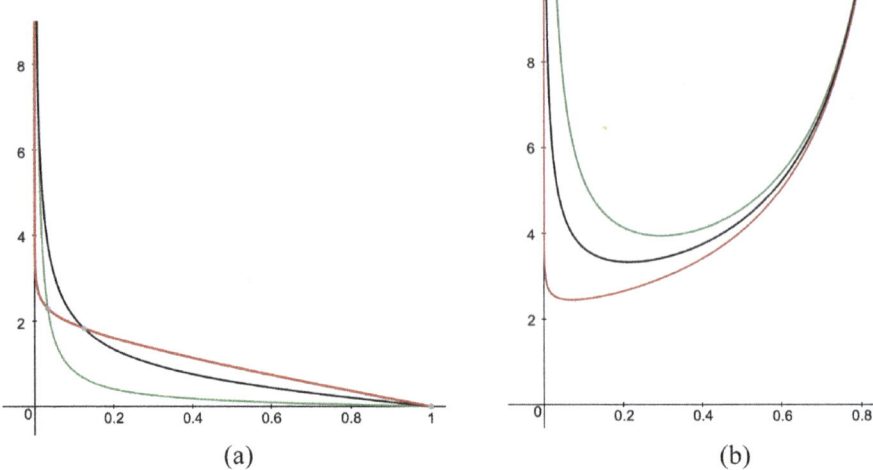

Fig. 3.11 a Shows different shapes of Topp-Leone density function, whereas **b** shows different shapes of the Topp-Leone failure rate function. In both **a** and **b**, the green curves are for $\alpha = 0.1$, black are for $\alpha = 0.5$, whereas the red curves are drawn for $\alpha = 0.9$ taking the scale parameter $b = 1$

Incidentally, this is also the Bayes' estimate of α in case of a quadratic loss function. The estimator is weakly consistent and its finite-sample properties are yet to be studied.

Giles (2021) reports use of the Topp-Leone (TL) distribution to life spans of 18 electronic devices (values divided by 1000) and to data on 23 measures of unit capability factors (studied by Mazumder and Gaver, 1984).

TL-Exponential (TL-Ex) has the density function

$$f(x; \alpha, \lambda) = \alpha \lambda e^{-\lambda x} \left(1 - e^{-\lambda x}\right)^{\alpha - 1}, \ x > 0, \ \alpha > 0, \ \lambda > 0.$$

The failure rate is increasing if $\alpha > 1$ and is decreasing if $\alpha < 1$.

Rather than being used as a probability model for lifetime data representation, the Topp-Leone distribution has been used as a generator to work out certain new models, starting with some baseline distribution function G and deriving the Topp-Leone generated distribution with distribution function given by

$$F(x) = G^\alpha(x)(2 - G(x))^\alpha, \ x \geqslant 0, \ \alpha > 0.$$

In fact, one can even replace $G(x)$ by a continuous function $w(G(x))$ satisfying certain conditions to derive a more general family of distributions. A commonly used model is the TL-Ex distribution as given above. Muhammad (2017) reviewed several extensions of the Topp-Leone generated distributions and offered a new extension.

3.8 Extensions and Generalizations

During the last few decades and specially since the beginning of this century, many of the existing distributions used to represent lifetime or related data have been generalized or modified or extended, to accommodate varying reliability properties revealed by data sets in real life. In some cases, the form of the existing distribution or density function has been modified, usually introducing an additional parameter. Exponentiating the existing distribution and/or density function has been quite often used to change or modify some properties of the existing distribution and expand the horizon of its applications.

Shaw and Buckley (2009) introduced the concept of the rank transmutation map (RTM) as the composition of the distribution function of one distribution with the inverse of distribution function (or of the quantile) of another distribution to come up with a generalized version of the baseline distribution and generate a new family of distributions. To be specific, if F and G are two distribution functions, then the RTM is defined as $F(G^{-1}(u))$ for $u \in (0, 1)$. One of the motivations behind this approach was to introduce skewness in a symmetrical distribution. The quadratic rank transmutation map ($QRTM$) is defined by $(1 + \lambda)u - \lambda u^2$, where $-1 \leqslant \lambda \leqslant 1$. This means that

$$F(G^{-1}(u)) = (1 + \lambda)u - \lambda u^2.$$

Thus, starting with a baseline distribution $u = G(x)$ and using the $QRTM$, a new (transmuted) distribution has a distribution function (cf. Aryal and Tsokos, 2011)

$$F(x) = (1 + \lambda)G(x) - \lambda G^2(x).$$

Different choices of $G(x)$ will yield transmuted versions of several different distributions of interest in reliability studies. This concept has been extended to yield a generalized transmuted-G (GT-G) family of distributions with cdf given by

$$F(x; a, b, \lambda) = (1 + \lambda)G^a(x) - \lambda G^b(x). \tag{3.8.2}$$

Here a and b are shape parameters. Starting with a baseline distribution, transmutations result in a general family of distributions, by adding some new parameters (usually shape) and the transmuting parameter. Several families of distributions using cubic and quartic transmutation have been suggested. In the case of cubic transmutation, we have

$$F(x) = \lambda_1 G(x) + (\lambda_2 - \lambda_1)G^2(x) + (1 - \lambda_2)G^3(x).$$

Clearly, on using $QRTM$, transmuted exponential distribution will have the density function

$$f(x) = \alpha(1 - \lambda)e^{-\alpha x} + 2\lambda\alpha e^{-2\alpha x}, \ x \geqslant 0, \ \alpha > 0.$$

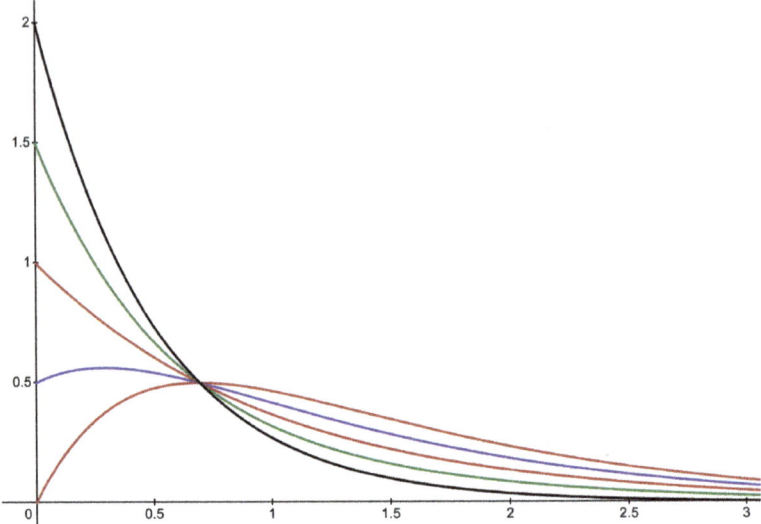

Fig. 3.12 Shapes of transmuted exponential density function for different values of λ with $\alpha = 1$

The failure rate function has the form

$$r(x) = \alpha \left[\frac{1 - \lambda + 2\lambda e^{-\alpha x}}{\lambda e^{-\alpha x} + 1 - \lambda} \right].$$

The density takes various shapes for different values of the parameter λ with $\alpha = 1$ as shown in Fig. 3.12. The distribution could be unimodal with a median given by

$$Me = \left(\frac{1}{\alpha} \right) \left[-\ln \left(\frac{\lambda - 1 + \sqrt{1 + \lambda^2}}{2\lambda} \right) \right].$$

The r-th moment comes out simply as $E(X^r) = \alpha^{-r} r! [1 - \lambda + \lambda 2^{-r}]$. The mean works out simply as $E(X) = \frac{2-\lambda}{2\alpha}$.

Performance rating of the transmuted exponential distribution compared to that of other distributions in analysing failure-time data has been attempted by Owoloko et al. (2015).

Hussain et al. (2018) have suggested a transmuted Size-biased exponential distribution and discussed its versatile reliability properties.

There is a huge literature on transmuted distributions. Rahman et al. (2020) have mentioned more than six dozens of different transmuted distributions in their review work.

Rasekhi et al. (2017) proposed a modified general family of distributions, starting from a baseline distribution with distribution function $G(\cdot)$ and introducing three additional parameters α, β and λ to yield the distribution function

3.8 Extensions and Generalizations

$$F(x) = 1 - [1 - G^\lambda(x)]^\alpha \left[1 - \frac{\alpha\beta}{\alpha + \beta} \ln\left(1 - G^\lambda(x)\right) \right], \ \alpha, \lambda > 0, \ \beta \geqslant 0.$$

Considering the exponential model with parameter δ as the baseline distribution, we get the modified exponential distribution with the distribution function given by

$$F(x) = 1 - \left[1 - \left(1 - e^{-\delta x}\right)^\lambda\right]^\alpha \left\{ 1 - \frac{\alpha\beta}{\alpha + \beta} \ln\left[1 - \left(1 - e^{-\delta x}\right)^\lambda\right] \right\}.$$

This can also be obtained by compounding the Gomez-Deniz and the Kundu-Gupta model. The failure rate of this distribution is either increasing or sigmoid. With $\beta = 0$, we get the Kumaraswamy distribution, while the case $\beta = 0$ and $\alpha = 1$ corresponds to the generalized exponential distribution and the choice $\beta = 0$, $\alpha = \lambda = 1$ gives the exponential distribution.

Aryal and Yousof (2017) have referred to more than a score of such generalizations. Korkmaz and Yousof (2017) mention a similar number of extensions and generalizations of the exponential and the Weibull distributions. They also refer to the Odd-Lindley model along with a few modified forms of baseline distributions.

Intervened versions of some existing distributions arose in the context of clinical data analysis (cf. Shanmugam et al., 2002). Further, some general families of distributions have been suggested by appropriately introducing an additional parameter to develop a family of distribution functions. Some research workers have even used the underlying techniques like exponentiation, transmutation or modification or intervention and the like in combinations. Most of the generalizations or extensions were motivated by the need to accommodate different shapes of failure rate or mean remaining life or some such property of interest in reliability and survival analysis.

Exponentiation has been used to extend the applicability of some well-known distributions by removing some property of the original model that limits its applicability, e.g., the exponentiated normal distribution accommodates non-symmetrical data. Instead of starting with an existing distribution and exponentiating a component of the density function or the distribution function, several modifications, generalizations and extensions of the existing or traditional distributions have been offered by a whole host of research workers in the domain of reliability and survival analysis. It must be admitted, however, that some of these generalizations or modifications or extensions are not user-friendly and have reliability properties which are mathematically involved and cumbersome.

3.8.1 Generalizations of the Exponential Model

In all fairness, we should start with such modifications, extensions and generalizations of the one-parameter exponential distribution and briefly consider their reliability properties. Some of these new distributions include, as special cases, some of the existing distributions and also some of the extensions or generalizations of some other existing distributions. These usually involve at least two parameters. Mention may be made of the extended exponential (Gómez et al., 2014), modified exponential (Rasekhi et al., 2017), slashed generalized exponential (Astorga et al., 2017), gamma-exponentiated exponential (Ristic and Balakrishnan, 2012), among a few others. Andrade (2017) proposed a four-parameter exponentiated generalized Extended exponential distribution which can exhibit both bathtub failure rate (Shanmugam et al., 2002) and inverted bathtub failure rate properties.

The extended exponential distribution due to Gómez et al. (2014), sometimes referred to as the weighted exponential distribution, has the density function f given by

$$f(x) = \frac{\alpha^2}{\alpha+\beta}(1+\beta x)e^{-\alpha x}, \ \alpha, \beta > 0, \ x \geqslant 0$$

with a mean life $E(X) = (\alpha + 2\beta)/[\alpha(\alpha + \beta)]$ and a failure rate

$$r(x) = \frac{\alpha^2(1+\beta x)}{\beta + \alpha(1+\beta x)},$$

which is increasing and bounded between $\alpha^2/(\alpha + \beta)$ and α. The shapes of the density function are shown in Fig. 3.13. This distribution can be regarded as the mixture of exponential distribution with failure rate α and gamma distribution with scale parameter α and shape parameter 2 with the respective weights $\alpha/(\alpha+\beta)$ and $\beta/(\alpha+\beta)$. This reduces to the exponential distribution when $\beta = 0$. The Intervened exponential distribution proposed in the context of neurological studies has the density function

$$f(x;\rho,\theta,\tau) = \begin{cases} \frac{e^{-(x-\tau)/(\rho\theta)} - e^{-(x-\tau)/\theta}}{(\rho-1)\theta}, & \rho \neq 1 \\ \frac{x-\tau}{\theta^2}e^{-(x-\tau)/\theta}, & \rho = 1, \end{cases}$$

$x > \tau \geqslant 0$, and $\theta > 0$. Here $\rho > 0$ is the intervention parameter. The mean of the distribution (taking $\tau = 0$) works out as $(\rho+1)\theta$ and the failure rate is increasing. For some more properties of intervened exponential distribution one may refer to Bhat and Pundir (2022).

Nadarajah and Haghighi (2011) came up with a new two-parameter extension of the exponential distribution having the density

$$f(x) = \alpha\beta(1+\alpha x)^{\beta-1}e^{1-(1+\alpha x)^\beta}, \ \alpha, \beta > 0, \ x \geqslant 0,$$

reducing to the exponential distribution when $\beta = 1$. The failure rate function works out as $r(x) = \alpha\beta(1+\alpha x)^{\beta-1}$ which is increasing if $\beta > 1$. In fact, $\beta = 2$ yields a linearly increasing failure rate.

3.8 Extensions and Generalizations

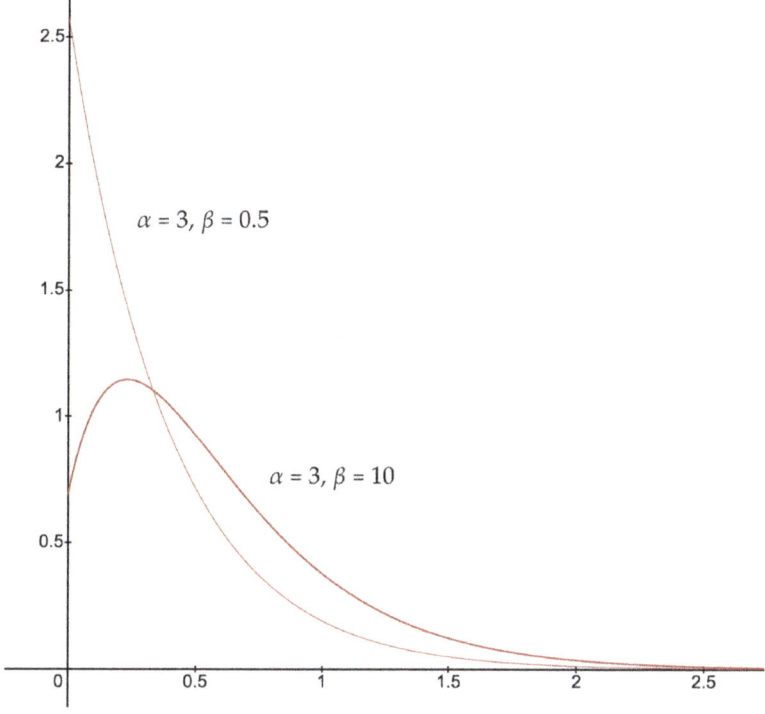

Fig. 3.13 Different shapes of extended exponential distribution

Dias et al. (2018) discussed the beta Nadarajah-Haghighi distribution with a density function given by

$$f(x; \alpha, \lambda, a, b) = \frac{\alpha\lambda}{B(a,b)}(1+\lambda x)^{\alpha-1}\left(1 - e^{(1-(1+\lambda x)^\alpha)}\right)^{a-1} e^{b(1-(1+\lambda x)^\alpha)}, \ x \geqslant 0, \ \lambda > 0, \ \alpha > 0, a, b > 0.$$

This is derived from the relation between two random variables X and Y given by

$$X = \frac{1}{\lambda}\left[(1 - \ln(1-Y))^{1/\alpha} - 1\right],$$

where Y is assumed to follow a beta distribution with parameters a and b. This general distribution includes quite a few other traditional as well as generalized models as particular cases.

A relatively often used generalization was suggested by Gupta and Kundu (1999) as a special case of the exponentiated Weibull distribution studied by Mudholkar et al. (1995) with a zero location parameter. The density function of the generalized exponential distribution has the form

$$f(x; \alpha, \lambda) = \alpha \lambda e^{-\lambda x} \left(1 - e^{-\lambda x}\right)^{\alpha - 1}, \quad \alpha, \lambda > 0, \ x \geq 0,$$

defining a right-skewed, unimodal curve with a monotone failure rate function increasing for $\alpha > 1$ and decreasing for $\alpha < 1$. This is actually an exponentiated exponential distribution which can be interpreted as the distribution of failure time of a parallel system with α (if an integer) components, each having an exponential distribution with parameter λ. Mean life for the distribution is given by

$$E(X) = \psi(\alpha + 1) - \psi(1),$$

where ψ is the di-gamma function. This distribution can be used in place of Weibull, gamma and log-normal distributions.

Astorga et al. (2017) proposed the slashed generalized exponential distribution which is robust against influential observations and is an extension of the generalized exponential. A random variable $S = Z/U^{1/q}$ follows a standard Slashed distribution with shape parameter q if Z has a standard normal distribution and U has a standard Uniform distribution independently of Z. This will have heavier tails than the normal. In case X has a generalized exponential distribution with parameters α and λ, and Y has a beta distribution with parameters $(q, 1)$ then $Z = X/Y$ has the slashed generalized exponential distribution with parameters (α, λ, q) and the density function is given by

$$f(z; \alpha, \lambda, q) = \frac{\alpha q}{\lambda^q z^{q+1}} J_{(\alpha, q)} \left(1 - e^{-\lambda z}\right), \quad z \geq 0, \ \lambda, \alpha, q > 0,$$

where $J_{(\alpha, q)}(t) = \int_0^t \ln^q[1/(1-u)] u^{\alpha-1} du$. Putting $\alpha = \lambda = q = 1$ we get the canonical slashed generalized exponential distribution with the density function given by

$$f(z) = \frac{1}{z^2} \left(1 - e^{-z} - z e^{-z}\right), \quad z > 0.$$

We have $E(Z) = q d_1 / [\lambda (q - 1)]$, $q > 1$ where $d_1 = \psi(\alpha + 1) - \psi(1)$, ψ being the di-gamma function and $\psi(1) = \gamma$, the Euler constant. The distribution has given good fits to data on failure times of airborne communication transceiver as well as data on remission times of bladder cancer patients in survival analysis, which has the cdf involving the confluent hyper-geometric function.

The failure rate function is unexpectedly cumbersome and can exhibit both bathtub and inverted bathtub shapes. This distribution can also be derived as a mixture of Erlang distributions.

Chaudhary and Kumar (2021) proposed the Arctan exponential distribution and illustrated its applications. The survival function of this distribution is given by

$$\bar{F}(x) = \frac{\arctan\left(\alpha e^{-\lambda x}\right)}{\arctan \alpha}, \quad x \geq 0, \ \lambda > 0, \ \alpha > 0.$$

3.8.2 Generalizations of the Weibull Model

In view of the fact that models to represent lifetime data with bathtub failure rate, as proposed by Hjorth (1980), Rajarshi and Rajarshi (1988), Haupt and Schäbe (1992) and several others are not attractive in practice, new families of models based on the Weibull distribution have come up recently to fit into such situations. Among these, mention may be made of the exponentiated Weibull introduced by Mudholkar et al. (1995), the Additive Weibull distribution by Xie and Lai (1995), the Extended Weibull distribution by Xie et al. (2002) and the Modified Weibull introduced by Lai et al. (2003). Pham and Lai (2007) provide a good review of recent generalizations of the Weibull distribution.

Bhattacharjee et al. (2013a) have characterized several families of Weibull distributions in terms of the ageing intensity function. These families are listed in Table 3.2.

Elbatal (2011) considers a generalization of the Weibull model, called the exponentiated Modified Weibull distribution ($EMWD$) with a four-parameter density function given by

$$f(t; \alpha, \beta, \lambda, \delta) = \delta \left(\alpha + \lambda\beta t^{\lambda-1}\right) e^{-(\alpha t + \beta t^\lambda)} \left(1 - e^{-(\alpha t + \beta t^\lambda)}\right)^{\delta-1}, \; t \geq 0, \; \alpha, \beta, \lambda, \delta > 0.$$

Here (α, β) are the scale parameters of the distribution, while (λ, δ) are the shape parameters. As expected, the failure rate function has a complicated form. This distribution takes different shapes including bi-modal one as shown in Fig. 3.14.

The Modified Weibull distribution (cf. Sarhan and Zaindin, 2009) has the distribution function

$$F(x; \alpha, \beta, \lambda) = 1 - e^{-\alpha x - \beta x^\lambda}, \; \alpha + \beta > 0,$$

Table 3.2 Different survival functions along with the notations used in Bhattacharjee et al. (2013b)

Survival function	Notation used
$\bar{F}_X(t) = \exp\left(-at^b\right)$, $a, b > 0$, $t \geq 0$	$X \sim W_2(a, b)$ (Weibull, 1951)
$\bar{F}_X(t) = \exp\left(-at^b e^{\lambda t}\right)$, $a, b > 0$, $\lambda \geq 0$, $t \geq 0$	$X \sim W_3(a, b, \lambda)$ (Lai et al., 2003)
$\bar{F}_X(t) = \exp\left[-\lambda \left(\frac{t-a}{b-t}\right)^\beta\right]$, $0 \leq a < t < b$, $\lambda, \beta > 0$	$X \sim W_4(a, b, \lambda, \beta)$ (Kies, 1958)
$\bar{F}_X(t) = \exp\left[-\lambda \left(\frac{(t-a)^\beta}{(b-t)^\gamma}\right)\right]$, $0 \leq a < t < b$, $\lambda, \beta, \gamma > 0$	$X \sim W_5(a, b, \lambda, \beta, \gamma)$ (Phani, 1987)
$\bar{F}_X(t) = \exp\left[\frac{\theta}{\alpha}(1 - \exp(\alpha t))\right]$, $\theta > 0$, $\alpha \in \mathbb{R}$, $t \geq 0$	$X \sim W^*(\theta, \alpha)$ (Gompertz, 1825)
$\bar{F}_X(t) = \exp\left(-(at)^b - (at)^{1/b} - h_0 t\right)$, $h_0 > 0$, $a \geq 0$, $b > 1$, $t \geq 0$	$X \sim RAW(a, b, h_0)$ (Xie and Lai, 1995)

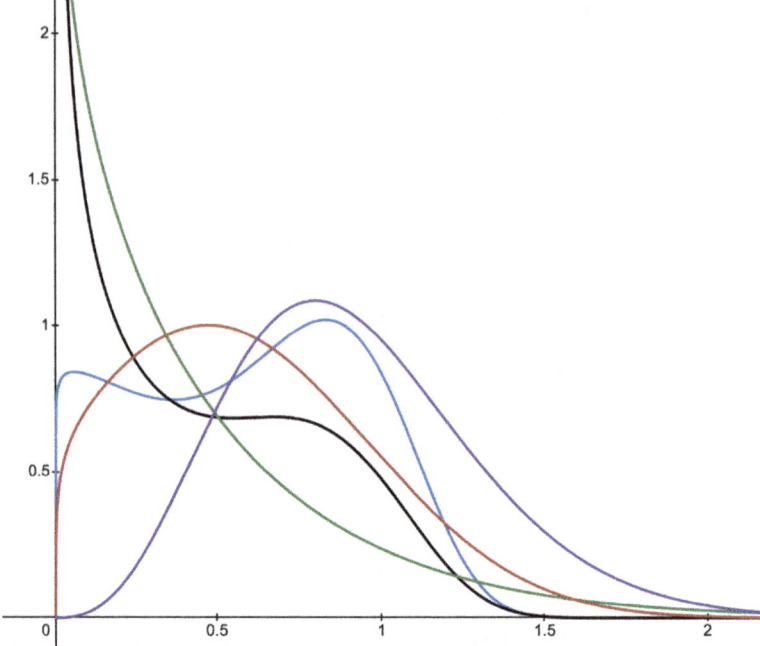

Fig. 3.14 Shapes of exponentiated modified Weibull distribution for different values of the shape parameters

λ being the shape parameter. The failure rate function works out as

$$r(x) = \alpha + \beta\lambda x^{\lambda-1}.$$

This model reduces to exponential model if $\beta = 0$ or $\lambda = 1$ and to the Rayleigh distribution if $\alpha = 0$ and $\lambda = 2$,

Starting with this, a Transmuted Modified Weibull distribution with a somewhat clumsy expression for the failure rate and the quantile functions has been shown to fit various uncommon types of failure data (cf. Khan and King, 2013).

As mentioned earlier, the $EMWD$ includes several life distributions as special cases, some of which are themselves quite general in the sense of covering as special cases some of the simpler and more widely used univariate life distributions. An interesting list is given in Table 3.3.

A new four-parameter Weibull model studied by Yousof et al. (2018) really yields a general family of distributions based on an initial distribution function G and having the distribution function F given by

$$F(x; a, b, \lambda, \delta) = \left[1 - (1 - G(x; \delta))^a\right]^b \left[1 + \lambda - \lambda(1 - (1 - G(x; \delta)))^a\right]^b$$

3.8 Extensions and Generalizations

Table 3.3 Some particular cases of $EMWD$ distribution

Parameter values	Resulting distribution
$\delta = 1$	Modified Weibull distribution
$\alpha = 0$	Exponentiated Weibull distribution
$\delta = 1, \alpha = 0$	Weibull distribution
$\alpha = 0, \lambda = 1$	Exponentiated Exponential distribution
$\alpha = 0, \lambda = 1, \delta = 1$	Exponential distribution
$\lambda = 2$	Generalized Linear Failure Rate distribution
$\lambda = 2, \delta = 0$	Linear Failure Rate distribution
$\alpha = 0, \lambda = 2$	Generalized Rayleigh distribution
$\alpha = 0, \lambda = 2, \delta = 0$	Rayleigh distribution

Table 3.4 Some special cases of four-parameter Weibull distribution

Parameter values	Distribution
$a = b = 1$	Transmuted class (Shaw and Buckley, 2009)
$\lambda = 0, b = 1$	Exponentiated G-family (Gupta et al., 1998)
$\lambda = 0$	Exponentiated Generalized G-family (Cordeiro et al., 2013)
$a = b = 1, \lambda = 0$	Baseline distribution

with the transmutation parameter λ, the exponentiation parameter a and the generalization parameter b, δ denoting the parameter in the baseline distribution. All the three additional parameters determine the shape of the resulting distribution. Special cases of this family of distributions are given in Table 3.4. The five-parameter Weibull model can then be easily written, but will have somewhat cumbersome expressions for reliability properties including the failure rate.

3.8.3 Other Generalizations

Stacy (1962) proposed the Generalized Gamma (GG) distribution as a family that includes as special cases some of the more commonly used life distributions. It has been subsequently discussed by quite a few authors and even further generalized. The pdf for this model involving two shape parameters β and λ and a scale parameter α can be written as

$$f(x; \alpha, \beta, \lambda) = \frac{\lambda \alpha^{-\beta}}{\Gamma\left(\frac{\beta}{\lambda}\right)} x^{\beta-1} e^{-(x/\alpha)^\lambda}, \ x \geqslant 0, \alpha, \beta, \lambda > 0. \quad (3.8.3)$$

This is denoted by $GG(\alpha, \beta, \lambda)$. The model has been re-parameterized in different ways by different authors and some considered a two-parameter version by taking the scale parameter as unity. The model can be regarded as a generalization of the gamma distribution, involving an exponentiation of the exponent factor in the latter. With no restriction on the parameters, this model becomes the same as the Amoroso distribution (cf. Johnson et al., 1994) used to graduate income distribution.

The GG distribution is closed under scale and power transformations of the variable. If X has the $GG(\alpha, \beta, \lambda)$ then, for some constant $c > 0$, $cX \sim GG(c\alpha, \beta, \lambda)$ and

$$X^k \sim GG\left(\alpha^k, \frac{\beta}{k}, \frac{\lambda}{k}\right)$$

with

$$E(X^r) = \alpha^r \frac{\Gamma\left(\frac{\beta+r}{\lambda}\right)}{\Gamma\left(\frac{\beta}{\lambda}\right)}.$$

Thus, the mean life for GG distribution is given by

$$E(X) = \frac{\alpha \Gamma\left(\frac{\beta+1}{\lambda}\right)}{\Gamma(\beta/\lambda)}.$$

The distribution is unimodal with the mode located at

$$\text{Mode} = \alpha \left(\frac{\beta - 1}{\lambda}\right)^{1/\lambda}$$

if $\beta > 1$ and is zero otherwise.

The following distributions come out as special cases of GG distribution.

Parameter values	Distribution
$\alpha = 1, \beta = \lambda > 0$	Weibull distribution
$\lambda = 1$	Gamma distribution
$\lambda = \alpha = 1$	Exponential distribution
$\lambda = 2, \beta = 1/2$	Half-normal distribution

Considering the problem of fitting this model, it may be pointed out that equating the sample mean, its r-th power and the ratio between the sample variance and n times the sample mean squared to their respective expectations, we can derive consistent estimates of the three parameters involved. Large sample properties of the estimators so derived have been studied by Mead et al. (2019).

Among some general distributions for bounded variables, mention may be made of the two-parameter Kumaraswamy (1980) distribution with the probability density function as

3.8 Extensions and Generalizations

$$f(x; a, b) = abx^{a-1}(1-x^a)^{b-1}, \ 0 \leqslant x \leqslant 1, a, b > 0,$$

where a and b are the shape parameters. Motivated by this form of a distribution, Cordeiro and de Castro (2011) defined the Kumaraswamy-G distribution with the distribution function specified by

$$F(x; a, b) = \left(1 - G^a(x)\right)^{b-1}$$

for a given distribution function $G(\cdot)$. Again, starting with different life distributions with some distribution function $G(\cdot)$ we get a general family of distributions.

The Kumaraswamy distribution which was proposed in the context of doubly bounded stochastic processes representing hydrological problems, is better suited to represent a finite-range life distribution (discussed in Chap. 6). To do away with this restriction, Lemonte et al. (2013) introduced the Log-exponentiated Kumaraswamy distribution to model usual lifetime data. It can be shown that $Y = -\ln(1-X)$ will follow the Log-Kumaraswamy distribution with the density function given by

$$f(x; a, b) = abe^{-x}(1-e^{-x})^{a-1}[1-(1-e^{-x})^a]^{b-1}, \ x \geqslant 0, a, b > 0.$$

3.8.4 Transformed-Transformer Distribution

A more recent development in generating new family of distributions is the Transformed-Transformer family of distributions studied by Alzaatreh et al. (2013a,c), who call it T-X family of distributions. This family of distributions is generated by transformation of a random variable T through another random variable X using weight function ω of the cumulative distribution function of X. This distribution has been called by Hazra et al. (2018b) as 'Transformed-Transformer' family of distributions. Different choices of T, X and ω lead to different families of distributions.

Let T be an absolutely continuous random variable with support $[a, b]$, where $-\infty < a < b < \infty$ and let X be another random variable with support $[c, d]$, where $-\infty < c < d < \infty$. Further, let $\omega_1 : [0, 1) \to [a, b]$ be a continuous function such that

(i) $\omega_1(\cdot)$ is differentiable and monotonically increasing;
(ii) $\omega_1(0) = a$ and $\lim_{x \to 1-} \omega_1(x) = b$.

For a random variable Z, we denote the probability density function of Z by f_Z with distributive function F_Z and survival function $\bar{F}_Z$. Then the distribution function $F(\cdot)$ of the Transformed-Transformer family of distributions is defined, for $x \in [c, d]$, as

$$F(x) = \int_a^{\omega_1(F_X(x))} f_T(u)du = F_T[\omega_1(F_X(x))]. \tag{3.8.4}$$

Let the corresponding random variable be denoted by R. For $\omega_2(x) = \omega_1(1-x)$, the reliability function of R is given by

$$\bar{F}(x) = \bar{F}_T[\omega_2(\bar{F}_X(x))].$$

Alzaatreh et al. (2013a) obtained different distributions for different choices of the distributions of T and X based on different weight function $\omega(\cdot)$. It is interesting to note (cf. Alzaatreh et al., 2014a) that, for any random variable with support in (a, b), $\omega(\cdot)$ can be taken as the quantile function of the distribution of that random variable.

One may notice that all the possible probability distributions will appear as special cases of the Transformed-Transformer family of distributions given in (3.8.4) for different choice of f_T, F_X and ω. Hazra et al. (2018b) have shown that if X and T are IFR, then R is IFR provided $x\omega_2'(x)$ is increasing in $x \in (0, 1]$. Using this result and different $\omega_2(x)$, one can generate large number of IFR distributions taking different T and X having IFR property. Table 3.5 shows some such cases.

Following counterexample shows that the condition on increasingness of $x\omega_2'(x)$ cannot be dropped.

Counterexample 3.8.1 *Take $\omega_2(x) = (1-x^3)/\sqrt{x}$, $x \in (0, 1]$ which is not monotone. Suppose X has the distribution function given by $F_X(x) = 1 - e^{-25x^2/4}$, $x \geqslant 0$ and T has the distribution $F_T(x) = 1 - e^{-2x}$, $x \geqslant 0$. Clearly X and T both are IFR. Then one can easily see that the failure rate of R is not monotone.* □

Consider the gamma-Weibull distribution generated by using $\omega_2(x) = -\ln x$ with density function given in Table 3.5. Then putting $\lambda = 1$ we get the generalized gamma distribution discussed by Stacy (1962) (also see Khodabin and Ahmadabadi, 2010);

Table 3.5 Gamma-Weibull family of distributions with IFR property

$\omega_2(x)$	Pdf of Gamma-Weibull family of distributions with $\alpha \geq 1, k \geq 1$
$-\ln x$	$\dfrac{k}{\Gamma(\alpha)\lambda^\alpha \beta}\left(\dfrac{x}{\beta}\right)^{k\alpha-1} e^{-\frac{1}{\lambda}\left(\frac{x}{\beta}\right)^k}$
$(1-x)/x$	$\dfrac{k}{\Gamma(\alpha)\lambda^\alpha \beta} e^{\left(\frac{x}{\beta}\right)^k} \left(\dfrac{x}{\beta}\right)^{k-1} \left(e^{\left(\frac{x}{\beta}\right)^k} - 1\right)^{\alpha-1} e^{-\frac{1}{\lambda}\left(e^{\left(\frac{x}{\beta}\right)^k}-1\right)}$
$(1-x)^2/x$	$\dfrac{k\left(2-e^{-\left(\frac{x}{\beta}\right)^k}\right)}{\Gamma(\alpha)\lambda^\alpha \beta}\left(\dfrac{x}{\beta}\right)^{k-1}\left(\dfrac{1-e^{-\left(\frac{x}{\beta}\right)^k}}{\sqrt{e^{-\left(\frac{x}{\beta}\right)^k}}}\right)^{2(\alpha-1)} e^{-\frac{1}{\lambda}\left(\frac{1-e^{-\left(\frac{x}{\beta}\right)^k}}{\sqrt{e^{-\left(\frac{x}{\beta}\right)^k}}}\right)^2}$
$(1-x^2)/x$	$\dfrac{k\left(e^{\left(\frac{x}{\beta}\right)^k}+e^{-\left(\frac{x}{\beta}\right)^k}\right)}{\Gamma(\alpha)\lambda^\alpha \beta}\left(\dfrac{x}{\beta}\right)^{k-1}\left(e^{\left(\frac{x}{\beta}\right)^k}-e^{-\left(\frac{x}{\beta}\right)^k}\right)^{\alpha-1} e^{-\frac{1}{\lambda}\left(e^{\left(\frac{x}{\beta}\right)^k}-e^{-\left(\frac{x}{\beta}\right)^k}\right)}$

for $\beta = k = 1$ or $\lambda = k = 1$, we get gamma distribution; for $k = 2$, $\alpha = 1/2$, $\lambda = 1$, setting $\beta^2 = 2\sigma^2$ we get Half-normal distribution; for $k = 2$, $\alpha = \lambda = 1$, setting $\beta^2 = 2\sigma^2$ we get Rayleigh distribution.

The following theorem may be obtained in Hazra et al. (2018b). A life distribution or equivalently its corresponding random variable with survival function $\bar{F}$ is said to be New Better than Used (NBU) if $\bar{F}(x+t) \leq \bar{F}(x).\bar{F}(t)$ for all $x, t \geq 0$.

Theorem 3.8.1 *Let $\omega_2(xy) \geq \omega_2(x) + \omega_2(y)$ for all $x, y \in (0, 1]$. If X and T are NBU then R is NBU.* □

Suppose a random variable R_1 is derived as above using two random variables T_1 and X_1, and R_2 is generated using T_2 and X_2. Then it is not difficult to see that if T_2 dominates T_1 and X_2 dominates X_1 both in usual stochastic order, then R_2 also dominates R_1 in the usual stochastic ordering.

Following theorem gives the conditions for the above result to generalize to hazard rate order (cf. Hazra et al., 2018b). Different stochastic orders are discussed in Chap. 7.

Theorem 3.8.2 *Suppose $x\omega_2'(x)$ is increasing in $x \in (0, 1]$. Let the conditions given below are also satisfied.*

(a) $T_1 \leq_{hr} T_2$.
(b) At least one of T_1 and T_2 is IFR.
(c) $X_1 \leq_{hr} X_2$.

Then $R_1 \leq_{hr} R_2$. □

3.9 Compound Distributions

Mixtures of lifetime distributions have been widely considered to represent some real-life situations. In an electronic device, components made with same material and by the same process may be exposed to the same but unknown level of environmental effects. It is also possible that use or deployment conditions vary from one component to another. Components coming out from different machines are likely to encounter varying environmental effects. In survival analysis, when the population under study is heterogeneous, one can consider the population as a mixture of individuals with different risks and modelled by a frailty model. Random effect models, over-dispersion models and some competing risk models also involve mixture distributions.

The Lindley distribution can be looked upon as a mixture of an exponential distribution and a gamma distribution with shape parameter 2. The mixture of an exponential distribution with a gamma distribution with shape parameter 3 gives the x-gamma distribution which is quite flexible, as cited in Sen et al. (2016). In a similar vein, the Hamza distribution due to Aijaz et al. (2020) with the density function given by

$$f(x) = \left[\frac{\beta^6}{\beta^5\alpha + 120}\right]\left(\alpha + \frac{\beta x^6}{6}\right)e^{-\beta x}, \ x \geqslant 0, \alpha > 0, \beta > 0$$

can be looked upon as the mixture of an exponential distribution with hazard rate β and a gamma distribution with scale parameter β and shape parameter 7, and the mixture proportion as $\alpha\beta^5/(\alpha\beta^5 + 120)$ and $120/(\alpha\beta^5 + 120)$, respectively. The mean life comes out as

$$E(X) = \frac{\alpha\beta^5 + 840}{\beta(\alpha\beta^5 + 120)}.$$

The failure rate function has a lengthy expression and the corresponding curve is more or less flat. Enogwe et al. (2022) have proposed a Power Hamza distribution with applications in lifetime data analysis.

Maiti and Pramanik (2019) introduced a wider class of distributions called the Odds x-gamma-G family by taking the odds function of a baseline distribution function and

$$f(t, \lambda) = \frac{\lambda^2}{1+\lambda}\left(1 + \frac{\lambda t^2}{2}\right)$$

as the generator. Reliability properties of several special distributions with uniform, exponential, Burr Type XII as the baseline distributions have been studied comprehensively by them. Expressions involved are, however, not so user-friendly.

Parameters in a life distribution may depend on some 'environmental' factors characterizing the stress and the related factors operating on the equipment or product whose life varies randomly from unit to unit, because of inherent variations in strength built into them essentially during manufacture. These environmental factors are likely to vary across use environments, causing random variations in the life distribution parameters. To get at the overall behaviour of life (time-to-failure), we may assume some probability distribution of the parameter(s) and derive what is usually known as a compound distribution. Otherwise also, if we are interested in studying failure rate and other properties of products coming out of several production lines in a manufacturing organization where the different production lines are not expected to be performing exactly similarly, and we consider the entire production by mixing the products coming out of all the production lines, the mixture distribution can also be regarded as a compound distribution.

Compound univariate and bivariate or, in general, multivariate life distributions and their reliability properties as also properties of series and parallel systems made of components which are either independently distributed or which follow a joint distribution have been studied by a whole host of authors, allowing parameters in the component life distributions as well as the association parameters to vary randomly.

Suppose that $F(\cdot|\phi)$ represents the cdf of a random variable T given ϕ and that $G(\cdot)$ represents the cdf of the random parameter ϕ. [We can simply extend the discussion to the case to a random vector and a vector parameter.] Then the function $H(\cdot)$ defined as $H(t) = \int F(t|\phi)\,dG(\phi)$ which, in some sense the unconditional or marginal distribution of T, was called by Fisher (1936) compound distribution of

3.9 Compound Distributions

F and G. Teicher (1961) called H a mixture, and $F(t|\phi)$ is called the kernel and G the mixing distribution. We may have, as particular cases, countable mixtures or even finite mixtures. In such cases, H can be expressed as a finite or an infinite sum and values of $G(\phi_j)$ can be taken as mixing proportions p_j. Al-Hussaini and Sultan (2001) have provided a comprehensive account of mixture models both when components belong to the same family as also when these belong to different families. They have included detailed discussion on estimation of parameters, besides reliability properties of mixtures.

Many applications of finite mixtures of component distributions in real world have been cited in Titterington et al. (1985). Kao (1958) proposed the use of a Weibull-exponential mixture in fitting the life distribution of electron tubes. One of the components in the mixture represented sudden or catastrophic failure and the other represented delayed or wear-out failure. Since wear-out failure usually occurs after sudden failure, a location parameter representing the delay time in the second component of the mixture may be introduced.

A normal-exponential model was used by Davis (1952) to represent random variations in bus motor failure times. Failure was either abrupt, in which some part broke and the motor would not function or when the maximum power produced, as measured by a dynamometer, fell short of a fixed percentage of the rated normal value. Failures of motor accessories were not included in the data analysed by Davis. In some cases it was found that the data could be approximately fitted by a normal distribution, while, in some other cases, an exponential distribution was found to provide a satisfactory fit. In a few cases neither of these two distributions worked, but a mixture of these two components looked better. AL-Hussaini et al. (1997) estimated stress-strength reliability $R = P(Y < X)$ in both parametric and non-parametric approaches when random variables X and Y are independent and each of which is a mixture of two log-normal components.

In a compound distribution or in a countable mixture, we can define the density function as $h(t) = \int f(t|\phi)\,dG(\phi)$ or as $h(t) = \sum p_j f_j(t)$.

Barlow and Proschan (1981) defined the failure rate function for a mixture and showed that the failure rate is concave. The mixture is DFR if each $F(t|\phi)$ is DFR and is $DFRA$ if each F is $DFRA$. They also pointed out that mixtures of IFR ($IFRA$) distributions are not necessarily IFR ($IFRA$), as mentioned in Sect. 2.4.1. Monotonicity of failure rate in a variety of mixture distributions has been examined by several authors. Vaupel and Yashin (1985) showed graphically that the failure rate function of a mixture of two distinct exponential distributions is strictly decreasing over the entire range. Vaupel and Yashin (1985) and Wang et al. (1998) have shown graphically that the failure rate functions of some two-component mixtures is strictly decreasing on $[0, \infty)$, even when all the distributions mixed have strictly increasing failure rate functions. Gurland and Sethuraman (1994) have given examples where the mixture had a strictly decreasing failure rate although each of the distributions being mixed had a non-decreasing failure rate. For example, a truncated Extreme-value distribution (with a strictly increasing failure rate) when slightly mixed (5%) with an exponential distribution gives rise to a mixture distribution with a strictly decreasing failure rate over the entire region. Rajarshi and Rajarshi (1988) and Gupta and Akman (1995) have described some mixture models characterized by bathtub failure rate functions. All this does not necessarily imply that all mixtures of IFR distributions will have strict DFR property over some region. In this direction, the following two results by Lynch (1999) are worth reporting.

Theorem 3.9.1 *If the survival function of the mixture R_M is given by $\int R(t|\phi)\, dG(\phi)$ and (i) $R(t|\phi)$ is log-concave[2] in (t, ϕ) and (ii) the mixing distribution G has a log-concave density, then the survival function of the mixture is log-concave and hence has the IFR property.* □

The second result due to Lynch states the following.

Theorem 3.9.2 *If (i) $R(t|\phi)$ is log-concave in (t, ϕ) and is non-decreasing in ϕ for each t, and (ii) the mixing distribution is IFR, then the mixture distribution is IFR.* □

Gurland and Sethuraman (1995) obtained necessary and sufficient conditions for a mixture of two IFR distributions to be DFR. They also studied the special case where the component distributions have proportional hazard rate functions. They also showed that mixing exponentials with various other distributions lead to failure rates that are ultimately decreasing on regions of the form (t_0, ∞) for some $t_0 < \infty$. It should be noted here that in case the mixture represents variations in life of an item then the probability that the life ever reaches t_0 is rather small. Thus, the results due to Gurland and Sethuraman or those due to Block and Savits (1976) need not be that important if our interest lies around the average lifetime of the item.

A class of mixtures is identifiable if a unique solution of G in the integral defining H can be found or, in other words, if the mapping of the class of G onto the class of H is one-to-one. Identifiability issue is quite important for estimating parameters in G based on observations from the mixture. Thus, any meaningful application of mixture distributions is constrained by the identifiability condition. It was shown by Ahmed (1988) that a finite mixture of Weibull distributions (and hence of exponential and Rayleigh components) is identifiable. Similarly, Teicher (1963) showed that a finite mixture of normal distributions is identifiable. Further, a finite mixture of inverse-Gaussian components is identifiable. This is true even for the multivariate inverse-Gaussian distribution mixtures.

For a mixture of two-component distributions with failure rate $r_i(t)$ and mean remaining life functions $m_i(t)$, we have the following results for the mixture.

$$r(t) = A(t)r_1(t) + [1 - A(t)]r_2(t),$$
$$m(t) = A(t)m_1(t) + [1 - A(t)]m_2(t),$$

---

[2] Let $g : \mathbb{R}^n \to \mathbb{R}$ be a twice partially differentiable function. Then g is convex if the matrix $\left(\left(\frac{\partial^2 g(x)}{\partial x_i \partial x_j}\right)\right)$ is positive semi-definite for all $x = (x_1, x_2, \ldots, x_n)$. If the matrix is negative semi-definite, the function g is called concave.

where

$$A(t) = \frac{p\bar{F}_1(t)}{p\bar{F}_1(t) + (1-p)\bar{F}_2(t)} = (1 - B(t))^{-1},$$

$$B(t) = \frac{(1-p)\bar{F}_2(t)}{p\bar{F}_1(t)}.$$

These results indicate that a mixture of two IFR distributions need not be IFR, though if the components are DFR then the mixture necessarily is DFR. Similar remarks apply to the mean remaining life functions. In fact, a finite mixture of two exponentials has a DFR distribution, although the components are having constant failure rates.

The inverse-Gaussian and the Weibull mixture is a peculiar case where the inverse-Gaussian distribution has an upside down bathtub failure rate ($UBTFR$) or increasing failure rate and the Weibull has a decreasing, constant or increasing failure rate, thus giving rise to six combinations of failure rate patterns for the mixture, viz. (IFR, DFR), (IFR, CFR), (IFR, IFR), ($UBTFR$, DFR), ($UBTFR$, CFR) and ($UBTFR$, IFR). Behaviour of the mean remaining life function of this mixture was also studied by Al-Hussaini and Al-Hussaini and Abd-El-Hakim (1989).

3.10 Models with Non-monotone Failure Rates

So far we have discussed distributions with monotone failure rates which result in two important non-parametric classes, $viz.$, IFR and DFR classes. It is important to note that all distributions do not have either the IFR or the DFR property. There exists distributions having failure rate non-monotone. Some such distributions will be discussed in this section.

3.10.1 Inverse Rayleigh and Inverse Weibull Models

The inverse-Rayleigh distribution is found to represent a non-monotonic failure rate function, initially increasing and decreasing thereafter. If a random variable Y has a Rayleigh distribution, then the reciprocal $X = 1/Y$ has the inverse Rayleigh distribution with survival function given by

$$\bar{F}(x) = 1 - e^{-\lambda/x^2}, x > 0, \lambda > 0,$$

and hence, the failure rate function is obtained as

$$r(x) = \frac{2\lambda}{x^3} \cdot \frac{e^{-\lambda/x^2}}{1 - e^{-\lambda/x^2}}.$$

It is found that this distribution shows three segments with properties (IFR, IFRA) followed by (DFR, IFRA) and (DFR, $DFRA$). Taking $\lambda = 0.1$, the graphs for failure rate and average failure rate are drawn in Fig. 3.15 which shows that, for $x \in (0, 0.33821)$ approximately, both the curves are increasing, for $x \in (0.33821, 0.75639)$ approximately, the failure rate function is decreasing but average failure rate function is increasing, and both the functions are decreasing thereafter.

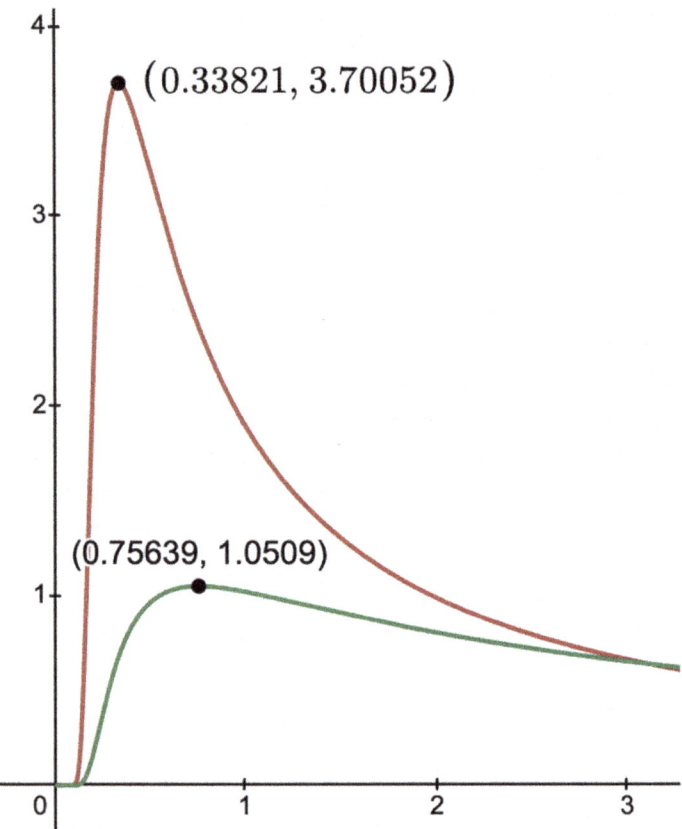

Fig. 3.15 The curves for failure rate (in red) and average failure rate (in green) of the inverse Rayleigh distribution, which show a portion where the failure rate is decreasing but average failure rate is increasing

3.10 Models with Non-monotone Failure Rates

The three-parameter generalized inverse Weibull model (de de Gusmão et al., 2011) which is virtually the exponentiated inverse Weibull model with the distribution function given by

$$F(x) = \exp\left[-\lambda \left(\frac{\alpha}{x}\right)^\beta\right], \ x > 0, \ \lambda > 0$$

has a unimodal failure rate function which rises to its peak at time t^* such that

$$\lambda\left(\frac{\alpha}{t^*}\right)\left[1 - \exp\left\{-\lambda\left(\frac{\alpha}{t^*}\right)^\beta\right\}\right]^{-1} = 1 + \frac{1}{\beta}.$$

3.10.2 Log-Logistic Distribution

Unlike the Weibull distribution with a monotonic failure rate, the log-logistic distribution has a non-monotonic failure rate function depending on the value of the shape parameter.

The two-parameter basic log-logistic distribution has the distribution function

$$F(x) = \frac{x^p}{b^p + x^p},$$

where b is the median, irrespective of the value of p. The failure rate function has the form

$$r(x) = \frac{p x^{p-1}}{b^p + x^p},$$

which is monotonic decreasing for $0 < p \leq 1$, as mentioned in Sect. 3.5.3. For $p > 1$, $r(x)$ is increasing to reach a maximum at $x = b(p-1)^{1/p}$ and then is decreasing. This distribution having a heavy tail is mathematically more tractable when dealing with incomplete (censored) data.

There have been many extensions and generalizations of the log-logistic distribution on the same lines in which several other distributions have been generalized or extended. Mention may be made of the exponentiated log-logistic (Rosaiah et al., 2006) and Beta log-logistic (cf. Lemonte, 2014) with the density function given by

$$f(x) = \frac{1}{B(a,b)} \cdot \frac{(\beta/\alpha)(x/\alpha)^{a\beta-1}}{[1 + (x/\alpha)^\beta]^{a+b}}, \ x > 0, a, b, \alpha, \beta > 0.$$

If X has the above pdf, we write $X \sim BLL(a, b, \alpha, \beta)$. Different shapes of BLL distribution are given in Lemonte (2014). When $a, b \in \mathbb{N}$, the above pdf is the pdf of $X_{a:a+b-1}$, the ath-order statistic from a sample of size $a + b - 1$ from log-logistic distribution. If V is a beta random variable having parameters a and b then

$$X = \alpha \left(\frac{V}{1-V} \right)^{1/\beta} \sim BLL(a, b, \alpha, \beta).$$

Further to note that the BLL distribution is closed under scale transformation. This means that if $X \sim BLL(a, b, \alpha, \beta)$ then, for any $k > 0$, $kX \sim BLL(a, b, k\alpha, \beta)$.

3.11 Distributions with Bathtub Failure Rate

Distributions with failure rate functions appearing like a bathtub i.e., something like the U-shape have engaged the attention of reliability analysts for representing failure time over the three distinct phases of life of an engineered product, $viz.$, early life, useful life and wear-out. A bathtub failure rate (BT) distribution is one in which the failure rate $r(t)$ is characterized by the following: it is strictly decreasing over the interval $[0, t_1]$, strictly increasing over the interval $[t_2, U]$ and remains constant over the interval $[t_1, t_2]$ where U is the upper limit to failure time (see Sect. 2.4.1). This definition has been modified by Mitra and Basu (1995) to require $r(t)$ to be non-increasing over $[0, t_0]$ and non-decreasing thereafter. The second definition does not guarantee a flat portion in the middle, but accommodates the presence of more than two change points in $r(t)$.

Rajarshi and Rajarshi (1988) and Jiang et al. (2001) provide comprehensive accounts of BT distributions, give out properties of such distributions and even suggest methods of constructing such distributions. It may be noted that very few BT distributions having an infinite range and with a flat middle portion (to justify the bathtub shape) have been cited in the relevant literature and one such model cited by Jiang et al. (2001) is somewhat contrived. However, several finite-range life distributions have been found to exhibit this pattern for $r(t)$. Thus, most BT distributions have usually a decreasing followed by an increasing failure rate, with one change point only. Several models reported in the previous section do exhibit a bathtub shape or an inverted bathtub shape for the failure rate function.

In an attempt to obtain a single reliability distribution which will have such a bathtub property, Hjorth (1980) obtained the survival function given by

$$\bar{F}(x) = (1 + \beta x)^{-\theta/\beta} e^{-\delta x^2/2}, \quad x \geq 0, \; \theta, \beta, \delta > 0$$

with failure rate function given by

$$r(x) = \delta x + \frac{\theta}{1 + \beta x}.$$

Although this represents an IFR distribution for $\theta \to 0$, a CFR distribution as $\delta \to 0$ and $\beta \to 0$, and a DFR distribution as $\delta \to 0$, and non-monotonic failure rate distribution for $0 < \delta < \theta \beta$, it cannot simultaneously represent the decreasing, constant and increasing (DCI) failure rates in a single choice of the parameters for

3.11 Distributions with Bathtub Failure Rate

early, middle and late ages. The constant part of the bathtub curve remains uncovered in Hjorth's formulation, although it is known as increasing, decreasing and bathtub (IDB) curve.

Mukherjee and Roy (1993) consider a non-negative random variable X with pdf $f(\cdot)$ given by

$$f(x) = \frac{\delta(|x-a|+|x-b|)}{b-a} \cdot \exp\left\{-\frac{\delta a^2}{b-a} - \delta x + (-1)^{k(x-a,\,|x-a|)}(b-a)\frac{\delta}{4}\left(\frac{|x-a|+|x-b|}{b-a}-1\right)^2\right\},$$

for $x \geq 0$, $0 < a < b < \infty$, $\delta > 0$ and $k(x, y)$ takes value 1 or 0 according as $x = y$ or $x \neq y$. The corresponding survival function comes out as

$$\bar{F}(x) = \exp\left\{-\frac{\delta a^2}{b-a} - \delta x + (-1)^{k(x-a,\,|x-a|)} \cdot \frac{\delta(b-a)}{4}\left(\frac{|x-a|+|x-b|}{b-a}-1\right)^2\right\},$$

whereas the failure rate is calculated as

$$r(x) = \begin{cases} \frac{\delta(b+a)}{b-a}, & \text{if } x = 0 \\ \frac{\delta(a+b-2x)}{b-a}, & \text{if } 0 < x \leq a \\ \delta, & \text{if } a < x \leq b \\ \frac{\delta(2x-a-b)}{b-a}, & \text{if } x > b, \end{cases}$$

which starts from $\delta(b+a)/(b-a)$ at $x = 0$, then decreases monotonically to δ at $x = a$ which is continued to remain same up to the point $x = b$ and then increases monotonically to infinity, making the failure rate curve bathtub-shaped (see Fig. 3.16).

Some basic properties of BT distributions have been pointed out in Lai et al. (2001). These include the following:

(*i*)   The convolution of BT distributions is not necessarily BT.
(*ii*) A mixture of BT distributions need not be BT.
(*iii*) A parallel system of two independent components with BT lifetimes need not have a BT life distribution.

Among the BT distributions which are essentially showing a decreasing followed by an increasing failure rate over an infinite range, mention may be made of the competing risk model (Lai et al., 2001) with the failure rate

$$r(t) = \frac{\alpha}{1+\beta t} + \lambda \delta t^{\delta-1}, \quad \alpha, \beta, \lambda > 0, \; \delta > 2.$$

This can be regarded as corresponding to a risk model involving a Lomax distribution and a Weibull distribution. The case $\delta = 2$ has been considered by Hjorth (1980) yielding a flexible family.

Fig. 3.16 The failure rate function of Mukherjee-Roy distribution

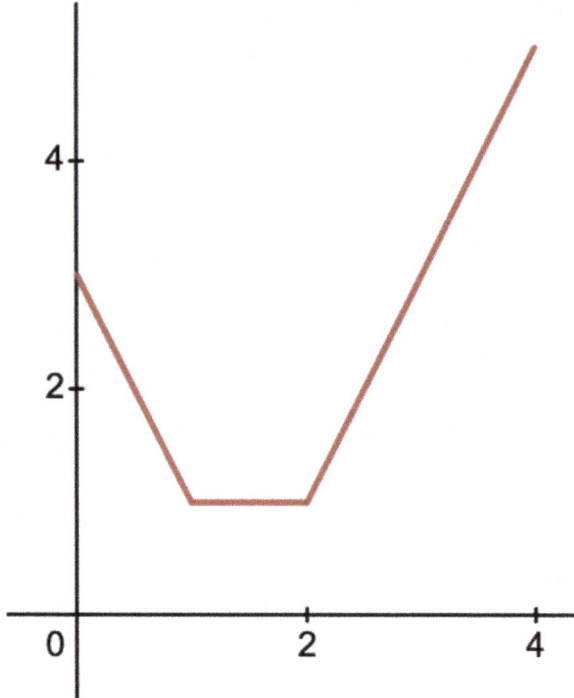

3.12 Other Failure Rate Patterns

The power hazard rate function defined as $r(x) = ax^k, x > 0$ with $a > 0$ and $k > -1$ induces the Power-hazard rate distribution with density function given by

$$f(x) = ax^k \exp\left[-\frac{ax^{k+1}}{k+1}\right], \quad x > 0.$$

Combining the linear hazard rate and the power hazard rate functions to get the general form $r(x) = ax^k + bx$ with $a + b > 0$ and $k > -1$ (which can include constant, decreasing, increasing and even bathtub-shaped cases), the distribution function of the P-LHRD (Power Linear Hazard Rate distribution) works out as

$$F(x) = 1 - \exp\left[-\frac{b}{2x^2} + \frac{ax^{k+1}}{k+1}\right].$$

The failure rate function has a bathtub shape when $a, b > 0$ and $k < 0$. As expected, this distribution covers, as special cases, exponential, Rayleigh, linear hazard rate, Quadratic Hazard Rate and Power Hazard Rate distributions.

3.12 Other Failure Rate Patterns

Also interesting is the use of a quadratic failure rate. Xie and Lai (1995) and Jiang and Murthy (1997) considered a competing risk model involving two Weibull distributions resulting in a failure rate given by

$$r(t) = \beta_1 \lambda_1^{-\beta_1} t^{\beta_1-1} + \beta_2 \lambda_2^{-\beta_2} t^{\beta_2-1},$$

which has a bathtub shape when $\beta_1 < 1$ and $\beta_2 > 1$. The change point is given by

$$t_0 = \left(\frac{\beta_1(1-\beta_1)\lambda_2^{\beta_2}}{\beta_2(\beta_2-1)\lambda_1^{\beta_1}} \right)^{1/(\beta_2-\beta_1)}.$$

Also $r(0) = 0$ and r is monotonic increasing beyond t_0.

Dagum (1977) proposed a three-parameter unimodal distribution in place of the log-normal and the Pareto distributions to model data on personal income, wealth etc. which was later used to represent meteorological and reliability data. This distribution can also be looked upon as Inverted Burr Type XII distribution and is also referred to as the Mielke Beta-kappa distribution. It has a density function given by

$$f(t; \beta, \lambda, \delta) = \beta \lambda \delta t^{-\delta-1}(1 + \lambda t^{-\delta})^{-\beta-1}, \ t > 0,$$

with $\lambda > 0$ as the scale parameter and $\beta, \delta > 0$ as the two shape parameters. It corresponds to a decreasing (resp. unimodal) distribution if $\beta\delta \leqslant 1$ (resp. $\beta\delta > 1$). The distribution function being

$$F(t) = \left(1 + \lambda t^{-\delta}\right)^{-\beta},$$

the failure rate function works out as

$$r(t) = \frac{\beta \lambda \delta t^{-(\delta+1)}(1 + \lambda t^{-\delta})^{-(\beta+1)}}{1 - (1 + \lambda t^{-\delta})^{-\beta}}.$$

Different shapes of the failure rate (taking scale parameter $\lambda = 1$) are shown in Fig. 3.17. The median or shelf-life is given by

$$Me = \left[\frac{1}{\lambda}(2^{1/\beta} - 1) \right]^{-1/\delta}.$$

On the other hand, the mean life has the value

$$E(T) = \beta \lambda^{1/\delta} B\left(1 - \frac{1}{\delta}, \beta + \frac{1}{\delta}\right).$$

It should be noted that the Dagum distribution (cf. Dey et al., 2017) has been used in analysis of failure-time data. The authors considered 100 observations on breaking

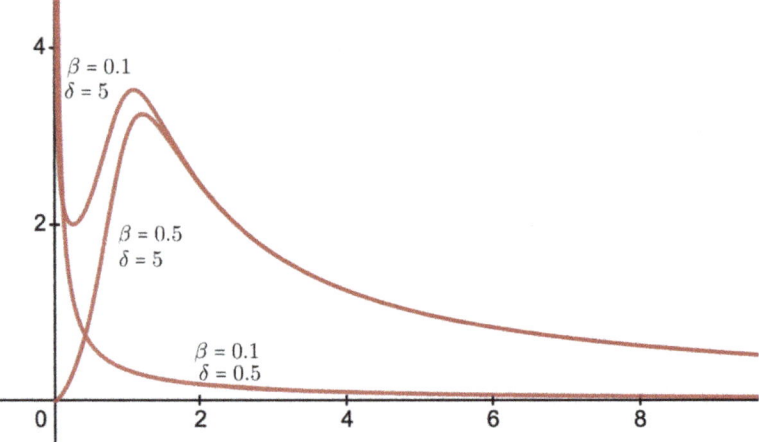

Fig. 3.17 Different shapes of Dagum distribution for different values of the parameters

stress of carbon filters and found that this distribution gave a better fit compared to the Lomax distribution and the Marshall-Olkin extended Lomax distribution. The failure rate for this distribution can be decreasing, bathtub-shaped or inverted bathtub-shaped.

Domma and Cordeiro (2013) proposed the five-parameter beta-Dagum family of distributions. Among several articles on Dagum and related distributions, reference may be made to the one by Dey et al. (2017).

A flexible distribution, called the Extended Quadratic Hazard Rate ($EQHR$) distribution, has been studied by Bhatti et al. (2018). Bain (1974) proposed the distribution with a quadratic hazard rate function in terms of the distribution function given by

$$F(x) = 1 - \exp\left[-\left(\alpha x + \frac{\beta x^2}{2} + \frac{\lambda x^3}{3}\right)\right], \; x > 0, \; \alpha > 0, \; \lambda > 0, \; \beta > -2\sqrt{\alpha\lambda}.$$

Sarhan (2009) introduced the Generalized Quadratic Hazard Rate distribution. Various modifications and extensions of the Quadratic Hazard Rate distributions have been considered by a host of authors. Use of these distributions in modelling various failure data of components as also of assemblies like aircraft windshield has been examined by several authors.

Bhatti et al. (2018) used data on failure times of 84 aircraft windshield to establish the superiority of the $EQHR$ model over Lomax, Rayleigh, Weibull, and linear failure rate models. Murthy et al. (2004) also analysed a sample of failure times of 50 components to find this model as quite an appropriate one.

Chapter 4
Some Discrete Distributions

4.1 Introduction

Reliability analysis is in some cases based on some discrete performance parameter like number of successful operation cycles before (the first) failure of a non-repairable product or the number of such cycles between successive failures of a repairable equipment or the number of units in a group put on test which fail before a specified time. Further, even when the reliability parameter is time to failure or time between successive failures, and hence measurable on a continuous scale, observations as recorded may invite discrete probability models for analysis. Quite often, we get lifetime data grouped into class intervals, requiring some discrete models as appropriate. Hence, the application of discrete probability models has been growing over the years. Apart from the well-known discrete probability models, $viz.$ the binomial, the negative binomial, the geometric and the Poisson distributions, less frequently used discrete distributions like those derived from the Mittag-Leffler functions and modifications, extensions and generalizations of the common discrete distributions to remove some inherent deficiencies in accommodating some special features of the data like over-dispersion, bi-modality or non-monotonicity of failure rate, etc., have also been briefly discussed. Some discrete distributions including some of the classical ones can be derived in terms of solutions to some difference equations—basically hinted at by Pearson himself connecting probability masses at consecutive points.

Discretized versions of some useful continuous distributions have also been reported in the literature on reliability analysis. In fact, several schemes for deriving discretized versions of some well-known and frequently used continuous probability models have been suggested by researchers and reviewed here. A very comprehensive survey has been provided by Chakraborty (2015) and various issues concerning discretization of continuous probability models have been discussed by Lai (2013b).

A brief review of probability models appropriate for count or categorical (including binary) data has been attempted in this chapter. A section is devoted to a discussion on the different approaches adopted for discretization. Reliability properties of discrete distributions have been discussed along with those for continuous distributions in an earlier chapter and some of these in respect of a few discrete models have been reported. A few discrete distributions over finite support have been considered in a subsequent chapter dealing with finite-range distributions.

Published materials on discrete distributions rarely report any applications to model lifetime or failure data, except indirectly in terms of such data recorded as integers. In fact, discretized versions of continuous distributions used in the reliability regime may be used to deal with data on continuous variables x recorded as discrete like $[x]$, the largest integer contained in x. Count data arise in the context of life-testing and prototype trials. Some applications in dealing with count data arising in bio-medical contexts are, of course, available. For example, Cragg (1971) talks of hurdle model to analyse count data, especially in case of binary count data. The fixed-effects Poisson regression model is sometimes a natural choice to analyse count data in economic applications, $viz.$, the number of occurrences of an event within a given time interval (cf. Du et al., 2012). However, for such a model, the mean and the variance being equal, may not be suited for modelling over-dispersed data. Thus, for modelling over-dispersed data, one may think of negative binomial model. For more discussion on the use of count data models, one may refer to Hausman et al. (1984), Cameron and Trivedi (1986), Mullahy (1986) to name a few.

While classical methods based on moments and on likelihood are most often used to estimate discrete distribution parameters, equating the observed relative frequency (proportion) of observations in the 'zero' cell to the corresponding probability has been tried out in some cases.

4.2 Common Discrete Distributions

First, let us define different reliability functions for discrete random variables. Let X be a discrete random variable having pmf $p(\cdot)$ defined as $p(k) = P(X = k)$ and cdf given by $P(k) = \sum_{i \leqslant k} p(i)$. The corresponding survival function is given by

$$\bar{P}(k) = \sum_{i \geqslant k} p(i). \qquad (4.2.1)$$

It is to be noted that some authors define survival function as $\bar{P}(k) = 1 - P(k)$. However, we shall use the convention as mentioned in (4.2.1). Then the failure rate function becomes $r(k) = p(k)/\bar{P}(k)$ for all k such that $\bar{P}(k) \neq 0$. The MRL function is given by

4.2 Common Discrete Distributions

$$m(k) = E(X - k | X \geqslant k) = \frac{\sum_{i \geqslant k}(i-k)p(i)}{\bar{P}(k)} = \frac{\sum_{i \geqslant k+1} \bar{P}(i)}{\bar{P}(k)}.$$

Note that the failure rate function and the mean residual life function are related as

$$r(k) = 1 - \frac{m(k)}{1 + m(k+1)}.$$

4.2.1 Binomial and Related Distributions

Coming to count data analysis, let us consider the experiment where the random variable of interest is the number X of failed units out of n units put on (life) test. Evidently, X will follow a binomial distribution with probability mass function given by

$$b(x; n, p) = P(X = x) = \binom{n}{x} p^x (1-p)^{n-x}, \; x = 0, 1, \ldots, n, \; 0 < p < 1,$$

where p is the probability that a unit will fail (by a pre-specified test time). This implies that $p = 1 - R$, where R is the reliability or probability of survival up to or beyond the test time. The test time may equal the mission time up to which the unit is expected to function satisfactorily or a shorter time in case testing is costly or a decision based on the test results has to be reached early. It is known that $E(X) = np$ and $Var(X) = npq < np$, where $q = 1 - p$. While n is provided by the experimenter, p can be easily estimated from the mean number of failures. This form of a life-test is less efficient compared to one in which the times at which the different units failed are to be noted. The attribute test is quite simple to operate.

It must be noted that the above model assumes a constant probability of success (reliability) for all the units and that results in the n units to be independent of one another. And an inevitable consequence is under-dispersion in the sense that $V(X) < E(X)$. In case the life-test is repeated on sets of n units each and the sets come from different product lines which conform to the same design requirements to different unknown extents (causing random variations in the corresponding probabilities of success), the observed values of X will not follow the Binomial model. In fact, a more applicable model in such a case is the beta-binomial distribution where the probability of success p is assumed to follow a beta distribution with parameters (α, β) to give

$$P(X = x) = \binom{n}{x} \frac{B(x + \alpha, n - x + \beta)}{B(\alpha, \beta)}, \; x = 0, 1, 2, \ldots, n,$$

where B is the usual beta function. This distribution is known as beta-binomial distribution and we write this as $X \sim BB(n; \alpha, \beta)$. This can be regarded as a compound distribution with the two-parameter beta distribution as the mixing distribution. The mean and the variance of this distribution work out, respectively, as

$$E(X) = \frac{n\alpha}{\alpha + \beta} \quad \text{and} \quad V(X) = \frac{n\alpha\beta(n + \alpha + \beta)}{(\alpha + \beta)^2 (1 + \alpha + \beta)}.$$

Expressions for the third and the fourth moments are pretty lengthy. It can be conveniently shown that this distribution—unlike the usual binomial—can take care of over-dispersion. Further, the shape of the distribution can also vary depending on the choice of the parameters α and β.

Fitting the beta-binomial is facilitated by using moment estimates of the parameters. In fact, denoting the first and the second order raw sample moments by m_1 and m_2, respectively, the estimates of α and β come out as

$$\widehat{\alpha} = \frac{nm_1 - m_2}{n\left(\frac{m_2}{m_1} - m_1 - 1\right) + m_1}$$

and

$$\widehat{\beta} = \frac{(n - m_1)\left(n - \frac{m_2}{m_1}\right)}{n\left(\frac{m_2}{m_1} - m_1 - 1\right) + m_1}.$$

The beta-binomial distribution with parameters $(n, 1, 1)$ is the discrete uniform distribution.

In prototype development exercise, we have to note the number of failures until the experiment is terminated or observing a specified number r of failures have taken place. Here we have to begin with a random number of units on test and the number of units working follows a negative binomial distribution. The negative binomial distribution has a probability mass function

$$P(X = x) = \binom{x + r - 1}{x} p^x (1 - p)^r$$

with the distribution function given by $1 - I_p(x + 1, r)$, where I is the incomplete beta function. The mean for this model is $rp/(1 - p)$ and the variance is $rp/(1 - p)^2$ which is greater than the mean. While r (a positive integer) is specified by the experimenter, the other parameter p is estimated simply as $1 - E(X)/V(X)$.

The Fisher information for this distribution is given by $r/[p(1 - p)^2]$. It follows that, as $n \to \infty$, the beta-binomial with parameters $(n, \alpha, n(1 - p)/p)$, where p is taken as $p = \alpha/(\alpha + \beta)$, becomes the negative binomial model with parameters α and β.

The geometric distribution corresponds to the special case $r = 1$. Here, p is the probability of failure by (survival till) the test time and is determined by the distribution of the underlying time-to-failure and the test time. This way the geometric distribution can be conveniently derived from the exponential distribution by considering $p = 1 - e^{-\lambda}$. It should be borne in mind that the test time and the design life of an item may differ and the choice of test time is influenced by resource (time and cost) considerations. One generalization of geometric distribution that has potential applications in life-testing can arise from the trinomial distribution, with three classes of events be recognized, $viz.$ failures, partial successes, and successes where success means survival up to the pre-specified test time and partial success corresponds to survival with less than full efficiency or performance.

4.2.2 Poisson and Related Distributions

The Poisson distribution, also known as the distribution of rare events, finds an obvious place in analysing variations in number of failures in situations where reliability is quite high and the number of possible failures is also large. In fact, for a group of repairable units put on test, required to function intermittently during a specified number n (assumed large) of cycles of operation, the number of units showing 0 or 1 or 2 or higher number of unsuccessful cycles is expected to follow a Poisson distribution, unless the units are unreliable in terms of their design and manufacturing/testing problems.

The distribution has been used rather widely in software reliability studies as also in the context of shock models where the number of external shocks received by a system is often assumed to follow a Poisson distribution with parameter λ which is both the mean and the variance of the distribution. Use of zero-truncated distributions has also been advocated by reliability analysts. And properties of zero-truncated discrete distributions and their applications have been discussed by a host of authors.

One deficiency of the Poisson model is the fact that its mean and variance are equal. This has led to the development of several models able to represent over-dispersion and under-dispersion besides this equi-dispersion. Examples include the Hyper-Poisson (Bardwell and Crow, 1964), Generalized Poisson (Consul, 1995), Double-Poisson of Efron (1986), Poisson polynomial of Conway and Maxwell (1962), Weighted Poisson of Castillo and Perez-Casany (1998), and the like. Pillai and Jayakumar (1995), and a few others proposed a Mittag-Leffler model, while Chakraborty and Ong (2017) have developed a Mittag-Leffler function distribution which includes most of the previously mentioned models as particular cases. For this model, reliability properties as also stochastic ordering have been studied by Chakraborty and Ong (2017).

The probability mass function of the Hyper-Poisson probability model is given by

$$p(k) = \frac{\Gamma(\beta)}{\Gamma(k+\beta)} \cdot \frac{\lambda^k}{{}_1F_1(1, \beta; \lambda)},$$

where

$$_1F_1(1, \beta; \lambda) = \sum_{i=0}^{\infty} \frac{\lambda^i}{(\beta)_i} = 1 + \sum_{i=1}^{\infty} \frac{\lambda^i}{(\beta)_i}$$

is a particular case of confluent hyper-geometric function defined as

$$_pF_q(\alpha_1, \alpha_2, \ldots, \alpha_p; \beta_1, \beta_2, \ldots, \beta_q; z) = \sum_{i=0}^{\infty} \frac{(\alpha_1)_i (\alpha_2)_i \ldots (\alpha_p)_i}{(\beta_1)_i (\beta_2)_i \ldots (\beta_q)_i} \cdot \frac{z^i}{i!},$$

where $(\alpha)_i = \alpha(\alpha+1)(\alpha+2)\ldots(\alpha+i-1) = \Gamma(\alpha+i)/\Gamma(\alpha)$ with $(\alpha)_0 = 1$. It can be seen that this series is everywhere convergent if $p \leq q$ and nowhere convergent for $p \geq q + 2$ whereas, for $p = q + 1$, it is convergent for $|z| < 1$. For details, one may refer to Buchholz (1969). It may be noted that the above distribution reduces to Poisson distribution for $\beta = 1$. It is known as super-Poisson distribution if $\beta > 1$ and sub-Poisson distribution if $\beta < 1$. It reduces to Displaced Poisson(λ, t) if $\beta = t + 1$, where t is a positive integer (cf. Staff, 1967). We must mention here that $_1F_1(a, b, 0) = 1$.

The probability mass function satisfies the relation $(\beta + k) p(k+1) = \lambda p(k)$. A three-parameter form of Hyper-Poisson distribution, suggested by Kumar and Nair (2014), has the pmf

$$p(k) = P(X = k) = \frac{_1F_1(1+k, \beta+k; \alpha)}{_1F_1(1, \beta; \lambda+\alpha)} \cdot \frac{\lambda^k}{(\beta)_k}.$$

This reduces to the Hyper-Poisson distribution for $\alpha = 0$.

The discrete Pareto distribution (also called the Zipf's law or the Riemann Zeta distribution) has the probability mass function given by

$$P(X = x) = \frac{1}{H_n} \cdot \frac{1}{x}, \quad x = 1, 2, \ldots, N,$$

where $H_N = \sum_{k=1}^{N} \frac{1}{k}$.

Gómez-Déniz (2010) offered a generalization of the geometric distribution by using the Marshall-Olkin approach and introducing an additional parameter. They assumed a variable X having the exponential distribution with survival function $\bar{F}(x) = e^{-\theta x} = q^x$ and then define the distribution of a discrete variable Y as

$$P(Y = k) = \frac{\alpha q^k (1-q)}{\left(1 - (1-\alpha)q^{k+1}\right)\left(1 - (1-\alpha)q^k\right)}, \quad \alpha > 0, \ k = 0, 1, 2, \ldots$$

This is a unimodal distribution with a monotonic failure rate. The distribution can be derived alternatively through the Marshall-Olkin scheme directly applied to the geometric distribution. The distribution (which reduces to the geometric distribution for $\alpha = 1$) has been found to provide satisfactory fits to both over-dispersed and under-dispersed data sets.

4.2.3 Distribution Using Mittag-Leffler Function

The special function

$$E_\alpha(u) = \sum_{k=0}^{\infty} \frac{u^k}{\Gamma(1+\alpha k)}, \quad u \in (0, \infty)$$

or its generalized form

$$E_{\alpha,\beta}(u) = \sum_{k=0}^{\infty} \frac{u^k}{\Gamma(\beta + \alpha k)},$$

$\alpha, \beta > 0$, was introduced in the context of a fractional differential equation. The Laplace transform of $E_\alpha(-x^\alpha)$, $0 < \alpha \leqslant 1$, is

$$\frac{\lambda^{\alpha-1}}{1+\lambda^\alpha}, \quad \lambda \geqslant 0.$$

Pillai (1990) showed that $1 - E_\alpha(-x^\alpha)$ is a distribution function for $0 < \alpha \leqslant 1$. Pillai and Jayakumar (1995) proposed a class of discrete Mittag-Leffler distribution which arises as a compounding of the Poisson distribution with parameter $\theta\lambda$, where θ is a constant, and λ follows the Mittag-Leffler distribution with cdf given by

$$F(x;\alpha) = \sum_{k=1}^{\infty} \frac{(-1)^{k-1} x^{k\alpha}}{\Gamma(\alpha k + 1)}, \quad 0 < \alpha \leqslant 1, x > 0. \tag{4.2.2}$$

Jose and Abraham (2011) introduced another generalization of the Poisson distribution by replacing the exponential waiting time in a Poisson process by the Mittag-Leffler distribution. This distribution has the pmf

$$P(X = k) = \sum_{i=k}^{\infty} \frac{\binom{i}{k}(-1)^{i-k} k^{i\alpha}}{\Gamma(\alpha i + 1)},$$

$k = 0, 1, 2, \ldots; 0 < \alpha \leqslant 1$. This distribution has a more flexible shape and can fit both under- and over-dispersed data.

Chakraborty and Ong (2017) proposed a Mittag-Leffler function distribution ($MLFD$) with pmf given by

$$P(X = k) = \frac{\lambda^k}{\Gamma(\alpha k + \beta) E_{\alpha,\beta}(\lambda)},$$

$k = 0, 1, 2 \ldots; \alpha, \beta, \lambda > 0$. $MLFD$ satisfies the recurrence relation

$$\Gamma(\alpha k + \alpha + \beta) P(X = k+1) = \lambda \Gamma(\alpha k + \beta) P(X = k),$$

$k \geqslant 1$, which reduces to the corresponding relation in the Hyper-Poisson distribution for $\alpha = 1$.

Consider the sequence of independent Bernoulli trials where the probability of success in the kth trial is α/k (likely to be true in case of prototype improvement trials), $0 < \alpha < 1, k = 1, 2, 3, \ldots$ and let the number of trials N follow the geometric law with $P(N = k) = q^k p$. Then $X_1 + X_2 + \ldots + X_N$ has the Mittag-Leffler distribution.

$MLFD$ covers a number of discrete distributions as special cases, $viz.$

Parameter values	Distribution
$\alpha = \beta = 1$	Poisson (λ)
$\alpha = 0, \beta \neq 1$	Geometric $(\lambda < 1)$
$\alpha = 1, \beta \neq 1$	Hyper-Poisson (α, β)
$\alpha = 1, \beta \in \mathbb{N}$	Displaced Poisson

$MLFD$ exhibits a wide range of shapes, depending on the choice of parameter values. It can be unimodal with a non-zero mode or can have two non-zero modes or be non-increasing with the mode at zero. It has an increasing failure rate.

4.3 Discretization of Continuous Distributions

There are different ways of discretizing a continuous distribution. However, these discretization methods can be broadly classified into three categories.

(a) First, consider a characteristic property of a continuous distribution and get the corresponding discrete counterpart of this property, and use that discrete counterpart to get the corresponding discrete distribution.

(b) If a continuous distribution is derived under some initial conditions, one may replace the corresponding discrete counterparts of the initial conditions and then derive the corresponding discrete distribution.

(c) Consider some integer points from the support of the continuous random variables, and redefine the probabilities at those discrete points properly.

There are different characteristic properties of a continuous distribution and each such property may lead to a discrete distribution on using the method described in (a) above. Similarly, there are different ways to get the probabilities at the integer points from the support and each one may lead to a different discrete distribution.

Below we discuss different discretization procedures. We must mention here that we are mentioning some commonly used discretization methods and there are quite a few other different ways of discretization. For example, the discretized version of a continuous life distribution derived by mixing two component (continuous) distributions using specified mixing proportions can be obtained through the corresponding mixtures of some discterized versions of the component distributions. For a more comprehensive study of discretization methods available in the literature, one may refer to the review articles by Bracquemond and Gaudoin (2003) and Chakraborty (2015).

Let X be a continuous random variable having survival function $\bar{F}$ defined as $\bar{F}(x) = P(X \geq x)$ and density function f.

4.3.1 Discretization-I

Define $Y = \lfloor X \rfloor + 1$, where $\lfloor X \rfloor$ is the integral part of X. Then the pmf of Y is given by

$$P(Y = k) = P(k - 1 \leq X < k)$$
$$= F(k) - F(k - 1).$$

Note that

$$F_Y(k) = P(Y \leq k)$$
$$= P(\lfloor X \rfloor \leq k - 1)$$
$$= P(X < k)$$
$$= P(X \leq k)$$
$$= F_X(k).$$

If $r_Z(\cdot)$ is the failure rate function of a random variable Z, then we have

$$r_Y(k) = \frac{P(Y = k)}{P(Y \geq k)}$$
$$= \frac{P(Y > k - 1) - P(Y > k)}{P(Y > k - 1)}$$
$$= 1 - \frac{\bar{F}_X(k)}{\bar{F}_X(k - 1)}.$$

It is known that

$$\bar{F}(x) = e^{-\int_0^x r(t)\,dt}.$$

So, we have

$$r_Y(k) = 1 - e^{-\int_{k-1}^k r_X(u)\,du}.$$

Thus, r_Y and r_X are monotonic in the same direction.

It can be seen that if X follows Exponential distribution with failure rate λ, then Y follows geometric distribution with probability of success $p = P(X \leq 1)$. Since exponential (resp. geometric) distribution is the only continuous (resp. discrete) distribution having the memoryless property, we can conclude that the memoryless property is transmitted through the above discretization process.

4.3.2 Discretization-2

To describe Method (b) of generating discrete distribution, one may consider the method used by Carver (1919, 1921) who solved the difference equation corresponding to the differential equation used to derive Pearsonian system of frequency curves (cf. Elderton and Johnson, 1969) given by

$$\frac{\Delta y_x}{\Delta x} = \frac{y_x(a-x)}{f(x)}$$

under the assumptions that

1. y_x has a single mode at $x = a$;
2. $f(x)$ can be expanded in a power series;
3. $y_x \to 0$, as $x \to \pm\infty$.

4.3.3 Discretization-3

Let X have the probability density function $f : \mathbb{R} \to [0, \infty)$. Then the corresponding discretized random variable Y is defined as

$$P(Y = k) = \frac{f(k)}{\sum_{k \in \mathbb{Z}} f(k)}, \quad k \in \mathbb{Z}.$$

The discrete distribution studied by Good (1953) can be seen as a discretized version of gamma distribution. This was later studied by Kulasekara and Tokyn (1992). The discretized version of normal distribution (called discrete normal distribution) was studied by Kemp (1997).

If X has support $S \subseteq \mathbb{R}$, one may consider, for any fixed $n \in \mathbb{N}$, the points $y_1, y_2, \ldots, y_n \in S$ and define the corresponding discrete random variable Y as

$$P(Y = y_i) = \frac{f(y_i)}{\sum_{j=1}^n f(y_j)}.$$

The details of this kind of discrete distribution may be obtained in Barbiero (2010).

4.3.4 Discretization-4

Roy and Dasgupta (2001) have defined a discretized version Y of a continuous random variable X as

$$P(Y = k) = F(k + \delta) - F(k - (1 - \delta)),$$

for $0 \leqslant \delta \leqslant 1$, where δ to be so chosen that $E(X^r) \cong E(Y^r)$, for $r = 1, 2$. Since X is a continuous random variable, taking $\delta = 1$, we get

$$P(Y = k) = F(k + 1) - F(k) = P(k < X \leqslant k + 1) = P(k \leqslant X < k + 1)$$

so that $Y \stackrel{D}{=} \lfloor X \rfloor$. Since positive power of any distribution function is a distribution function, for any given $\alpha > 0$, one may consider a discretized version of X as

$$P(Y = k) = F^\alpha(k + 1) - F^\alpha(k).$$

Again, we may write

$$P(Y = k) = P(k \leqslant X < k + 1) = \bar{F}(k) - \bar{F}(k + 1).$$

Further, since positive power of any reliability function is a reliability function, people have also considered

$$P(Y = k) = \bar{F}^\alpha(k) - \bar{F}^\alpha(k + 1).$$

Taking X as a normal random variable, Roy and Dasgupta (2001) have obtained optimum value of δ as 0.5. This means that

$$Y(Y = k) = F(k + 0.5) - F(k - 0.5) = P(k - 0.5 < X \leqslant k + 0.5).$$

They have studied different reliability properties of Y. For example, if X is IFR (resp. $IFRA$), then Y will also be IFR (resp. $IFRA$).

4.3.5 Discretization-5

If X is a continuous random variable having $r(\cdot)$ as the failure rate function with $r(0) = 0$ then, for any fixed $m \in \mathbb{N}$, Roy and Ghosh (2009) have considered a discrete random variable Y as

$$P(Y \geqslant k) = (1 - r(1))(1 - r(2)) \ldots (1 - r(k - 1)).$$

4.4 Some Discretized Versions

One of the early exercises to use a discrete analogue of a continuous life distribution was by Sato (1999) proposing the discrete exponential distribution in terms of the mass function

$$P(Y = k) = (1 - e^{-\lambda})e^{-\lambda k}, \quad k = 0, 1, 2 \ldots,$$

which comes out to be the probability function of the geometric distribution with parameter $p = 1 - e^{-\lambda}$. In a similar manner, it can be shown that the gamma distribution admits of a discretized version which is the same as the negative binomial distribution. Nekoukhou et al. (2012) proposed a discrete analogue of the generalized exponential distribution of Gupta and Kundu (1999) with the probability function defined as

$$P(Y = k) = c p^{k-1} \left(1 - p^k\right)^{\alpha - 1}, \quad k = 1, 2, 3, \ldots,$$

where

$$c^{-1} = \sum_{j=0}^{\infty} \binom{\alpha - 1}{j} \frac{(-1)^j p^j}{1 - p^{1+j}}, \quad p = e^{-\lambda}.$$

In view of the wide applicability of the Weibull distribution, several discrete analogues of the Weibull distributions and of its extensions/modifications have been proposed by research workers. Barbiero (2013) has considered three such forms, a few others have been mentioned by Chakraborty (2015). Nakagawa and Osaki (1975) first proposed the discrete Weibull distribution (preserving the form of the survival function of the continuous distribution) in terms of the mass function

$$P(Y = k) = q^{k^\alpha} - q^{(k+1)^\alpha}, \quad x = 0, 1, 2 \ldots, \ \alpha > 0,$$

where $q = e^{-\lambda}$. It can be easily shown that if X_i, $i = 1, 2, \ldots, n$, follow discrete Weibull distribution independently, then $\min\{X_1, X_2, \ldots, X_n\}$ follows discrete Weibull distribution. Three distinct forms for discrete Weibull distribution have been studied, the first preserving the survival function form (suggested by Nakagawa and Osaki) and a second one mentioned by Barbiero (2013) preserving the failure rate function introduced by Stein and Dattero (1984). They assumed a failure rate function given by

$$r(x) = \begin{cases} \left(\frac{x}{m}\right)^{\beta - 1}, & x = 1, 2, \ldots, m \\ 0, & x = 0, \text{ or } x > m. \end{cases}$$

This yields the mass function

$$p(x) = \left(\frac{x}{m}\right)^{\beta - 1} \prod_{i=0}^{x-1} \left[1 - \left(\frac{i}{m}\right)^{\beta - 1}\right], \quad x = 1, 2, \ldots, m.$$

4.4 Some Discretized Versions

The third type proposed by Padgett and Spurrier (1985) from a different stand point. This has the survival function defined as

$$\bar{P}(x) = e^{-c \sum_{j=1}^{x} j^{\beta}}, \; x = 1, 2, \ldots$$

and probability function given by

$$P(X = x) = e^{-c \sum_{j=1}^{x} j^{\beta}} \left(1 - e^{-c(x+1)^{\beta}}\right), \; x = 1, 2, \ldots, \; c > 0,$$

with $P(X = 0) = 1 - e^{-c}$. The hazard rate function comes out simply as

$$r(x) = 1 - e^{-c(x+1)^{\beta}}.$$

The complex expression for the probability function has hindered the diffusion of this model. The first and the second moments can be expressed only in terms of infinite series. Besides the method of maximum likelihood, a simple method based on proportions was suggested by Khan et al. (1989) for estimating the two parameters in this model. Let y denote the number of zeroes in a sample of size n. We can get an estimate of c readily in terms of $-\ln(1 - y/n)$. Similarly, denoting by z/n the proportion of 1's in the sample, an estimate of β can be worked out. While this method of using proportions suffers from the weakness that no feasible estimate of β will emerge if there are no zeroes in the sample, it does yield an estimate close to the MLE if the zero cell has a moderate frequency. Moreover, the estimates of the two parameters are based on the numbers of zeroes and 1's only. Theoretically, implausible estimates may also arise from this procedure, when

$$\frac{z}{n} < \left(1 - \frac{y}{n}\right) - \left(1 - \frac{y}{n}\right)^{1/2}.$$

4.4.1 Discrete Lindley Distribution

Al-Babtain et al. (2020) introduced a new discrete analogue of the one-parameter Lindley distribution without using any known discretization scheme. They considered a mixture of the Geometric and the negative binomial distributions for the purpose. The pmf for this discrete distribution was given as

$$p(x) = \frac{\theta^2 (2 + x)(1 - \theta)^x}{1 + \theta}, \; x = 0, 1, 2, \ldots; \; 0 < \theta < 1.$$

This is a unimodal distribution with a failure rate function given by

$$r(x) = \frac{\theta^2 (2 + x)}{1 + \theta + \theta x},$$

which is increasing in x. The mean remaining life function has the form

$$m(x) = \frac{(1-\theta)r(x)}{\theta^2} + \frac{(1-\theta)(2-\theta)}{\theta(1+\theta+\theta x)}$$
$$= \frac{(1-\theta)(2+\theta+\theta x)}{\theta(1+\theta+\theta x)},$$

which is decreasing in x. The maximum likelihood estimate of the parameter θ works out as

$$\frac{1}{2}\left(1 + \frac{8}{1+\bar{x}}\right)^{1/2} - \frac{1}{2},$$

which, incidentally, is the same as the moment estimate of θ, where $\bar{x}$ is the sample mean.

4.5 Telescopic Distribution

A discrete random variable X has a telescopic distribution (Roknabadi, 2000, 2006), denoted by $X \sim T(q, k_\theta)$, if its mass function has the form

$$f(x; q, k_\theta) = q^{k_\theta(x)} - q^{k_\theta(x+1)}, \quad x = 0, 1, 2, \ldots,$$

where $0 < q < 1$ and $k_\theta(x)$ is strictly increasing in x with $k_\theta(0) = 0$ and $k_\theta(x) \to \infty$ as $x \to \infty$.

The geometric and the discrete Weibull families belong to this class. Each member of this family has a continuous analogue with similar properties. In fact, if Y is a continuous random variable with distribution function given by

$$G(y) = 1 - e^{\alpha k_\theta(y)}, \quad y > 0,$$

where $\alpha > 0$ and θ is a parameter vector (which may include α) and $k_\theta(y)$ behaves like $k_\theta(x)$ above, then $X \sim T(q, k_\theta)$, where $X = [Y]$ and $q = e^{-\alpha}$. The class of distributions corresponding to $G(y)$ is called the extended exponential class that contains exponential, Rayleigh, Weibull, linear-exponential, Gompertz, and some distributions describing wear-out failures.

4.5 Telescopic Distribution

For the telescopic family, the following are generic expressions for some of the important reliability properties:

Reliability $R(x)$	=	$q^{k_\theta(x)}$
Failure rate $r(x)$	=	$1 - q^{k_\theta(x+1)-k_\theta(x)}$
Reversed hazard rate $\mu(x)$	=	$\dfrac{q^{k_\theta(x)-k_\theta(x+1)}}{1-q^{k_\theta(x+1)}}$

Mean residual life functions come out as partial sums.

A telescopic distribution belongs to the IFR class if one of the following four conditions holds.

(i) $k_\theta(x+1) - k_\theta(x)$ is increasing in x.
(ii) $q^{k_\theta(i+x)} - q^{k_\theta(x)}$ is decreasing in $i = 0, 1, 2, \ldots$, for every x.
(iii) For all $j_1, j_2, k_1, k_2 \in \{0, 1, 2, \ldots\}$ such that $j_1 < j_2$, $k_1 < k_2$,

$$k_\theta(j_1 - k_1) + k_\theta(j_2 - k_2) \leq k_\theta(j_1 - k_2) + k_\theta(j_2 - k_1).$$

(iv) The sequence $\{k_\theta(x)\}_{x \geq 0}$ is convex.

A further result in this connection given by Roknabadi et al. (2009) is the following.

Theorem 4.5.1 *Let*

$$T_\theta(x) = \frac{1}{2}\left(2k_\theta(x+1) - k_\theta(x) - k_\theta(x+2)\right).$$

Then X is IFR (DFR) if and only if $T_\theta(x) > (<)0$ for all $x \geq 0$, and X is CFR if and only if $T_\theta(x) = 0$ for all $x \geq 0$ (with X having geometric distribution). □

The distributions (i) Discrete Exponential, (ii) Discrete Rayleigh, (iii) Discrete Weibull, (iv) Discrete Linear-Exponential, and (v) Discrete Gompertz belong to this family.

Al-Masoud (2013) obtained the following particular cases.

Name of the Distribution	Particular Value of $k_\theta(x)$
Discrete Modified Weibull Extension	$k_\theta(x) = e^{(x/\theta)^\beta} - 1$
Discrete Modified Weibull Type I	$k_\theta(x) = (\delta/\alpha)x + x^\beta$
Discrete Modified Weibull Type II	$k_\theta(x) = e^{\alpha x} x^\beta$

Particular choices of α and/or β values have given rise to distributions studied by several investigators. The failure rate function for the case $\alpha = 1$ (Almalki, 2014) in Type II is increasing as well as bathtub-shaped.

Chapter 5
Distributions Derived from the Parent Distribution

5.1 Introduction

Associated with the lifetime of a living or a non-living object (a random variable) and its distribution are the distribution of residual or remaining life at some age (time) t, the equilibrium distribution and the distribution of the failure rate or the reversed failure rate (looked upon as a random function of t, the age or time survived). Linked up with the concept of residual life is that of inactivity time noted at some time t given that the unit had failed prior to t. These distributions and their properties along with inter-relations among the latter have been useful in determining the original life distributions uniquely, in characterizing specific lifetime distributions, in studying ageing properties of such distributions, in classifying life distributions as also in testing for departure of the underlying life distribution from exponentiality or from other class property.

One finds a good account of residual life distributions with explicit asymptotic expressions given for the mean, the variance and the quantiles for distributions associated with Gamma and Weibull families of life distributions in Siddiqui and Çağlar (1994). In fact, a whole lot of research in this area has resulted in many articles published in a wide spectrum of periodicals. Beyond the variance of residual life providing a measure of uncertainty about remaining life, different measures of uncertainty like entropy have also been worked out. The widely used proportional hazards model has been extended by Zahedi (1991) to the proportional mean residual life model, replacing the hazard rate function by the mean residual life function.

Equilibrium distributions have been extensively studied in the context of reliability and survival analysis. Equilibrium distributions were introduced as limiting distributions of forward or backward recurrence times in the context of Renewal Processes. Thus, they can be regarded as limiting distributions of used life or residual life (up to or beyond a given point) in Survival/Reliability Analysis.

We can also look upon these distributions as particular cases of weighted distributions. Let X be a non-negative random variable with pdf f. Let the realization of the variable value x of X enter the study with probability proportional to a weight $w(x)$. Then the recorded variable Y has the pdf given by

$$f_Y(x) = \frac{w(x)f_X(x)}{E[w(X)]}.$$

The choice of $w(x) = x$ leads to the length-biased distribution with pdf given by $f_Y(x) = xf_X(x)/E(X)$. If we take $w(x) = x^\alpha$, with $\alpha \geqslant 0$, the distribution is called size-biased distribution. Since length-biased distribution is obtained from size-biased distribution by taking $\alpha = 1$, the length-biased distribution is also called size-biased distribution by some authors. The choice $w(x) = \bar{F}_X(x)/f_X(x) = 1/r_X(x)$ gives the equilibrium distribution with pdf given by $f_Y(x) = \bar{F}_X(x)/E(X)$. If we choose $w(x) = f_X(x+t)/f_X(x)$ we get the remaining (residual) life distribution. Analogous to the remaining or residual life of a unit beyond a certain time up to which it has survived, the time from the point a unit failed to some specified time (when, say, the failure was noted), called inactivity time, and the properties of its distribution have attracted the attention of reliability analysts in recent times.

Inactivity time distribution and its properties have been investigated by several research workers to determine life distributions uniquely, to characterize some such distributions, and to relate properties of the inactivity time distribution with those defining distribution classes based on ageing properties.

We can look upon the failure rate as a transform of the original lifetime variable and work out its distribution conveniently. In fact, we may even consider the distribution of the reversed hazard rate as well. A few studies have been carried out to characterize some life distributions in terms of distributions of their failure rate transforms and to establish comparisons between the parent and the failure rate distributions.

Introduced primarily in the context of life-testing, total-time-on-test (TTT) transform has been somewhat more widely used as a plotting technique to identify the underlying failure mode. Subsequently, this transform and its stochastic properties have been studied in relation to classes of life distributions.

This chapter is devoted to a treatment of these distributions in relation to the parent distributions where these are derived from and the various uses to which properties of these derived distributions have been applied. A few illustrative examples have been provided.

5.2 Remaining (Residual) Life Distribution

In reliability analysis of manufactured products or survival analysis of living beings, we are not always interested in complete life or time-to-failure of a new (unused) unit. Thus, in a life table, we compute complete (or curtate) expectation of life at some age x in respect of individuals who have survived up to that age, besides deriving the complete expectation of life at birth. Similarly, in the case of a product, several units of the product may be put on test and analysis may focus on the fraction of units which survive a certain age t. For a patient entering a treatment protocol at age t, we remain interested in the remaining life of the patient.

Residual life at age t has been defined as $X_t = [X - t | X \geqslant t], t \geqslant 0$. Provided $E(X)$ is finite, mean residual (remaining) life MRL at age t, $m(t) = E(X_t) = E(X | X \geqslant t) - t$ exists and is given by

$$E(X_t) = \frac{1}{\bar{F}_X(t)} \int_t^\infty \bar{F}_X(x)\,dx.$$

Some discussions on MRL function have already been done in Sect. 2.4.2. The variance of residual life can be conveniently defined as

$$Var(X_t) = \frac{2}{\bar{F}(t))} \int_t^\infty \int_y^\infty \bar{F}(u)\,du\,dy - m^2(t),$$

which appears in the formula for variance of the estimated mean residual life at age t. Karlin (1982) studied montonicity of this variance when lifetime density is log-convex (log-concave). Defining the coefficient of variation of residual life at age t by $C_X(t)$, Gupta (1987) has characterized the monotonic behaviour of variance of residual life in terms of $C_X(t)$.

The MRL function, the reliability function, and the failure rate function are all related one to the other in terms of the relations $m'(t) = r(t)m(t) - 1$ where $r(t)$ is the failure rate and

$$\bar{F}(t) = \frac{m(0)}{m(t)} \exp\left\{-\int_0^t \frac{dx}{m(x)}\right\}.$$

It may be pointed out that the slope of the MRL curve is greater than or equal to -1. The coefficient of variation of residual (remaining) life, which can be explicitly obtained in many cases, also finds many applications in the study of orderings among life distributions.

Starting with the differential equation connecting the MRL, the reliability, and the failure rate functions, Dersin (2018) obtained a class of life distributions with a mean remaining life linear in time (age). This class has been found useful in prognostics and health management. This class is characterized by the reliability function

$$\bar{F}(k, t) = \left(\frac{\mu}{\mu - kt}\right)^{1 - 1/k}$$

and a failure rate function

$$r(k, t) = \frac{1 - k}{\mu - kt},$$

defined over the range $0 \leqslant t < \mu/k$, $0 < k \leqslant 1$. Allowing $k \to 0$ yields the Exponential distribution. With $k = 1$, we get the Dirac distribution having MRL at age t as $\mu - t$. If $k = 1/2$, we get the Uniform distribution over the range $(0, 2\mu)$ with the mean remaining life as $m(t) = \mu - t/2$. It is noted that $\mu = m(0)$ is mean life or mean time-to-failure. For this Uniform distribution, failure rate becomes $r(1/2, t) = 1/(2\mu - t)$ and the reliability function is $\bar{F}(t) = 1 - t/(2\mu)$.

It can be easily shown that the ageing rate k can be expressed as

$$k = \frac{1 - C^2(X)}{1 + C^2(X)},$$

where $C(X)$ is the coefficient of variation of X. While the MRL function has been pretty widely used by many authors for classification and characterization of life distributions, it may be of some interest to mention the median residual life at age t. In fact, an attempt towards this direction was made as early as the sixteenth century by the English businessman John Graunt (1620–1674), recognized as the first demographer, who analysed the weekly London Bills of Mortality, constructed what may be called a rudimentary life table and also computed, for a person of age t, a further lifetime x_0 such that there was an even chance of survival up to age $t + x_0$. In other words, x_0 can be regarded as the median residual life at age t. The great astronomer Halley (1656–1742) also contributed to the early development of life tables, though Farr (1860) is considered with the first construction of modern life tables.

As is well known, residual life at any age t for Exponential lifetime distribution is independent of age t and has the same Exponential distribution. Suppose we consider the Weibull distribution with parameters α (shape) and λ (scale) with a survival function given by $\bar{F}(x) = e^{-\lambda x^\alpha}$, $x > 0$. In this case, residual life at age t has the survival function given by

$$P(X_t \geq x) = \frac{\bar{F}_X(x+t)}{\bar{F}_X(t)} = e^{\lambda t^\alpha - \lambda(x+t)^\alpha}$$

and the median x_0 of residual life comes out as the solution of the equation

$$\lambda(t + x_0)^\alpha - \lambda t^\alpha = \ln 2 \cong 0.30103.$$

It can be easily seen that with $\alpha = 1$ (Exponential distribution), $x_0 = (1/\lambda) \ln 2$. With $\alpha = 2$ (Rayleigh distribution) and a scale parameter $\lambda = 1$, it can be shown that median works out as the root of a quadratic equation having a value 0.115 approximately.

For the finite-range distribution corresponding to the Exponentiated Rectangular distribution with the survival function given by

$$\bar{F}(x) = \left(1 - \frac{x}{a}\right)^p, \quad 0 < x < a, \ p > 0,$$

5.2 Remaining (Residual) Life Distribution

the residual life at age t has the survival function as

$$\bar{F}_{X_t}(x) = \left(\frac{a-t-x}{a-t}\right)^p,$$

which can be equated to $1/2$ to get the median residual life at any age t. For example, with $a = 4$ and $p = 2$ we get the median x_0 as the root of the quadratic equation

$$(4-t)^2 - 4x_0(4-t) + 2x_0^2 = 0$$

which has the roots $(4-t)/(2+\sqrt{2})$ and $(4-t)/(2-\sqrt{2})$. Taking $a = 4$ and $p = 2$, the median at age 3 is 0.3 approximately and at age 2 the figure is 0.6 approximately. Here we have taken the first root because the second root gives the median values as approximately 1.7 and 3.4, respectively, which are not tenable in this case.

Schmittlein and Morrison (1981) pointed out several shortcomings of the MRL function. The empirical mean cannot be obtained when the data are truncated. Even with complete data, estimated MRL will be unstable owing to its strong dependence on the very large durations. And in such cases, the median of residual life may be used with some advantage.

Going beyond the median, Haines and Singpurwalla (1974) defined the α-th percentile of remaining life distribution and proposed the increasing (decreasing) α-percentile residual life class IPR-α (DPR-α). The issue of determining the original lifetime distribution uniquely in terms of the percentiles of residual life was examined by Gupta and Langford (1984). Using Schröder's equation, which states that

$$F(\delta(t)) = sF(t), \quad 0 \leqslant t < \infty,$$

where $0 < s < 1$ and δ is a continuous, strictly increasing function on $[0, \infty)$ with $\delta(t) > t$ for every $t \geqslant 0$ and its solution, they could show that a single percentile cannot uniquely determine the original survival function. However, a knowledge of two different percentiles can lead to a unique determination.

Just as expectation dominance has been used to establish an ordering between two distributions, we can also define MRL dominance between any two life distributions. Ahmed (1988) showed that the mean remaining life ordering is closed under convolutions, mixtures, and weak convergence. Coming to ageing properties and corresponding classes of the parent and the residual life distributions, one notes the following results related to dominance of original life (distribution) over remaining life (distribution). For definitions and properties of different stochastic orders mentioned below one may consult Chap. 7.

(a) X dominates X_t in likelihood ratio order if and only if X is PF_2.
(b) X dominates X_t in failure rate order if and only if X is IFR.
(c) X dominates X_t in failure rate average if and only if X is NBU.

(d) X stochastically dominates X_t if and only if X is NBU.
(e) X dominates X_t in expectation if and only if F is $NBUE$[1]
(f) X dominates X_t in MRL order if and only if X is $DMRL$.
(g) X dominates X_t in VRL order if and only if X is $DVRL$[2]

5.3 Equilibrium Distribution

As mentioned earlier, equilibrium distributions were introduced as limiting distributions of forward or backward recurrence times in the context of Renewal Processes. Thus, they can be regarded as limiting distributions of used life or residual life (up to or beyond a given point) in Survival/Reliability Analysis.

The pdf of any equilibrium distribution diminishes monotonically from $1/\mu$ at $x = 0$ to 0 as $x \to \infty$. In general, $E(Y) = (\sigma^2 + \mu^2)/(2\mu)$ where $E(X) = \mu$ and $V(X) = \sigma^2$. Thus, the equilibrium distribution dominates the original distribution in expectation if $\sigma \geqslant \mu$ or $C(X) \geqslant 1$. Some discussion of equilibrium distribution may be available at Sect. 7.5.

It is interesting to note that the original and the equilibrium distributions are identical only for the Exponential distribution. It can be shown that the failure rate $r_Y(t)$ of the equilibrium distribution is the reciprocal of the mean residual life $m_X(t)$ of the original distribution.

Just to illustrate some dominance relations between a life distribution and the corresponding equilibrium distribution, let us take the somewhat trivial case of the Uniform distribution. Let X be distributed as $U(0, \beta)$ with pdf given by $f_X(x, \beta) = 1/\beta$, survival (reliability) function as $\bar{F}_X(x) = 1 - x/\beta$, $E(X) = \beta/2$ and failure rate $r_X(x) = 1/(\beta - x)$ which is increasing in x. The pdf of the random variable Y (i.e, of the corresponding equilibrium distribution) is given by $f_Y(y) = 2(\beta - y)/\beta^2$ so that the survival function works out as $\bar{F}_Y(y) = (1 - y/\beta)^2$. Further, $E(Y) = \beta/3 < \beta/2 = E(X)$. The failure rate function for the equilibrium distribution becomes $r_Y(y) = 2/(\beta - y)$ which is increasing in y and $r_Y(t)$ is strictly greater than $r_X(t)$ for all t. Thus, X dominates Y in hazard rate order.

The likelihood ratio between densities f_X and f_Y becomes $f_X(x)/f_Y(x) = \beta/(2(\beta - x))$ which increases monotonically with x. Thus, the original life distribution of X is dominated by the equilibrium distribution in likelihood ratio order (and hence in usual stochastic order, in expectation order, and in the failure rate order).

---

[1] A non-negative random variable X is said to be $NBUE$ (new better than used in expectation) if $E(X_t) \leqslant E(X)$ for all $t \geqslant 0$. If this inequality is reversed, we say that X is $NWUE$ (new worse than used in expectation).

[2] A random variable X is said to belong to the $DVRL$ (decreasing in variance of residual life) class if $V(X_t)$ is decreasing in t. If $V(X_t)$ is increasing, the variable is said to belong to the $IVRL$ (increasing in variance of residual life) class.

5.3 Equilibrium Distribution

It can be easily shown that the Lorenz function of X is given by

$$L_X(p) = \frac{\int_0^p F_X^{-1}(u)\,du}{\int_0^1 F_X^{-1}(u)\,du} = p^2$$

for every $p \in [0, 1]$. Note that Lorenz function $L(\cdot)$ (commonly known as Lorenz curve) is convex having a bow-like structure with $L(0) = 0$ and $L(1) = 1$. It always lies below the line joining the points $(0, 0)$ and $(1, 1)$, which easily follows from the convexity of the curve.

To find the Lorenz function of Y, note that $F_Y^{-1}(u) = \beta\left(1 - \sqrt{1-u}\right)$ so that

$$\begin{aligned}
L_Y(p) &= \frac{\int_0^p F_Y^{-1}(u)\,du}{\int_0^1 F_Y^{-1}(u)\,du} \\
&= \frac{\int_0^p \left(1 - \sqrt{1-u}\right) du}{\int_0^1 \left(1 - \sqrt{1-u}\right) du} \\
&= 3p + 2(1-p)^{3/2} - 2 \\
&< p^2 = L_X(p).
\end{aligned}$$

Thus, the Lorenz curve of X dominates that of Y for all $p \in [0, 1]$.

For the Weibull distribution with scale parameter unity and shape parameter α, we have $E(X) = \Gamma\left(\frac{1}{\alpha} + 1\right)$, and

$$\begin{aligned}
E(Y) &= \frac{1}{E(X)} \int_0^\infty x \bar{F}_X(x)\,dx \\
&= \frac{1}{2E(X)} \int_0^\infty x^2 f_X(x)\,dx \quad (5.3.1) \\
&= \frac{\alpha}{2E(X)} \int_0^\infty x^{\alpha+1} e^{-x^\alpha}\,dx \\
&= \frac{1}{2E(X)} \int_0^\infty t^{2/\alpha} e^{-t}\,dt \\
&= \frac{\Gamma(2/\alpha)}{\Gamma(1/\alpha)}.
\end{aligned}$$

From (5.3.1) we note that

$$E(Y) = \frac{\Gamma(2/\alpha)}{\Gamma(1/\alpha)} \begin{cases} > E(X), & \text{if } \alpha < 1 \\ < E(X), & \text{if } \alpha > 1. \end{cases}$$

We can easily see that $f_X(x)/f_Y(x)$ increases (decreases) in x if $\alpha > (<)1$, i.e., for a *DFR* Weibull distribution, the equilibrium distribution dominates the original distribution in likelihood ratio order (and hence, in failure rate order, stochastic order, and expectation order). Some general results regarding equilibrium distributions are discussed in Theorem 7.8.18.

Some stochastic dominance relations between the original random variable and its equilibrium random variable are discussed below while some more are discussed in Sect. 7.8.3 but before that let us define a few related concepts.

Write $\phi = F_Y^{-1} F_X$. It may be mentioned here that if F_X and F_Y are two distribution functions, we say that F_Y is convex ordered w.r.t. F_X or, equivalently, Y dominates X in convex transform order (written as $X \leqslant_c Y$) if $\phi(x)$ is convex, i.e., $\phi(\lambda x + (1 - \lambda)y) \leqslant \lambda \phi(x) + (1 - \lambda)\phi(y)$ for $x, y \geqslant 0$ and for all $\lambda \in [0, 1]$. Again, F_Y is star shaped w.r.t. F_X (denoted by $X \leqslant_* Y$) if $\phi(x)$ is star shaped, i.e., $\phi(\lambda x) \leqslant \lambda \phi(x)$ for $x \geqslant 0$ and for all $\lambda \in [0, 1]$. This is equivalent to saying that $\phi(x)/x$ is increasing in $x > 0$. Similarly, F_Y is super-additive in relation to F_X, denoted as $X \leqslant_{sa} Y$, if ϕ is super-additive, i.e., if $\phi(x + y) \geqslant \phi(x) + \phi(y)$ for all $x, y \geqslant 0$.

It is well known that every convex function is star shaped and every star-shaped function is super-additive (cf. Bruckner and Ostrow, 1962). If the inequalities in above are reversed, the function is called concave, anti-star shaped, and sub-additive, respectively.

As earlier, below we take Y as the equilibrium random variable corresponding to X. It can be shown that

(a) F_Y is convex ordered in relation to F_X implies F_X is IFR;
(b) F_Y is star shaped in relation to F_X implies F_X is $IFRA$;
(c) F_Y is super-additive in relation to F_X implies F_X is $NBUE$.

5.4 Distribution of Failure Rate Transform

Analysis of failure rate $r(x)$ has been usually made in the 'time domain' in terms of its behaviour with respect to time or age x. It is quite logical to treat $r(X)$ as a random function being a transform of X, and its probability distribution can be conveniently worked out to analyse failure rate in the 'frequency domain'. Thus, one may be interested in the probability that the failure rate of a certain device or unit will never exceed a given value. While a lot has been reported on the time domain analysis of failure rate, not much has been done in respect of the frequency domain analysis. And the probability distribution of the failure rate transform has provided some interesting characterization results. Let us denote the failure rate transform as Y with a density function g and a distribution function G.

It is well known that if lifetime X follows an Exponential distribution, its failure rate follows a degenerate distribution, and conversely if $r(X)$ has a degenerate distribution then X has an Exponential distribution. In fact, one related result that can be easily proved by using the Cauchy-Schwarz inequality can be stated as follows.

Theorem 5.4.1 *For a non-negative random variable X, $E[r(X)] = 1/E(X)$ is a necessary and sufficient condition for X to have an exponential distribution.* □

5.4 Distribution of Failure Rate Transform

Just incidentally, one notes the fact that, for the Rayleigh distribution, $E(r(X)) = 2E(X)$.

Makino (1984) proved that, for any non-negative random variable X,

$$E\left(\frac{1}{r(X)}\right) \geq \frac{1}{E(r(X))}.$$

The equality in the above expression holds if and only if X follows an Exponential distribution. This has motivated Nanda (2010) to prove the following result.

Theorem 5.4.2 *Let X be a non-negative random variable with $E(X) = \mu$ and coefficient of variation c. Then*

(i) $E(Xr(X)) \geq \frac{2}{1+c^2}$, *where equality holds if and only if X follows Exponential distribution with mean μ and coefficient of variation c;*

(ii) $E\left(\frac{r(X)}{X}\right) \geq \frac{2}{\mu^2(1+c^2)}$, *where equality holds if and only if X follows Exponential distribution with mean μ and coefficient of variation c.* □

Similarly, considering $m(x) = E(X - x | X \geq x)$ in frequency domain as above, Nanda (2010) proved the following result.

Theorem 5.4.3 *For any non-negative random variable X,*

(a) $E\left(\frac{1}{m(X)}\right) \geq \frac{1}{E(m(X))}$. *Here the equality holds if and only if X follows as Exponential distribution.*

(b) $E\left(\frac{m(X)}{r(X)}\right) \geq \frac{1}{E\left(\frac{r(X)}{m(X)}\right)}$, *where equality holds if and only if X follows Exponential distribution.* □

A further generalization of this result is given in the following:

Theorem 5.4.4 *For a non-negative random variable X, if raw moment of order k is denoted by μ_k, then $E\left(X^{2\alpha}r(X)\right) \geq \frac{\mu_{\alpha+\beta}}{\mu_{2\beta+1}}$, with equality holding true if and only if $r(x)$ is proportional to $x^{\beta-\alpha}$.* □

Bhattacharjee et al. (2013b) have proved the following characterization result where $L(\cdot)$ is the corresponding ageing intensity function discussed in Sect. 2.4.6.

Theorem 5.4.5 *For any non-negative random variable X,*

$$E\left(\frac{L(X)}{r(X)}\right) \geq \frac{1}{E\left(\frac{r(X)}{L(X)}\right)}.$$

The equality holds above if and only if X has an Exponential distribution. □

For the Pearsonian Type XI distribution[3] with failure rate

$$r(x) = \frac{q}{a}\left(1 + \frac{x}{a}\right)^{-1}, \quad x > 0, a > 0, q > 1$$

the failure rate transform given by $r(X) = \frac{q}{a}\left(1 + \frac{X}{a}\right)^{-1}$ has the distribution function given by

$$G(y) = P\left(\frac{q}{a}\left(1 + \frac{X}{a}\right)^{-1} \leq y\right) = P\left(X \geq \frac{q - ay}{y}\right) = \left(\frac{a}{q}\right)^q y^q,$$

which is the distribution function of the Power distribution. While $E(X) = a/(q - 1)$, the expected failure rate works out as $q^2/[a(q+1)]$.

It is interesting to note that the random variable $Z^* = Xr(X)$ has an expectation that equals the shape parameter in the Weibull distribution of X. Thus the sample mean of Z^* can provide an unbiased estimator of the shape parameter. However, Z^* in this case cannot be directly observed and does involve estimation (usually non-parametric) of the failure rate.

There exist situations where the mean failure rate does not exist. For example, if X follows the Uniform distribution over $[0, \alpha]$ the failure rate function $r(X) = 1/(\alpha - X)$ increases from $1/\alpha$ at $x = 0$ to infinity at $x = \alpha$ with a survival function $1/(\alpha x)$, the mean failure rate does not exist.

The distribution of failure rate transform yields an important characterization (due to Mukherjee and Roy, 1987) of the Weibull distribution contained in the following result.

Theorem 5.4.6 *If a random variable X follows DFR Weibull distribution (or has a shape parameter less than unity) then its failure rate transform Y follows Inverse Weibull distribution, and conversely if X belongs to the class of DFR distributions with y tending to infinity as x tends to 0 then Y follows inverse Weibull distribution only if X is DRF Weibull.* □

A similar result is the following.

Theorem 5.4.7 *If a random variable X follows IFR Weibull distribution, then Y, the failure rate transform of X, also follows IFR Weibull distribution, and conversely if X belongs to the class of IFR distributions with y tending to 0 as x tends to 0, then Y follows IFR Weibull distribution only if X follows IFR Weibull distribution.*
Proof In this case

$$P(Y \geq y) = P\left[X \geq \left(\frac{y}{\lambda\alpha}\right)^{1/(\alpha-1)}\right] = e^{-\lambda' y^{\alpha'}},$$

where $\lambda' = \lambda(\lambda\alpha)^{-\alpha'}$ and $\alpha' = \alpha(\alpha - 1)^{-1}$. Observing that $\alpha' > 1$ and $\lambda' \geq 0$, we conclude that Y has IFR Weibull distribution. The proof of the converse is easy. □

---

[3] Some characterization results for this distribution are discussed in Sects. 9.6 and 9.8.

Coming to the convex transform ordering between X and Y (the failure rate transform), Mukherjee and Roy (1987) gave the following theorem.

Theorem 5.4.8 *If X follows the IFR Weibull distribution, then whenever $1 < \alpha < 2$, $F \geqslant_c G$ and hence*

(a) *$\bar{G}(x)$ crosses $\bar{F}(x)$ at most once from above as x increases from 0 to ∞;*
(b) *G is star shaped (and hence super-additive) with respect to F, i.e., $F \geqslant_\star G$ and $F \geqslant_{su} G$;*
(c) *$C(X) \geqslant C(Y)$.*

Further, when $\alpha > 2$, $F \leqslant_c G$ with corresponding implications $\bar{F}(x)$ crosses $\bar{G}(x)$ at most once from above, F is star shaped and super-additive with respect to G, and $C(Y) \geqslant C(X)$, i.e., coefficient of variation of Y is greater than that of X. □

It may be incidentally pointed out that the point of crossing of the two ogives is given by $x_0 = (\lambda \alpha)^{2-\alpha}$.

Examining expectation dominance of X over Y, we get some interesting results for X following the Weibull distribution. Similar results can be explored in respect of other distributions of X. We now provide one such result in respect of Weibull X.

Theorem 5.4.9 *For a random variable X following $W(\alpha, \lambda)$ distribution, $X \leqslant_E Y$ if $\lambda > \lambda_0(\alpha)$, where*

$$\lambda_0(\alpha) = \left(\frac{\Gamma\left(1+\frac{1}{\alpha}\right)}{\alpha \Gamma\left(2-\frac{1}{\alpha}\right)} \right)^{\alpha/2}.$$

Proof The proof follows from a direct comparison of $E(X)$ with $E(Y)$. □

It comes out that, for $\lambda > 1.1$, $X \leqslant_E Y$ irrespective of the choice of α. Also, λ_0 steadily increases in $\alpha \in (0.5, 0.76331)$ approximately and then steadily decreases (see Fig. 5.1).

5.5 Inactivity Time or Past Residual Life

The reversed residual lifetime, also called the inactivity time (cf. Ruiz and Navarro, 1994), is formally defined as the conditional random variable $_t X = (t - X | X \leqslant t)$. Properties of the distribution of $_t X$ have been used by various workers to characterize some life distributions, to establish stochastic ordering among distributions and even to delineate some life distribution classes. In a somewhat broader context of event-history analysis in medicine and related fields, the inactivity time, being the time between the point of time an event of interest took place to the current time, provides an important summary of developments during the interval that can help deciding on the appropriate action to be taken on the unit (in case it is repairable).

The expected or mean inactivity time EIT (also called expected stopped time, somewhat inappropriately) is defined as

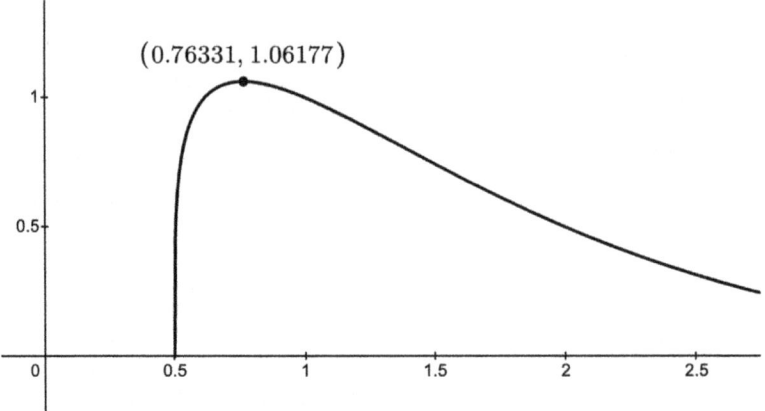

Fig. 5.1 Figure shows the shape of $\lambda_0(\alpha)$

$$\alpha(t) = E(_tX) = \frac{1}{F(t)} \int_0^t F(x)\,dx.$$

It can be easily shown that the reversed hazard rate $\mu(t) = (1 - \alpha'(t))/\alpha(t)$ where $\alpha'(t)$ is the derivative of $\alpha(t)$ with respect to time t. And this way, it follows that EIT, $\alpha(t)$, determines the life distribution uniquely. Kundu and Nanda (2010b) have used higher order moments of inactivity time to characterize some life distributions. We note from Kundu and Nanda that the rth-order moment of the inactivity time given by

$$\alpha_r(t) = E[(t - X)^r | X \leqslant t]$$

satisfies the relation

$$\alpha'_{r-1}(t) = (r-1)\alpha_{r-2}(t) + \left[\frac{\alpha'_r(t)}{\alpha_r(t)}\right] \alpha_{r-1}(t) - \frac{r\alpha^2_{r-1}(t)}{\alpha_r(t)}.$$

They have proved the following theorem to show that not only $\alpha(t)$ but also $\alpha_2(t)$ alone is enough to determine the distribution uniquely.

Theorem 5.5.1 *If X has a finite variance, then the distribution of X is uniquely determined by $\alpha_2(t)$.* □

The following theorem shows that the ratio of two consecutive moments determines the distribution uniquely.

Theorem 5.5.2 *For a given value of r, if $\alpha_r(t)/(r\alpha_{r-1}(t)) = g(t)$ is known, then the distribution of X is determined uniquely, provided $g(t)$ is at least r times differentiable with $g'(t) \neq 1$ for all t.* □

These authors provide the following characterization of a finite-range distribution based on the first two moments of inactivity time.

Theorem 5.5.3 *If*
$$\frac{\alpha_2(t)}{2\alpha(t)} = ct, \ 0 < t < 1, \ 0 < c < \frac{1}{2}$$
then the underlying distribution has the distribution $F(t) = t^{(1/c)-2}$ and conversely.
□

The case $c = 1/3$ corresponds to the $U(0, 1)$ distribution. In some sense, $F(t)$ can be looked upon as an Exponentiated Rectangular distribution, as discussed in a subsequent chapter.

In Theorem 5.5.2, if we take $r = 2$ then $g(t) = c(> 0)$, a constant, characterizes the distribution having cumulative distribution function
$$F(t) = e^{\frac{t-b}{b-\mu}}, \ t \in (-\infty, b),$$
where μ is the mean of the distribution. The same distribution has been characterized for $\alpha(t) = 1$ by Chandra and Roy (2001).

Defining increasing (decreasing) expected inactivity time ($IEIT$ ($DEIT$)) class as $E(_t X)$ increasing (decreasing), and increasing (decreasing) variance inactivity time ($IVIT$ ($DVIT$)) class as $V(_t X)$ increasing (decreasing), Nanda et al. (2003) relate these classes to the increasing (decreasing) reversed hazard rate class, earlier defined by Chandra and Roy (2001). In fact, they showed that the DRHR class is stronger than the IVIT class. They proceed further to prove that the $IEIT$ ($DEIT$) class is a sub-class of the $IVIT$ ($DVIT$) class.

For establishing stochastic orders between life distributions based on expected inactivity times, it is preferable to consider quantile EIT values computed at quantiles of specified orders for each of the distributions. This has been extensively studied by Nair et al. (2013). Some more discussions on inactivity time may be obtained in Sect. 7.7.

5.6 Total Time on Test

Epstein and Sobel (1954) introduced the Total Time on Test (TTT) statistic in the context of life-testing and reliability as a tool for testing exponentiality defined as
$$S_i = \sum_{j=0}^{i} (n - j + 1) \left(x_{(j)} - x_{(j-1)} \right), \ i = 1, 2, \ldots, n,$$
where $x_{(j)}$s are the ordered failure times out of n units put on test simultaneously.

Barlow and Campo (1975) considered the asymptotic form of TTT as the sample size n increases and defined the TTT transform of the parent life distribution, with distribution function F, as

$$T_u = \int_0^{F^{-1}(u)} \bar{F}(t)\,dt,\ 0 \leqslant u \leqslant 1.$$

They used this transform for model identification purpose. The scaled TTT transform T_u/T_1, $0 \leqslant u \leqslant 1$, where $T_1 = \mu = E(X)$, can be plotted on a unit square, consisting of linear segments starting at $(0, 0)$ and ending at $(1, 1)$.

It can be verified that T_u^{-1} is a distribution function on the support $[0, \mu]$, where μ is the mean of F. We can see that if X is distributed exponentially, T_u^{-1} (the transform) has a Rectangular distribution. Further, $F^{-1}T_u^{-1}(x)$ is convex for $0 \leqslant x \leqslant \mu$.

Barlow (1979) showed the following relations between life distributions and TTT transform distribution T_u^{-1}.

F	T_u^{-1}
Exponential	Linear
IFR	Convex
DFR	Concave
$IFRA$	Star shaped
$DFRA$	Anti-star shaped

Nair et al. (2013) give necessary and sufficient conditions for IFR, $IFRA$, $DMRL$, NBU, $NBUE$, $HNBUE$, $NBUFR$, and NBU-t_0 classes in terms of the TTT transform. They also indicate that the transform determines the original life distribution uniquely. A distribution F with mean μ ($< \infty$) is said to be Harmonic New Better than Used in Expectation ($HNBUE$) if

$$\int_t^\infty \bar{F}(x)\,dx \leqslant \mu e^{-t/\mu},\ t \geqslant 0. \tag{5.6.2}$$

If the inequality in (5.6.2) is reversed, the class is called Harmonic New Worse than Used in Expectation ($HNWUE$) (cf. Rolski, 1975; Klefsjö, 1981). An absolutely continuous non-negative random variable having failure rate r is said to be New Better than Used in Failure Rate ($NBUFR$) if $r(t) \leqslant r(0)$ for all $t \geqslant 0$. In case this inequality is reversed, we call this as New Worse than Used in Failure Rate $NWUFR$ (cf. Abouammoh and Ahmed, 1988). A life distribution F is said to be New Better than Used at t_0 (NBU-t_0) if, for all $x \geqslant 0$ (cf. Zahui and Xiaohu, 1998),

$$\bar{F}(x+y) \leqslant \bar{F}(x)\bar{F}(y),\ \text{for all } y \geqslant t_0 > 0.$$

The dual class of New Worse than Used at t_0 (NWU-t_0) is defined analogously by reversing the above inequality.

It can be easily shown that $T_u/T_1 = u$, for exponential distribution. This implies the equiangular line within the unit square. In fact, there is a one-one correspondence between the parent distribution F and its TTT transform. Many of the ageing properties of a life distribution are related to the corresponding properties for the transform.

5.7 Order Statistics and Record Values

Let $X_1, X_2, \ldots, X_n$ be a random sample of size n from a cdf F. Then $X_{1:n} \leqslant X_{2:n} \leqslant \ldots \leqslant X_{n:n}$ represent the order statistics corresponding to the random sample $X_1, X_2, \ldots, X_n$. Here $X_{i:n}$ is called the ith-order statistic. When there is no ambiguity about the sample size n, we generally write $X_{(i)}$ to mean $X_{i:n}$. It is to be noted that although $X_1, X_2, \ldots, X_n$ are independent, $X_{(1)} \leqslant X_{(2)} \leqslant \ldots \leqslant X_{(n)}$ are not. This is because $\{X_{(i)} \geqslant x\} \Rightarrow \{X_{(i+1)} \geqslant x\}$.

5.7.1 Distribution of Order Statistics

Below we discuss distribution of order statistics.

Theorem 5.7.1 *Let $X_{(1)} \leqslant X_{(2)} \leqslant \ldots \leqslant X_{(n)}$ represent the order statistics from a cdf F. The marginal cdf of $X_{(i)}$, $i = 1, 2, \ldots, n$, is given by*

$$F_{X_{(i)}}(x) = \sum_{j=i}^{n} \binom{n}{j} F^j(x)(1 - F(x))^{n-j}.$$

Proof For fixed x, let $Z_i = I_{(-\infty, x]}(X_i)$. Then

$$Y = \sum_{j=1}^{n} Z_j = \{Number\ of\ X_j \leqslant x\} \sim b(n, F(x)).$$

Note that

$$\{X_{(i)} \leqslant x\} \equiv \left\{ \sum_{j=1}^{n} Z_j \geqslant i \right\} \equiv \{Y \geqslant i\}.$$

Now,

$$\begin{aligned} F_{X_{(i)}}(x) &= P(X_{(i)} \leqslant x) \\ &= P(Y \geqslant i) \\ &= \sum_{j=i}^{n} \binom{n}{j} F^j(x)(1 - F(x))^{n-j}. \end{aligned}$$

Corollary 5.7.1 *Taking $i = 1$ and $i = n$ we get the distribution functions of $X_{(1)}$ and $X_{(n)}$ given, respectively, as*

(a) $F_{X_{(1)}}(x) = 1 - (1 - F(x))^n$;
(b) $F_{X_{(n)}}(x) = F^n(x)$. □

The density function $f_{X_{(i)}}$ of $X_{(i)}$ is obtained as

$$f_{X_{(i)}}(x) = \frac{d}{dx} F_{X_{(i)}}(x) = \frac{n!}{(i-1)!(n-i)!} F^{i-1}(x)(1-F(x))^{n-i} f(x).$$

The same can also be obtained as follows:

$$\begin{aligned}
f_{X_{(i)}}(x) &= \lim_{\Delta x \to 0} \frac{F_{X_{(i)}}(x+\Delta x) - F_{X_{(i)}}(x)}{\Delta x} \\
&= \lim_{\Delta x \to 0} \frac{P(x < X_{(i)} \leq x + \Delta x)}{\Delta x} \\
&= \lim_{\Delta x \to 0} \frac{1}{\Delta x} \big[P\{(i-1) \text{ of the } X_k \leq x; \text{ one } X_k \text{ in } (x, x+\Delta x]; \\
&\quad (n-i) \text{ of the } X_k > x + \Delta x\} \big] \\
&= \lim_{\Delta x \to 0} \frac{n!}{(i-1)!1!(n-i)!} \frac{F^{i-1}(x)(F(x+\Delta x) - F(x))(1 - F(x+\Delta x))^{n-i}}{\Delta x} \\
&= \frac{n!}{(i-1)!(n-i)!} F^{i-1}(x)(1-F(x))^{n-i} f(x).
\end{aligned}$$

The second last line follows from multinomial distribution.

Similarly, the joint density of $X_{(i)}$ and $X_{(j)}$, $1 \leq i < j \leq n$, can be obtained as follows. Let $x < y$. Then

$$\begin{aligned}
f_{X_{(i)}, X_{(j)}}(x, y) &= \lim_{\substack{\Delta x \to 0 \\ \Delta y \to 0}} \frac{1}{\Delta x . \Delta y} P[x < X_{(i)} \leq x + \Delta x, \ y < X_{(j)} \leq y + \Delta x] \\
&= \lim_{\substack{\Delta x \to 0 \\ \Delta y \to 0}} \frac{1}{\Delta x . \Delta y} P\big[(i-1) \text{ of the } X_k \leq x; \text{ one } X_k \text{ in } (x, x+\Delta x]; \ (j-i-1) \\
&\quad \text{of } X_k \text{ in } (x+\Delta x, y]; \text{ one } X_k \text{ in } (y, y+\Delta y]; \ (n-j) \text{ of the } X_k > y + \Delta y \big] \\
&= \frac{n!}{(i-1)!1!(j-i-1)!1!(n-j)!} \lim_{\substack{\Delta x \to 0 \\ \Delta y \to 0}} \frac{1}{\Delta x . \Delta y} \Big[F^{i-1}(x)(F(x+\Delta x) - F(x)) \\
&\quad (F(y) - F(x+\Delta x))^{j-i-1}(F(y+\Delta y) - F(y))(1 - F(y+\Delta y))^{n-j} \Big] \\
&= \frac{n!}{(i-1)!(j-i-1)!(n-j)!} F^{i-1}(x)(F(y) - F(x))^{j-i-1}(1 - F(y))^{n-j} f(x) f(y).
\end{aligned}$$

On using similar argument as above, the joint distribution of $X_{(1)}, X_{(2)}, \ldots, X_{(n)}$ can be obtained as

$$f_{X_{(1)}, X_{(2)}, \ldots, X_{(n)}}(x_1, \ldots, x_n) = n! f(x_1). f(x_2) \ldots f(x_n). \quad (5.7.3)$$

Any set of marginal distributions can be obtained from (5.7.3) by integrating out the remaining variables.

5.7.2 Record Values

The concept of records was first mathematically formulated by Chandler (1952) for *iid* observations. Since then people have tried to study different results related to records. However, in last seven decades, the results on records have not been so extensively studied in the literature as it should have been. The reason behind this is the difficulty level of the problems once we consider non-*iid* observations. A monograph in this direction is by Arnold et al. (1998).

For a sequence of *iid* random variables X_i, $i = 1, 2, \ldots$, having absolutely continuous distribution function F with finite mean, X_n is said to be an upper record value (or simply a record) if $X_n > X_i$, for every $i < n$. Clearly, X_1 is trivially a record. Let T_n, $n = 1, 2, \ldots$, be the random variables denoting the record times. Then we have $T_1 = 1$ and

$$T_n = \min\{j : X_j > X_{T_{n-1}}\},$$

for $n > 1$. The sequence of record values corresponding to X_n is then defined by $R_n^X = X_{T_n}$, $n > 1$. Here R_n^X is known as the nth upper record. A lower record can analogously be defined.

Dziubdziela and Kopociński (1976) have introduced a sequence of k-records as $T^{(k)}(1) = 1$, and

$$T^{(k)}(n+1) = \min\{j > T^{(k)}(n) : X_{j:j+k-1} > X_{T^{(k)}(n):T^{(k)}(n)+k-1}\},$$

for $n \geq 1$. Here $T^{(k)}(n)$ is known as nth k-record time of $\{X_i, i \geq 1\}$ and $R_{n:k}^X$, defined as

$$R_{n:k}^X = X_{T^{(k)}(n):T^{(k)}(n)+k-1},$$

is known as nth upper k-record. Clearly $R_{n:1}^X = R_n^X$. Kundu et al. (2009) have shown that, for $k \in \mathbb{N}$,

$$R_n^X \text{ is DRHR} \Rightarrow R_{n:k}^X \text{ is DRHR}$$
$$\Rightarrow R_{n-1:k}^X \text{ is DRHR}$$
$$\Rightarrow R_{n-1}^X \text{ is DRHR}.$$

Let us define, for $n \geq m$,

$$\Psi_{n:m}^{(k)}(X; x) = \left(R_{n:k}^X - x \mid R_{m:k}^X > x\right),$$

the generalized residual life of records and

$$\psi_{n:m}^{(k)}(x) = E\left(\Psi_{n:m}^{(k)}(X; x)\right),$$

which is known as generalized mean residual life of records. Kundu and Nanda (2010a) have shown the following result.

Theorem 5.7.2 *For any $x \geq 0$,*

(i) $\psi_{n:m}^{(k)}(x)$ *is increasing in n if m and k are fixed.*
(ii) $\psi_{n:m}^{(k)}(x)$ *is decreasing in m if n and k are fixed.*
(iii) $\psi_{n:m}^{(k)}(x)$ *is decreasing in k if m and n are fixed.* □

They have characterized the IFR class and the DFR class in terms of record values as follows.

Theorem 5.7.3 *A random variable X is IFR (resp. DFR) if and only if $\Psi_{n:1}^{(k)}(X;x)$ is stochastically decreasing (resp. increasing) in x.* □

The monotonicity of $\Psi_{n:m}^{(k)}(X;x)$ has also been studied by them as given below.

Theorem 5.7.4 *1. If X is IFR then $\Psi_{n:m}^{(k)}(X;x)$ is stochastically decreasing in x.*
2. If $\Psi_{n:m}^{(k)}(X;x)$ is stochastically increasing in x, then X is DFR.
3. If, for $x \geq 0$, $\Psi_{n:m}^{(k)}(X;x) \leq_{st} R_{n:k}^{X}$ then X is NBU.
4. If X is NWU then $\Psi_{n:m}^{(k)}(X;x) \geq_{st} R_{n:k}^{X}$, for $x \geq 0$. □

Let $R_{n:k}^{X}$ be the nth k-record corresponding to the random variable X and let $R_{n':k'}^{Y}$ be the n'-th k'-record corresponding to the random variable Y with $k' \leq k$. Let $X_t = (X - t | X > t)$ be the residual random variable corresponding to the random variable X. Kundu and Nanda (2010a) have shown that, for $t \geq 0$,

$$X_t \leq_{hr} Y_t \Rightarrow \Psi_{n:m}^{(k)}(X;t) \leq_{hr} \Psi_{n':m'}^{(k')}(Y;t),$$

whenever $n > m$, $n' > m'$ $n - 1 \leq n' - m'$.

Some more characterization results in terms of record values are discussed in Sect. 9.3.

Chapter 6
Finite-Range Life Distributions

6.1 Introduction

The 'mission time' that guides the design of a manufactured item as also its 'design life' which incorporates 'safety factors' to ensure that the item does not fail before the mission time, are always finite. In stress-strength analysis, both stress encountered by the product during use or deployment and strength built into the product through its design and conformance of manufacture to design do vary over finite ranges. Time to repair of a failed product or time between two consecutive repairs is also finite. Besides these continuous variables, number of successful operation cycles before failure or number of repairs before eventual disposal of a product and similar other discrete variables have finite ranges of variation.

Modelling such variables, we need probability distributions over finite range as distinct from the commonly used life distributions with infinite support. Life distributions are generally quite skewed, though repair time may follow a symmetrical distribution. One possibility is to truncate some existing life distribution (with an infinite support) beyond a certain high value of the underlying variable. Some finite-range life distributions have, of course, been proposed by a few authors. Among interesting features of such distributions, non-monotonicity of the failure rate function and its consequences are quite appealing. In fact, some of these distributions exhibit a bathtub shape of the failure rate function, not just like some other infinite-range distributions in which the failure rate is initially monotonically diminishing to reach a minimum and increases monotonically thereafter. In fact, developments in the area of bathtub-shaped failure rate distributions have incidentally led to certain finite-range life distributions.

This chapter is devoted to a discussion on finite-range life distributions—both continuous and discrete, their properties and associated issues of parameter estimation, of characterization, of generalizations, of application in reliability analysis, etc. It should be added that most of the material content of this chapter has laid more stress on two distributions studied by Mukherjee and his associates. Some reference has been made to estimation of stress-strength reliability as well. Use of existing life distributions truncated beyond a certain upper bound has simply been mentioned.

6.2 Some Finite-Range Continuous Distributions

Among the few finite-range continuous life distributions reported in reliability literature, we have the following distribution functions:

$$F_1(x) = \left(\frac{x}{a}\right)^p, \ 0 < x < a, \ p > 0 \tag{6.2.1}$$

and

$$F_2(x) = 1 - \left(1 - \frac{x}{a}\right)^p, \ 0 < x < a, \ p > 0. \tag{6.2.2}$$

The first distribution, suggested by Mukherjee and Islam (1983), can be regarded as the Exponentiated Rectangular distribution (sometimes referred to as the Power Law distribution), while the second, studied by Mukherjee and Roy (1986) (see also Mukherjee, 2017), besides some others, can be looked upon as a Beta-type distribution. The Rectangular distribution is a particular case of either distribution as $p = 1$. If a random variable X has the distribution given by (6.2.1) we denote it as $X \sim FR_1(a, p)$, whereas if X follows (6.2.2) we denote it by $X \sim FR_2(a, p)$. However, asymmetrical forms are also covered. For the first distribution, the failure rate becomes

$$r(t) = \frac{pt^{p-1}}{a^p - t^p}, \ 0 \leqslant t < a; \ p > 0$$

with $r(t) \to \infty$ as $t \to a$. Moore and Lai (1994) proposed another finite-range distribution having a bathtub-shaped failure rate function given by

$$r(t) = c(t + p)^{a-1}(q - t)^{b-1}, \ 0 < a, b < 1, \ 0 \leqslant t \leqslant q, c > 0, \ p \geqslant 0$$

with $r(0) = cp^{a-1}q^{b-1}$ and $r(t) \to \infty$ as $t \to q$.

Lai and Moore (1998) considered a lifetime distribution for which the cumulative failure rate is a beta function, but the failure rate is given by

$$r(t) = t^{a-1}(1 - t)^{b-1}[a - (a + b)t], \ 0 < t < 1, \ a > 0, \ b < 0.$$

It is obvious that $r(t) \to \infty$ as $t \to 0$ provided $0 < a < 1$. A minor extension yields

$$R(t) = \int_0^t r(x)dx = ct^a(1 - dt)^b, \ b < 0, \ 0 < a \leqslant 1, \ 0 < t < 1/d.$$

6.2 Some Finite-Range Continuous Distributions

It may be noted that the above expressions for the hazard rate correspond to life distributions which have complicated probability density functions.

Govindarajulu (1977) defined a class of finite-range distributions given by their quantiles in the form

$$x = \theta + \sigma\left[(\lambda+1)u^\lambda - \lambda u^{\lambda+1}\right],$$

where θ, λ, and σ are location, shape, and scale parameters, respectively, and $u = F(x; \theta, \sigma, \lambda)$ with $\theta \leqslant x \leqslant \theta + \sigma$. It can be easily shown that $\mu = 2/(\lambda+2)$ and $\sigma_Z^2 = (5\lambda+3)(2\lambda+1)^{-1}(2\lambda+3)^{-1} - 4(\lambda+2)^{-2}$ where $Z = (X-\theta)/\sigma$. The hazard rate of the standardized variate is given by

$$r(u) = \left[\lambda(\lambda+1)u^{\lambda-1}(1-u)^2\right]^{-1},$$

where $u = F(z; \lambda) = F(z)$ with $z = (\lambda+1)u^\lambda - u^{\lambda+1}$. It follows that

$$\lim_{z\to 0} r(z) = \lim_{u\to 0} r(z) = \begin{cases} 0, & \text{if } \lambda < 1 \\ 0.5, & \text{if } \lambda = 1 \\ \infty, & \text{if } \lambda > 1 \end{cases}$$

and $\lim_{z\to\lambda} r(z) = \lim_{u\to 1} r(z) \to \infty$. When u is close to 0, $u \cong [z/(\lambda+1)]^{1/\lambda}$ and when u is close to 1, we have $u \cong (z/\lambda)^{1/\lambda}$. This distribution with a bathtub-shaped or U-shaped failure rate has not found much of an application in real-life situations.

It does not have tractable forms of the density or the distribution function. Estimates of the parameters indicated by Govindarajulu are given by

$$\widehat{\theta} = X_{(1)},$$
$$\widehat{\sigma} = X_{(n)} - X_{(1)},$$
$$\widehat{\lambda} = 2(\widehat{\sigma} - \bar{X} + \widehat{\theta})/(\bar{X} - \widehat{\theta}).$$

Haupt and Schäbe (1992, 1994) proposed a finite-range distribution with failure rate function given by

$$r(t) = \frac{1+2\beta}{2T\sqrt{\beta^2+(1+2\beta)t/T}\left[1+\beta-\sqrt{\beta^2+(1+2\beta)t/T}\right]}, \quad 0 \leqslant t \leqslant T, \ \beta > -1/2,$$

where the support of the distribution is $[0, T]$ for some $T > 0$. It is to be noted that $r(t)$ is increasing in t for $\beta \in (-1/2, -1/3) \cup (1, \infty)$ and bathtub shaped for $\beta \in [-1/3, 1]$ with $r(0) = (1+2\beta)/(2T\beta)$ and $r(t) \to \infty$ as $t \to T$. The cumulative distribution function is given by

$$F(t) = \begin{cases} 0, & \text{if } t \leqslant T \\ -\beta + \sqrt{\beta^2+(1+2\beta)t/T}, & \text{if } 0 \leqslant t \leqslant T \\ 1, & \text{otherwise.} \end{cases}$$

The corresponding density is given by

$$f(t) = \frac{1+2\beta}{2T\sqrt{\beta^2 + (1+2\beta)t/T}}, \quad 0 \leqslant t \leqslant T.$$

For $\beta > 0$, the density is decreasing from $(1+2\beta)/(2\beta T)$ at $t = 0$ to $(1+2\beta)/(2(1+\beta)T)$ at $t = T$. Note that $\beta = 0$ gives the Mukherjee-Islam model with $p = 1/2$ in (6.2.1).

Sindhu et al. (2020) considered the Marshall-Olkin extension of the Exponentiated Rectangular distribution to have the three-parameter distribution function as

$$G(x) = \frac{x^p}{\lambda a^p + (1-\lambda)x^p}.$$

This is a unimodal distribution with its spread increasing with the parameter λ.

It will be of interest to consider distributions of system life with components having identical or different finite-range marginal distributions. With components functioning independently within simple coherent systems, this can be easily done. Otherwise, we proceed to work out multivariate (essentially bivariate) extensions of finite-range distributions and then work system life distributions. It must be admitted, however, that finite-range life distributions have not yet found more than a few applications. Some applications of finite-range distributions (specially uniform) may be obtained in Nanda et al. (2024). Sections 6.2.2 and 6.2.1 are devoted to discussions of various reliability properties of Mukherjee-Islam (1983) and Mukherjee-Roy (1986) distributions. Section 6.3 relates to stress-strength reliability with finite-range distributions of both stress encountered and strength built in. Section 6.4 considers a discrete finite-range distribution derived from the Mukherjee-Islam model. Extension to the Marshall-Olkin model, by considering the baseline distribution as Mukherjee-Islam model, has been discussed in Sect. 6.5. Estimation of the parameter of the model given in (6.2.1) is discussed in the last section.

6.2.1 Properties of Mukherjee-Islam Model

The distribution function of this model is given by

$$F(x) = \left(\frac{x}{a}\right)^p, \quad 0 < x < a, \ p > 0$$

having density

$$f(x) = \frac{p}{a}\left(\frac{x}{a}\right)^{p-1}, \quad 0 < x < a$$

6.2 Some Finite-Range Continuous Distributions

which immediately gives $E(X^r) = pa^r/(r+p)$ so that

$$E(X) = \frac{pa}{p+1}, \quad \text{and} \quad V(X) = \frac{pa^2}{(p+2)(p+1)^2}.$$

Thus, we have the coefficient of variation as $C^2(X) = 1/[p(p+2)] < 1$. We first consider the case $p > 1$. The failure rate function given by

$$r(x) = \frac{px^{p-1}}{a^p - x^p}$$

increases monotonically in x. The failure rate average $A(x)$ over $(0, x)$ is

$$A(x) = -\frac{1}{x} \ln\left(1 - \left(\frac{x}{a}\right)^p\right)$$

and this also is monotonic increasing. Graph of $r(x)$ appears in Fig. 6.1. Thus the ageing intensity function $L(\cdot)$ has the expression

$$L(x) = -\frac{px^p}{(a^p - x^p) \ln\left(1 - \left(\frac{x}{a}\right)^p\right)}. \tag{6.2.3}$$

The reversed hazard rate $\mu(x) = p/x$ decreases monotonically in x and, regarded as a random function, has the survival function given by

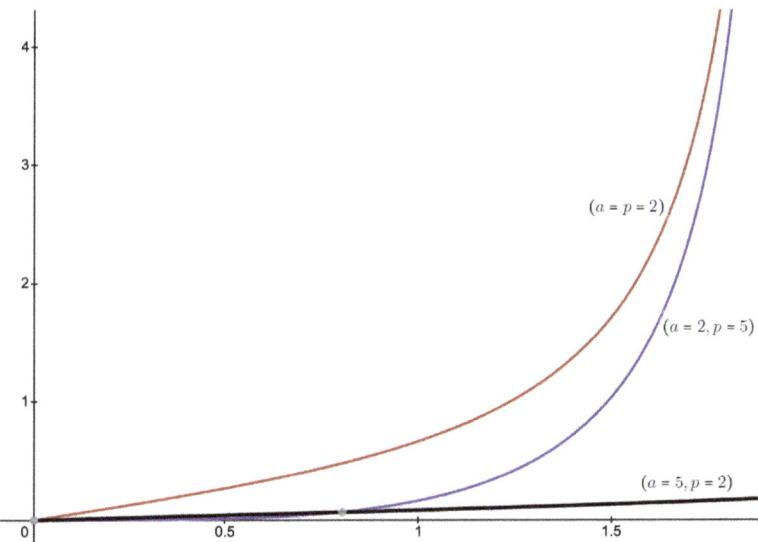

Fig. 6.1 The failure rate function of Mukherjee-Islam model when $p > 1$

$$P(\mu(X) \leqslant x) = 1 - \left(\frac{p}{ax}\right)^p. \qquad (6.2.4)$$

The equilibrium distribution has the density $g(x) = (1 - F(x))/E(X)$ with survival function

$$\bar{G}(x) = 1 - \frac{(p+1)x}{ap} + \frac{1}{p}\left(\frac{x}{a}\right)^{p+1}. \qquad (6.2.5)$$

It is well known that, for the equilibrium distribution corresponding to any distribution for a non-negative random variable,

$$E_g(X) = \frac{(C^2+1)E(X)}{2}, \qquad (6.2.6)$$

where C is the coefficient of variation of the parent distribution. Since we have $C < 1$, by (6.2.6), mean of the equilibrium distribution will be smaller than $E(X)$. The failure rate function is given by

$$r_g(x) = \frac{1}{MRL \text{ at age } x}.$$

The MRL works out as

$$m(u) = \frac{p}{p+1}\left[\frac{a^{p+1} - u^{p+1}}{a^p - u^p}\right] - u. \qquad (6.2.7)$$

It can be shown that the memory function corresponding to (6.2.7) is positive. Like the Mukherjee-Roy model, this distribution also has been used in income analysis. The Lorenz curve is given by

$$L_F(t) = \frac{1}{\mu}\int_0^p F^{-1}(t)\,dt = t^{\frac{1}{p}+1},$$

where $\mu = E(X)$ and hence the Generalized Lorenz Curve has the expression

$$GL_F(t) = \mu.L_F(t) = \frac{ap}{p+1}t^{\frac{1}{p}+1}.$$

The Bonferroni curve is obtained as $B_F(t) = L_F(t)/t = t^{1/p}$ so that the Bonferroni Index has the simple form

$$B = 1 - \int_0^1 B_F(t)\,dt = \frac{1}{p+1}.$$

6.2 Some Finite-Range Continuous Distributions

The Gini Index is thus obtained as

$$GI = 1 - 2\int_0^1 L_p(t)\,dt = \frac{1}{1+2p}.$$

We now consider the case $p < 1$. In this case the distribution exhibits a number of properties not shared by the Mukherjee-Roy model (to be discussed later) nor even by this distribution in case the shape parameter numerically exceeds unity. Its failure rate and average failure rate functions given below are non-monotonic.

$$r(x) = \frac{px^{p-1}}{a^p - x^p},\ 0 < x < a,\quad A(x) = -\frac{1}{x}\ln\left[1 - \left(\frac{x}{a}\right)^p\right].$$

It can be seen that failure rate initially decreases, remains almost flat over some range of life and subsequently increases. Thus the failure rate behaves like a bathtub (see Fig. 6.2). Further, the curves of $A(x)$ and $r(x)$ cross each other at only one point. This has been demonstrated in Fig. 6.4 by taking $a = 2$ and $p = 0.1$. In fact, the distribution is DFR in $[0, a(1-p)^{1/p}]$ and IFR over the rest of its range. Note that

$$\frac{1}{x}\int_0^x r(u)\,du = -\frac{1}{x}\ln \bar{F}(x) = -\frac{1}{x}\ln\left(1 - (x/a)^p\right)$$

which is increasing for all $x \in (0, a)$. Thus, the distribution is $IFRA$, though not IFR, over the entire range. This implies that the usual relationship among ageing properties of life distributions may not necessarily hold over a segment of the time domain when $r(x)$ is non-monotonic. It can be shown that the mean remaining life at age x as given in Eq. (6.2.7) has a change point at au_0 and the distribution is $IMRL$ in $(0, au_0)$ and $DMRL$ in (au_0, a). This is shown in Fig. 6.3. It may be noted that u_0 is quite small.

It thus appears that this distribution may be one among the very few life distributions with a finite range that have been reported to have a bathtub failure rate pattern. It may be incidentally mentioned that quite a few such distributions over an infinite range can be come across in literature and there are good reviews of such distributions (see, e.g., Rajarshi and Rajarshi, 1988).

Siddiqui et al. (2016a) introduced a generalization of the model which they called the Beta-extended Mukherjee-Islam (BEMI) distribution given by the distribution function

$$F(x) = \frac{1}{B(a,b)}\int_0^{(x/\theta)^p} w^{a-1}(1-w)^{b-1}dw,\ a, b, p, \theta > 0,\ x \leqslant \theta$$

which has a somewhat lengthy expression for the failure rate function. The authors provide tractable forms involving infinite sums of the density function when b is an integer as also when b is not an integer.

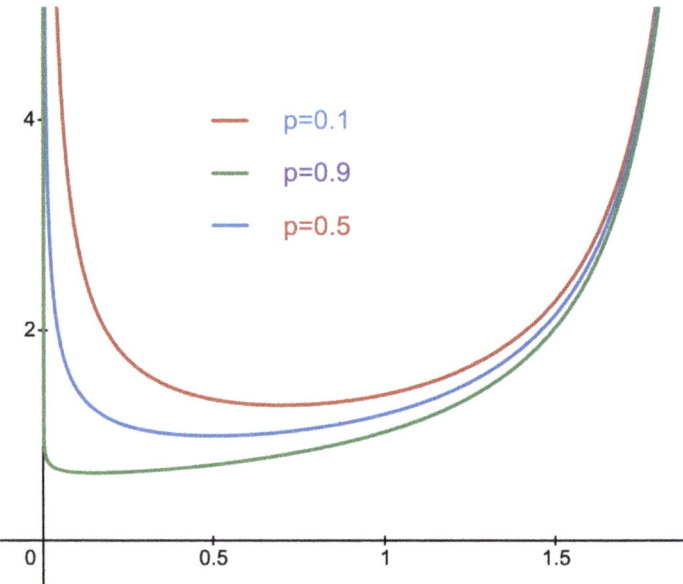

Fig. 6.2 The failure rate functions of Mukherjee-Islam model for different values of $p < 1$

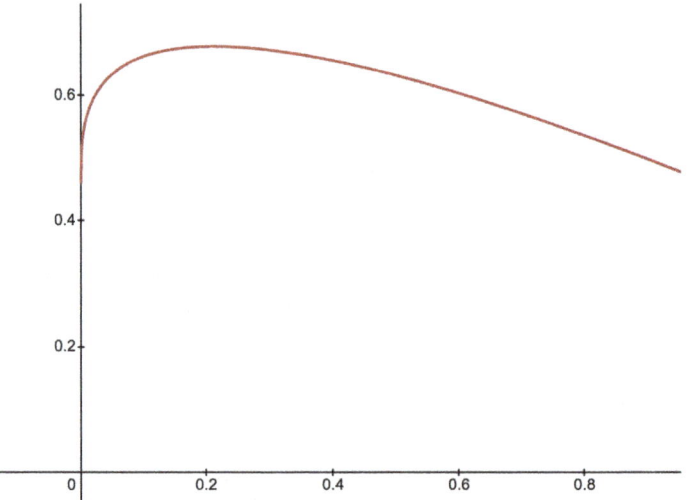

Fig. 6.3 The shape of the mean residual life function given in (6.2.7) taking $p = 0.3$

The first two moments work out as $\mu'_r = t_r \theta^r$ where

$$t_r = \sum_{j=0}^{\infty} \binom{b-1}{j}(-1)^j \frac{1}{\frac{r}{p} + (a+j)B(a,b)},$$

6.2 Some Finite-Range Continuous Distributions

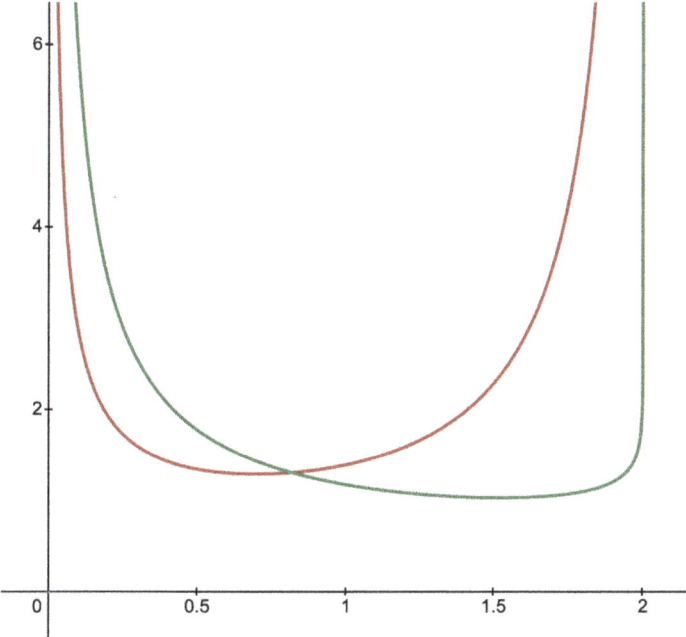

Fig. 6.4 This demonstrates the fact that the failure rate function and the average failure rate function of Mukherjee-Islam model cross each other once. Here we have taken $a = 2$ and $p = 0.1$. The red curve shows the failure rate whereas the green curve shows the average failure rate

for $r = 1$ or 2. Assuming a and b as known, maximum likelihood estimators of p and θ can be worked out numerically. The three shape parameters a, b, and p are related in terms of $p = 1/(1 + a - b)$.

6.2.2 Mukherjee-Roy Model and its Properties

The Beta-type distribution in (6.2.2) has the probability density

$$f(x; a, p) = \left(\frac{p}{a}\right)\left(1 - \frac{x}{a}\right)^{p-1}, \quad 0 < x < a, \ p > 0.$$

The rth moment is $E(X^r) = pa^r B(r+1, p)$. Thus

$$E(X) = \frac{a}{p+1} \quad \text{and} \quad V(X) = \frac{pa^2}{(p+1)^2(p+2)}$$

so that the coefficient of variation $C(X)$ is such that $C^2(X) = p/(p+2) < 1$. The failure rate $r(x) = p/(a - x)$ increases monotonically from p/a at $x = 0$ so that it is IFR. We can look upon failure rate as a random function and work out its probability distribution. In fact,

$$P(r(X) \geq t) = P\left(X \geq \frac{at - p}{t}\right) = \left(\frac{p}{at}\right)^p. \tag{6.2.8}$$

As expected, $E(1/r(X)) = a/(p+1)$. The mean remaining life at age x is

$$m(x) = E(X - x | X \geq x) = \frac{a - x}{p + 1}$$

and hence this distribution is DMRL. The variance of the remaining life at age x comes out as

$$V_X(x) = a^2 \frac{p(1 - x/a)^2}{(p+1)^2(p+2)}, \tag{6.2.9}$$

which implies that the coefficient of variation of remaining life, $C_X^2(x) = p/(p+2) = C^2$, independent of age x. This leads to a characterization of this finite-range life distribution (Mukherjee and Roy, 1986) which states that $C(x)$ is a constant less than 1 if and only if X follows the finite-range distribution given in (6.2.2). It was also pointed out that if $C_X(x)$ is a constant, then the distribution of X is DMRL. It can be easily verified that $r(x)m(x) = p/(p+1)$ is a constant less than unity. This provides another characterization of the finite-range distribution.

The equilibrium distribution has the probability density

$$g(x) = \frac{1 - F(x)}{E(X)} = \frac{p+1}{a}\left(1 - \frac{x}{a}\right)^p \tag{6.2.10}$$

with an expectation $E_g(X) = \frac{a}{p+2} < E(X)$. The survival function of the equilibrium distribution works out as

$$\bar{G}(x) = \left(1 - \frac{x}{a}\right)^{p+1} < \bar{F}(x).$$

The failure rate function (of the corresponding equilibrium distribution) takes the form $(p + 1)/(a - x)$ which is greater than the failure rate of the original distribution for all x. Further, the ratio between the two probability densities is

$$\frac{f(x)}{g(x)} = \frac{pa}{(p+1)(a-x)},$$

which increases monotonically with x. Thus, the original distribution dominates the equilibrium distribution in likelihood ratio (and hence in distribution, in expectation, and in failure rate) order.

6.2 Some Finite-Range Continuous Distributions

Coming back to the mean remaining life function $m(\cdot)$, we find that F has a positive memory (as discussed in Sect. 2.4.7) since

$$\frac{E(X) - m(x)}{x} = \frac{1}{p+1} > 0.$$

Incidentally, this distribution has also been used as an income distribution and to work out measures of income inequality and poverty, with the help of well-known indices which come out conveniently for this distribution. It can be shown that the Gini index for this distribution comes out to be $p/(2p+1)$. Some other properties of this distribution are quite useful. These are stated below.

Theorem 6.2.1 *Identical range finite-range distributions given in (6.2.2) are closed under series combinations (of independent components).*

Proof Let $X_i \sim FR_2(a, p_i)$, $i = 1, 2, \ldots, n$ independently and let $Z = \min\{X_1, X_2, \ldots, X_n\}$. Then

$$P(Z > x) = \prod_{i=1}^{n} P(X_i > x) = \left(1 - \frac{x}{a}\right)^{\sum_{i=1}^{n} p_i}$$

and this implies $Z \sim FR_2\left(a, \sum_{i=1}^{n} p_i\right)$.

Theorem 6.2.2 *The class of identical shape FR_2 distributions is closed under residual life transformation.*

Proof Let $X \sim FR_2(a, p)$. Then $P(X - t > x | X > t) = \left(1 - \frac{x}{a-t}\right)^p$, which is the survival function of $FR_2(a - t, p)$, implying $[X - t | X > t] \sim FR_2(a - t, p)$.

6.2.3 Some Characterization Results

We now consider a few results characterizing this finite-range distribution. Most of these results were proved by Roy (1988). Proofs are not reproduced here.

Theorem 6.2.3 *The MRL of an absolutely continuous non-negative random variable is proportional to the corresponding failure rate (constant of proportionality < 1) if and only if the original life has either FR_1 or FR_2 distribution.*

Theorem 6.2.4 *The coefficient of variation of the residual life of an absolutely continuous non-negative random variable is a constant less than unity if and only if the life has either FR_1 or FR_2 distribution.* □

The following theorem characterizes the FR_2 distribution.

Theorem 6.2.5 *Let X be an absolutely continuous random variable. We write $\bar{F}_a$ to denote the survival function of X if the support of X is $[0, a]$. Then*

(i) $\bar{F}_b(x) = \bar{F}_1(x/b)$, for all $b > 0$;
(ii) $\bar{F}_a(x+y) = \bar{F}_a(x) \cdot \bar{F}_{a-x}(y)$ for all $0 \leq x \leq a$ and for all y in the neighbourhood of 0,

if and only if $X \sim FR_2(a, p)$.

Proof Only if part: For $a = 1$, we get $\bar{F}_1(x+y) = \bar{F}_1(x) \cdot \bar{F}_{1-x}(y)$, which gives

$$g(x+y) = g(x) + g\left(\frac{y}{1-x}\right),$$

where $g(x) = \ln \bar{F}_1(x)$. Thus, $g(0) = 0$, and

$$g'(x) = \lim_{y \to 0} \frac{g\left(\frac{y}{1-x}\right) - g(0)}{y} = \frac{g'(0)}{1-x}.$$

Here the limit exists because the pdf of X exists. This being true for all x, we get

$$g(x) = -g'(0) \ln(1-x) + \ln A,$$

where $\ln A$ is the integration constant. Thus,

$$\bar{F}_1(x) = A(1-x)^{-g'(0)}.$$

Writing $p = -g'(0) = f_1(0) \geq 0$ and noting that $\bar{F}_1(0) = 1 = A$, we have $\bar{F}_1(x) = (1-x)^p$ giving $\bar{F}_a(x) = \left(1 - \frac{x}{a}\right)^p$ which proves the result. The 'If part' is easy to prove.

Theorem 6.2.6 Let the failure rate function $r(\cdot)$ of a random variable X be continuous with $r(x) > 0$. Then the failure rate at the arithmetic mean of two time points s and t is equal to the harmonic mean of the failure rates at s and t if and only if $X \sim FR_2$.

Proof If $X \sim FR_2$, the proof is simple. Now, assume that

$$r\left(\frac{s+t}{2}\right) = \frac{2}{\frac{1}{r(s)} + \frac{1}{r(t)}}. \tag{6.2.11}$$

Write

$$g(x) = \frac{1}{r(x)} - \frac{1}{r(0)} \tag{6.2.12}$$

so that $g(0) = 0$. Now we have

$$g\left(\frac{s+t}{2}\right) = \frac{1}{r\left(\frac{s+t}{2}\right)} - \frac{1}{r(0)}$$

6.3 Stress-Strength Reliability

$$\begin{aligned}
&= \frac{\frac{1}{r(s)} + \frac{1}{r(t)}}{2} - \frac{1}{r(0)}, \quad \text{from (6.2.11)} \\
&= \frac{1}{2}\left[\left(\frac{1}{r(s)} - \frac{1}{r(0)}\right) + \left(\frac{1}{r(t)} - \frac{1}{r(0)}\right)\right] \\
&= \frac{1}{2}(g(s) + g(t)).
\end{aligned} \quad (6.2.13)$$

Taking $s = 0$, we get

$$g\left(\frac{t}{2}\right) = \frac{1}{2} g(t). \quad (6.2.14)$$

From (6.2.13) and (6.2.14) we get

$$g(s+t) = g(s) + g(t)$$

which gives $g(x) = kx$ so that

$$r(x) = \frac{1}{kx + \frac{1}{r(0)}}.$$

Clearly, if $k < 0$, then $X \sim FR_2$.

6.3 Stress-Strength Reliability

We can conveniently find out $P(Y > X) = R$, assuming X and Y to represent stress and strength, respectively, having a distribution for both the variables of the form FR_1 or FR_2 with a common range (scale) but varying shape parameters.

The following theorem holds both for FR_1 and FR_2.

Theorem 6.3.1 *Let X and Y be two independent random variables denoting the stress and the strength, respectively, such that $X \sim FR_i(a, p_1)$ and $Y \sim FR_i(a, p_2)$. Then the product reliability is obtained as*

$$P(X < Y) = \frac{p_{3-i}}{p_1 + p_2},$$

for $i = 1, 2$. □

If we take X to follow FR_1 with parameter p_1 and Y to follow FR_2 with parameter p_2, and a common scale parameter a, we have

$$P(Y > X) = \int_0^a P(X < y) dF_2(y)$$

$$= \frac{p_2}{a}\int_0^a \left(\frac{x}{a}\right)^{p_1}\left(1-\frac{x}{a}\right)^{p_2-1}dx$$

$$= p_2\int_0^1 t^{p_1}(1-t)^{p_2-1}dt$$

$$= p_2.B(p_1+1, p_2).$$

Suppose X and Y are independently distributed as FR_2 with scale parameters a_1 and a_2, respectively, with common shape parameter p. First suppose $a_1 < a_2$. Then we have

$$P(X < Y) = \int_0^{a_2} P(X < x)dF_Y(x)$$

$$= \int_0^{a_1} P(X < x)dF_Y(x) + \int_{a_1}^{a_2} P(X < x)dF_Y(x)$$

$$= \frac{p}{a_2}\int_0^{a_1}\left(\frac{x}{a_1}\right)^p\left(\frac{x}{a_2}\right)^{p-1}dx + \frac{p}{a_2}\int_{a_1}^{a_2}\left(\frac{x}{a_2}\right)^{p-1}dx$$

$$= \frac{p}{a_1^p a_2^p}\int_0^{a_1} x^{2p-1}dx + p\int_{a_1/a_2}^1 x^{p-1}dx$$

$$= \frac{1}{2}\left(\frac{a_1}{a_2}\right)^p + 1 - \left(\frac{a_1}{a_2}\right)^p$$

$$= 1 - \frac{1}{2}\left(\frac{a_1}{a_2}\right)^p.$$

Now we consider $a_1 > a_2$. Then we have

$$P(X < Y) = \int_0^{a_2} P(X < x)dF_Y(x)$$

$$= \frac{p}{a_2}\int_0^{a_2}\left(\frac{x}{a_1}\right)^p\left(\frac{x}{a_2}\right)^{p-1}dx$$

$$= \frac{1}{2}\left(\frac{a_2}{a_1}\right)^p.$$

Theorem 6.3.2 *If $X \sim FR_2(a, p)$ and Y is the failure rate transform of X, then Y has probability density function given by $f_Y(y) = a(ay/p)^{-p-1}$.*

Proof $P(Y > y) = P(X > a - p/y) = (ay/p)^{-p}$ implying that Y has a Pareto distribution. □

R is found to be related to the Bhattacharyya distance between the two distributions. For two distributions f_1 and f_2, the Bhattacharyya distance is defined as

6.4 Finite-Range Discrete Distribution

$$-\ln \int_S \sqrt{f_1(x) f_2(x)} dx,$$

where S is the common support of f_1 and f_2. Thus, if X and Y both independently follow FR_1 with respective shape parameters p_1 and p_2 we have the Bhattacharyya distance as

$$-\ln \left[\frac{1}{a} \int_0^a \sqrt{p_1 p_2 \left(\frac{x}{a}\right)^{p_1+p_2-2}} dx \right] = -\ln \left[\sqrt{p_1 p_2} \int_0^1 t^{\frac{p_1+p_2}{2}-1} dt \right]$$

$$= -\ln \left[\frac{2\sqrt{p_1 p_2}}{p_1 + p_2} \right].$$

It is interesting to note that if X and Y both independently follow FR_2, then also the expression of the Bhattacharyya distance will be same as above.

We conclude this section by stating this easy-to-prove but important theorem in its own right.

Theorem 6.3.3 *If X is distributed as $FR_2(a, p)$, the probability density of X converges to that of exponential with failure rate μ as $p \to \infty$ and $a \to \infty$ such that $p/a \to \mu$ (a constant).*

6.4 Finite-Range Discrete Distribution

Discrete lifetime data arise in several different situations in life-testing and reliability analysis. As pointed out by Lai and Wang (1995), a device may be monitored only once per time period (e.g., an hour, a shift, a day, etc.) and we observe the number of such periods successfully completed before the device fails or the device operates in cycles or an experimenter discretizes the data or groups them. Literature search will reveal quite a few contributions on discrete failure time models and their properties. Starting with the Mukherjee-Islam model, Lai and Wang (1995) proposed a discrete distribution given by the following probability function:

$$f(i) = P(T = i) = \frac{i^\alpha}{\sum_{x=0}^N x^\alpha}, \quad i = 0, 1, 2, \ldots, N.$$

For $\alpha \in \mathbb{N}$, the distribution function works out as

$$F(t) = \frac{c(j, \alpha)}{c(N, \alpha)}, \quad j \leqslant t < j+1, \; j = 0, 1, 2, \ldots, N,$$

where $c(n, k) = [B_{k+1}(n+1) - B_{k+1}(n)]/(k+1)$ for any positive integer k, $B_i(x)$ being the ith Bernoulli polynomial defined by $B_m(x) = \sum_{j=0}^m B_j x^{m-j}$ with $B_i = B_i(0)$ and $B_0 = 1$. The mean and the variance for $\alpha = 1$ work out, respectively, as

$$\mu = \frac{2N+1}{3}, \quad \text{and} \quad \sigma^2 = \frac{1}{18}(5N^2 + 5N - 1).$$

For $\alpha = 2$, we have

$$\mu = \frac{3N(N+1)}{2(2N+1)} \quad \text{and} \quad \sigma^2 = \frac{1}{5}(3N^2 + 3N - 1) - \mu^2.$$

For $\alpha = -1$, $\mu = N/\sum_{x=1}^{N} \frac{1}{x}$, which is just the harmonic mean of the sequence $1, 2, \ldots, N$. It can be easily shown that the probability of survival till time i is given by

$$\bar{F}(i) = \frac{\sum_{x=i}^{N} x^\alpha}{c(N, \alpha)}$$

and the failure rate is given by $r(i) = f(i)/\bar{F}(i)$ with the convention $r(i) = 1$ for $i \geq N$. Lai and Wang prove that the distribution has a bathtub-shaped failure rate function for $\alpha < 0$. They also prove that

(i) The mean residual life $\{m(i), i \geq 1\}$ is a decreasing sequence for non-negative α.
(ii) $m(i)$ has an upside-down bathtub shape for $\alpha \leq -0.57$.
(iii) For $-0.57 < \alpha < 0$, the shape of the sequence $m(i)$ depends on the value of μ relative to the value of $f(1)$. The authors provide several numerical examples to illustrate this point.

6.5 Marshall-Olkin Extension

It is quite convenient to work out a Marshall-Olkin extension (discussed in Sect. 3.3) of the Mukherjee-Islam model (which is virtually an exponentiated rectangular distribution) to yield the distribution function as

$$F(x) = \frac{x^p}{qa^p + (1-q)x^p}, \quad 0 < x < a, \ p, q > 0$$

with a failure rate function given by

$$r(x; a, p, q) = \frac{pa^p x^{p-1}}{a^p - x^p} \cdot [qa^p + (1-q)x^p]^{-1},$$

which is bathtub shaped for $p < 1$. The pattern of the failure rate function may be seen in Fig. 6.5. The median comes out to be

$$Me = \left(\frac{qa^p}{1+q}\right)^{1/p}.$$

6.6 Estimation of Parameters in Mukherjee-Islam Model

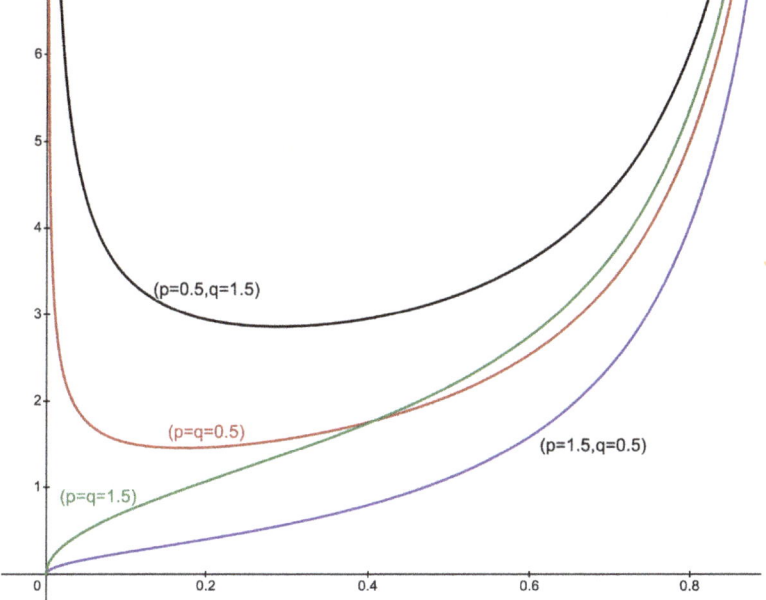

Fig. 6.5 Different shapes of the failure rate functions of Marshall-Olkin exponentiated rectangular distribution for different values of p and q taking $a = 1$.

Various properties of this Marshall-Olkin exponentiated rectangular distribution including its order statistics and estimation of its parameters have been studied by Sindhu et al. (2020).

6.6 Estimation of Parameters in Mukherjee-Islam Model

The parameters p and a can be estimated by the method of moments, by equating the square of the sample coefficient of variation C^2 to that of population coefficient of variation $1/[p(p+2)]$ to yield the estimate $\widehat{p}$ and then equating the sample mean $\bar{x}$ to $\widehat{p} a/(\widehat{p}+1)$ to get $\widehat{a}$. Large sample variances of these moment estimates can be routinely worked out.

Another method would be to consider two population quantiles y_1 and y_2, say, such that $F(y_1) = \alpha_1$ and $F(y_2) = \alpha_2$, and equate them to the corresponding sample quantiles x_1 and x_2, say, so that the estimate of p comes out as $\widehat{p} = (\ln \alpha_2 - \ln \alpha_1)/(\ln x_2 - \ln x_1)$. Subsequently, $\widehat{a}$ can be worked out as in the method of moments. Alternatively, a can be estimated as $\widehat{a} = x_1/\alpha_1^{1/\widehat{p}}$. We can even take the mean of these two estimates.

The choice of α_1 and α_2 can be simplified and taken as α and $1 - \alpha$, respectively, where the value of α can be so determined that the asymptotic variance of $\widehat{p}$ as a function of α is minimized. A slightly different procedure could be to consider half-life or median and a second quantile of order determined optimally. The best procedure would be to find α_1 and α_2 in such a way that either (i) asymptotic variance of $\hat{p}$ as a function of these two orders is a minimum or (ii) the generalized variance (the determinant of the asymptotic variance-covariance matrix of $\hat{p}$ and $\hat{a}$) is a minimum.

We can offer (sample median)$\times 2^{1/p}$ as the estimate of a, and this will reduce to $2\times$median in the case of a rectangular distribution. Of course, the estimate of the range parameter a can be simply taken as $\bar{x}(p+1)/p$ if p is known.

Chapter 7
Ordering Among Life Distributions

7.1 Introduction

In different fields of reliability, engineering, econometrics, biological sciences, operations research, and so on, different brands need to be compared in terms of their lifetimes. In comparing effectiveness of a drug, a group of patients taking drug need to be compared with the control groups. In order to decide which breed of animals to be preferred, we need to compare lifetimes of different breeds of animals. To the best of our knowledge, first such comparison was done in portfolio analysis (cf. Tobin, 1958) where the comparison was done in terms of moments of utility function. If the utility function is an nth-degree polynomial, we need to compare corresponding n moments. In particular, comparison in terms of a quadratic utility function needs comparison of the corresponding means and variances. Thus, in comparing two prospects having same variance, the one which has the higher mean value should be preferred (provided higher the mean better is the prospect) whereas, for two prospects having same mean value, the one with lower variance is to be preferred. If means and variances are all different then one way of comparison could be in terms of the corresponding coefficient of variations. However, if the utility function is a polynomial of degree more than 2 and all the corresponding moments are different, it is not possible to compare the prospects using the moments. Further, there are cases where moments may not exist, making this kind of comparison impossible. Here comes the role of partial orders. We may also be interested to compare the remaining life of a system at different ages to see whether and how the system is ageing with time.

Suppose we are interested to know which brand of a particular car (produced by two companies) is to be preferred. Let X and Y be the lifetimes of the two cars produced by two companies under consideration. If we see that the expected life of the cars by the first company is smaller than that of the second company (which we can mathematically write $E(X) < E(Y)$) we sometimes decide that the car produced by the latter company is better than that produced by the former. However, this is comparison of only one point taken from the whole right half of the real line. So, an immediate question about the effectiveness of such a comparison arises. Below we discuss some properties of different partial orders.

7.2 Usual Stochastic Order

The use of stochastic orders may be seen in Lehmann (1955) followed by Stoyan (1983), Shaked and Shanthikumar (1994, 2007) and others.

Definition 7.2.1 For two non-negative random variables X and Y, X is said to be smaller than Y in usual stochastic order, written as $X \leqslant_{st} Y$, if $P(X \geqslant x) \leqslant P(Y \geqslant x)$, for all $x \in (0, \infty)$. □

The above definition says that the survival function of X lies below that of Y. Now, integrating both sides of the above expression, we get that, if $X \leqslant_{st} Y$, then

$$\int_t^\infty \bar{F}_X(x)\,dx \leqslant \int_t^\infty \bar{F}_Y(x)\,dx$$

for all t. Taking $t \to 0$, we immediately get

$$X \leqslant_{st} Y \Rightarrow E(X) \leqslant E(Y).$$

The usual stochastic order tells that survival function of one distribution is dominated by that of another distribution. For two such distributions if we have the means same, then it is clear that the graphs of both the survival functions must overlap. This immediately tells that the two distributions are identical. Thus, we get the following.

Theorem 7.2.1 *For two non-negative random variables X and Y having same means (i.e., $E(X) = E(Y)$), if we have $X \leqslant_{st} Y$, then we must have $X =_{st} Y$.* □

It is well known (cf. Shaked and Shanthikumar, 2007) that stochastic order is closed under convolution, i.e., if $X_i, i = 1, 2, \ldots, n$, is a set of independent random variables and $Y_i, i = 1, 2, \ldots, n$, is another set of independent random variables such that $X_i \leqslant_{st} Y_i$ for all $i = 1, 2, \ldots, n$, then $\sum_{i=1}^n X_i \leqslant_{st} \sum_{i=1}^n Y_i$. However, it is interesting to note that (cf. Aubrun and Nechita, 2009) if $X_i, i = 1, 2, \ldots, n$, is a set of independent random variables and $Y_i, i = 1, 2, \ldots, n$, is another set of independent random variables then, in general,

$$\sum_{i=1}^n X_i \leqslant_{st} \sum_{i=1}^n Y_i \nRightarrow \sum_{i=1}^{n+1} X_i \leqslant_{st} \sum_{i=1}^{n+1} Y_i.$$

The following result is due to Balakrishnan and Iliopoulos (2009).

7.3 Hazard Rate Order

Theorem 7.2.2 *Let X, Θ, Z be three random variables such that*

(i) $[\Theta|Z = z] \leq_{st} [\Theta|Z = z']$ *for* $z \leq z'$,
(ii) $[X|\Theta = \theta, Z = z] \leq_{st} [X|\Theta = \theta', Z = z]$ *for* $\theta \leq \theta'$ *and for all z,*
(iii) $[X|\Theta = \theta, Z = z] \leq_{st} [X|\Theta = \theta, Z = z']$ *for* $z \leq z'$ *and for all θ.*

Then $[X|Z = z] \leq_{st} [X|Z = z']$ for $z \leq z'$. □

Suppose Θ is a random variable following the distribution F with support $\mathcal{X}$. Let $X(\theta)$ be a random variable having distribution G_θ. Then we define the random variable $X(\Theta)$ as the one having distribution H given by

$$H(y) = \int_{\mathcal{X}} G_\theta(y) dF(\theta).$$

Taking the random variable $[X|\Theta = \theta, Z = z]$ as independent of z in Theorem 7.2.2, we get the following corollary which may be available in Shaked and Shanthikumar (2007).

Corollary 7.2.1 *Let $X(\theta), \Theta_1, \Theta_2$ be the random variables such that*

(a) $X(\theta) \leq_{st} X(\theta')$ *for* $\theta \leq \theta'$,
(b) $\Theta_1 \leq_{st} \Theta_2$.

Then $X(\Theta_1) \leq_{st} X(\Theta_2)$.

7.3 Hazard Rate Order

Apart from the survival function, the failure rate (also known as hazard rate) plays a very important role in reliability and survival analysis. Suppose a system having lifetime X has survived for x units of time. For such a unit the probability that it fails on or before $x + \epsilon$, for some $\epsilon > 0$, is approximately given by $\epsilon.r_X(x)$, provided ϵ is very small, where $r_X(x) = f_X(x)/\bar{F}_X(x)$, for all $x \in \{t : \bar{F}_X(t) > 0\}$ (see Sect. 2.4.1). Thus, $r_X(x)$ represents the instantaneous failure rate of the random variable X at the point x. Note that

$$d\left(\ln \bar{F}_X(x)\right) = -r_X(x) dx$$

which gives, for a proper lifetime distribution,

$$\bar{F}(x) = e^{-\int_0^x r_X(u) du}. \qquad (7.3.1)$$

By taking limit, as $x \to \infty$, on both sides of the above expression, we immediately get

$$\int_0^\infty r_X(u) \, du = \infty \qquad (7.3.2)$$

From (7.3.1), we see that, for all $x > 0$,

$$\int_0^x r_X(u)\,du < \infty. \qquad (7.3.3)$$

Thus, by comparing (7.3.2) and (7.3.3), we get

$$\int_x^\infty r_X(u)\,du = \infty$$

for all $x > 0$. This immediately tells

$$\lim_{x \to \infty} r_X(x)\,dx = \infty$$

Clearly, smaller the value of failure rate better is the underlying random variable. Based on this idea, below we define another stochastic order, called failure rate (or hazard rate) order.

7.3.1 Definitions and Preliminaries

Let X and Y be two continuous (not necessarily non-negative) random variables with interval supports (l_X, u_X) and (l_Y, u_Y), respectively, where l_X, l_Y could be $-\infty$ and u_X, u_Y could be ∞, i.e.,

$$l_X = \inf\{x : F_X(x) > 0, x \in \mathbb{R}\};$$
$$u_X = \sup\{x : F_X(x) < 1, x \in \mathbb{R}\};$$
$$l_Y = \inf\{x : F_Y(x) > 0, x \in \mathbb{R}\};$$
$$u_X = \sup\{x : F_Y(x) < 1, x \in \mathbb{R}\}.$$

Definition 7.3.1 For two random variables X and Y as defined above, X is said to be smaller than Y in failure rate order, written as $X \leqslant_{fr} Y$, if $r_X(x) \geqslant r_Y(x)$, for all $x \in (-\infty, \max\{u_X, u_Y\})$. □

It is easy to see that

$$u_X \leqslant l_Y \Rightarrow X \leqslant_{fr} Y \quad \text{and} \quad X \leqslant_{fr} Y \Rightarrow (l_X \leqslant l_Y, u_X \leqslant u_Y).$$

Sometimes $X \leqslant_{fr} Y$ is also written as $X \leqslant_{hr} Y$, where hr stands for 'hazard rate'. In other words, failure rate order is sometimes called hazard rate order. In actuarial science, the hazard rate order is known as *mortality* order.

7.3 Hazard Rate Order

If the random variable is not absolutely continuous, then density does not exist. In this case, an equivalent definition of the above-discussed failure rate order can be given as $X \leq_{fr} Y$ if and only if $\bar{F}_X(x)/\bar{F}_Y(x)$ is decreasing in $x \in (-\infty, \max\{u_X, u_Y\})$. Thus, if $X \leq_{fr} Y$ we get

$$\frac{\bar{F}_X(x)}{\bar{F}_Y(x)} \leq \lim_{x \to -\infty} \frac{\bar{F}_X(x)}{\bar{F}_Y(x)} = 1$$

which is same as $X \leq_{st} Y$. Hence, we see that failure rate order is stronger than the usual stochastic order. Again, if $X \leq_{fr} Y$ we have, for all $x \geq 0$ and for all t,

$$\frac{\bar{F}_X(x+t)}{\bar{F}_Y(x+t)} \leq \frac{\bar{F}_X(t)}{\bar{F}_Y(t)}$$

which can equivalently be written as

$$\frac{\bar{F}_X(x+t)}{\bar{F}_X(t)} \leq \frac{\bar{F}_Y(x+t)}{\bar{F}_Y(t)}. \tag{7.3.4}$$

In case X and Y are discrete random variables having support $S_d = \{0, 1, 2 \ldots\}$ the failure rate order, written as $X \leq_{hr}^d Y$, is defined as

$$\frac{P(X=k)}{P(X \geq k)} \geq \frac{P(Y=k)}{P(Y \geq k)}$$

for all $k \in S_d$. It should be mentioned here that some authors define failure rate order, in case of discrete random variable, as

$$\frac{P(X=k)}{P(X > k)} \geq \frac{P(Y=k)}{P(Y > k)}$$

for all $k \in S_d$.

For $p \in [0, 1]$, let us consider a mixture random variable W_p having distribution function F_p given by

$$F_p(x) = p.F_X(x) + (1-p).F_Y(x),$$

which immediately gives the survival function $\bar{F}_p$ as

$$\bar{F}_p(x) = p.\bar{F}_X(x) + (1-p).\bar{F}_Y(x).$$

Clearly, $F_1 \equiv F_X$ and $F_0 \equiv F_Y$.

The following lemma will be used in sequel. The proof is simple and hence omitted.

Lemma 7.3.1 *Let* $a : [0, \infty) \to \mathbb{R}$ *be a differentiable decreasing function. Then*

$$\frac{p.a(x) + (1-p)}{p'a(x) + (1-p')} \text{ is increasing in } x,$$

for any $0 \leq p \leq p' \leq 1$. □

Following theorem generalizes Theorem 1.B.22 of Shaked and Shanthikumar (2007).

Theorem 7.3.1 *Let X and Y be two random variables having respective distribution functions F_X and F_Y such that $X \leq_{hr} Y$. For $p \in [0, 1]$, let W_p be as defined above. Then we have $W_{p'} \leq_{hr} W_p$, for $p \leq p'$.*

Proof Write $a(x) = \bar{F}_X(x)/\bar{F}_Y(x)$. Since $X \leq_{hr} Y$, $a(x)$ is decreasing in x. Note that, for $p \leq p'$,

$$\frac{\bar{F}_p(x)}{\bar{F}_{p'}(x)} = \frac{p.a(x) + (1-p)}{p'a(x) + (1-p')}$$

is increasing, by Lemma 7.3.1. So,

$$W_{p'} \leq_{hr} W_p$$

for $p \leq p'$. □

The following corollary is immediately obtained from Theorem 7.3.1.

Corollary 7.3.1 *Let X and Y be two random variables having respective distribution functions F_X and F_Y and let W_p be as defined above. Then, for $p \in [0, 1]$, $X \leq_{hr} Y \Rightarrow X \leq_{hr} W_p \leq_{hr} Y$.*

7.3.2 Component Redundancy Versus System Redundancy

In order to maximize the lifetime of a system, we sometimes allocate redundant (or spare) components in the system. There are mainly two kinds of redundancy, namely, active redundancy and standby redundancy. In active redundancy, the redundant component works simultaneously with the components, leading to the maximum of the random variables, whereas, in standby redundancy, the redundant component starts working once the original component fails, leading to the sum (convolution) of the random variables.

It is well known to the design engineers that redundancy at the component level is more effective than that at the system level (Barlow and Proschan, 1975). This actually tells that the life of a coherent system of independent components where an active spare is attached to each of its components is larger in usual stochastic order than that where the life of the system is enhanced by creating a duplicate system

7.3 Hazard Rate Order

to perform as a spare to the original system. To explain it mathematically, let $X = (X_1, X_2, \ldots, X_n)$ be the independent lifetimes of n components forming a coherent system with lifetime $\tau = \tau(X_1, X_2, \ldots, X_n)$. Suppose $Y = (Y_1, Y_2, \ldots, Y_n)$ be the independent lifetimes of n spares to be used to increase the system lifetime. Then

$$\tau(X \vee Y) \geqslant_{st} \tau(X) \vee \tau(Y),$$

where $X \vee Y = (\max(X_1, Y_1), \max(X_2, Y_2), \ldots, \max(X_n, Y_n))$. The above result cannot be extended to hazard rate order as the following example shows (cf. Boland and El-Neweihi, 1995).

Let us consider two iid components having lifetimes X_1, X_2 each having standard Exponential life distribution and two iid spares having lifetimes X_1^*, X_2^* following Exponential life distribution with hazard rate 2. Write

$$T_A = \tau(X) \vee \tau(Y) = \max[\min(X_1, X_2), \min(X_1^*, X_2^*)],$$
$$T_B = \tau(X \vee Y) = \min[\max(X_1, X_1^*), \min(X_2, X_2^*)].$$

Then we get

$$\begin{aligned}P(T_A > x) &= P\left(\max[\min(X_1, X_2), \min(X_1^*, X_2^*)] > x\right) \\ &= 1 - \{1 - P[\min(X_1, X_2) > x]\} \cdot \{1 - P[\min(X_1^*, X_2^*) > x]\} \\ &= 1 - \left[1 - P^2(X_1 > x)\right] \cdot \left[1 - P^2(X_1^* > x)\right] \\ &= 1 - \left(1 - e^{-2x}\right) \cdot \left(1 - e^{-4x}\right)\end{aligned}$$

and

$$\begin{aligned}P(T_B > x) &= P\left(\tau(X \vee Y) = \min[\max(X_1, X_1^*), \min(X_2, X_2^*)]\right) \\ &= \left[1 - P(X_1 \leqslant x).P(X_1^* \leqslant x)\right] \cdot \left[1 - P(X_2 \leqslant x).P(X_2^* \leqslant x)\right] \\ &= \left[1 - (1 - e^{-x})(1 - e^{-2x})\right]^2.\end{aligned}$$

Note that $T_A \leqslant_{hr} T_B$ if and only if

$$\frac{1 - \left(1 - e^{-2x}\right) \cdot \left(1 - e^{-4x}\right)}{\left[1 - (1 - e^{-x})(1 - e^{-2x})\right]^2} \text{ is decreasing in } x$$

which is equivalent to

$$a(t) \stackrel{def}{=} \frac{1 - (1 - t^2)(1 - t^4)}{[1 - (1 - t)(1 - t^2)]^2} \text{ is increasing in } t$$

which does not hold as Fig. 7.1 shows.

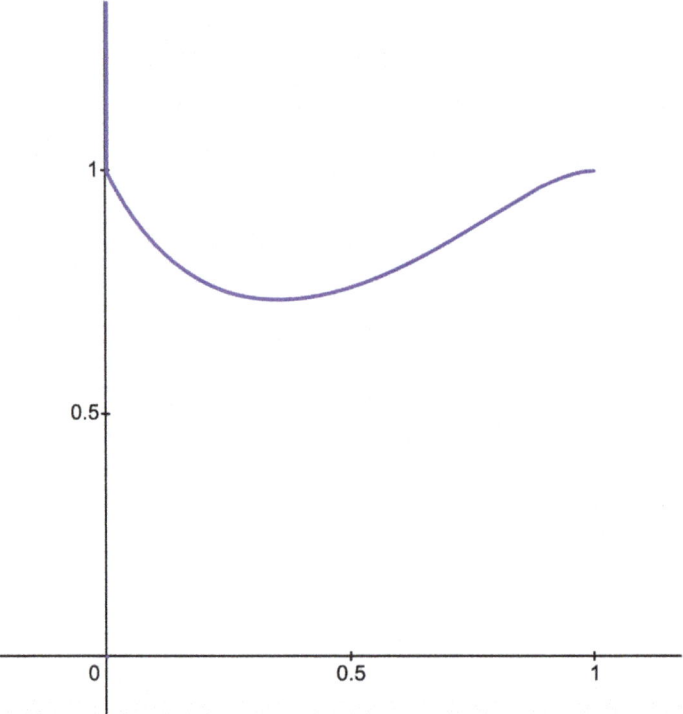

Fig. 7.1 Graph of $a(t)$ showing that the component redundancy is not superior to system redundancy in failure rate ordering

If X_i has same distribution as Y_i for all i, then the spares are called matching spares. Boland and El-Neweihi (1995) have shown that, in case of matching spares, under certain restrictions, the above result holds when the two series systems are compared in hazard rate ordering.

Theorem 7.3.2 *Suppose $X_1, X_2, \ldots, X_n$ denote the iid component lifetimes and $Y_1, Y_2, \ldots, Y_n$ denote the lifetimes of n iid matching spares. Further suppose τ denotes the lifetime of the coherent system formed from the n components with h as the system reliability function. If $ph'(p)/h(p)$ is increasing in p and $h(p) \leq p$ for all $p \in [0, 1]$, then*

$$\tau(X \vee Y) \geq_{hr} \tau(X) \vee \tau(Y)$$

where $h'(p) = dh(p)/dp$. □

Suppose different k-out-of-n sub-systems are put in series where at least one subsystem consists of single component, then such systems satisfy all the conditions of the above theorem.

7.4 Likelihood Ratio Order

In some of the cases, for example, for Gamma distribution, it is not possible to get an expression of distribution function (or equivalently of survival function). In such a case, comparing two such distributions in terms of either hazard rate order or reversed hazard rate order may be difficult. Below we define an order, known as likelihood ratio order, which may be considered as a sufficient condition for both hazard rate and reversed hazard rate orders to hold.

Definition 7.4.1 For two random variables X and Y having respective density functions f_X and f_Y, X is said to be smaller than Y in likelihood ratio order, written as $X \leqslant_{lr} Y$, if

$$\frac{f_X(x)}{f_Y(x)} \text{ is decreasing in } x. \tag{7.4.5}$$

Let us consider two measurable sets A and B such that $(x \in A, y \in B) \Rightarrow x \leqslant y$. In such cases, we write $A \leqslant B$. Note that (7.4.5) can equivalently be written as

$$\frac{f_X(x)}{f_Y(x)} \geqslant \frac{f_X(y)}{f_Y(y)} \text{ for all } x \leqslant y$$

which is same as

$$f_X(x) f_Y(y) \geqslant f_Y(x) f_X(y) \tag{7.4.6}$$

for all $x \leqslant y$. By integrating both sides of (7.4.6) over the sets A and B as defined above, we get

$$P(X \in A).P(Y \in B) \geqslant P(X \in B).P(Y \in A)$$

for all $A \leqslant B$. It can be noted that the above condition, which defines the likelihood ratio order, does not need existence of density functions. This representation of likelihood ratio order is given in Müller (1997) (see also Block et al., 1982).

Integrating both sides of (7.4.6) with respect to $x \in [t, \infty)$ and $y \in [s, \infty)$ with $t \leqslant s$, we see that $X \leqslant_{lr} Y \Rightarrow X \leqslant_{hr} Y$. Similarly integrating over $x \in (0, t]$ and $y \in (0, s]$ with $t \leqslant s$, we see that $X \leqslant_{lr} Y \Rightarrow X \leqslant_{rh} Y$. However, neither hazard rate order nor the reversed hazard rate order implies the other.

Taking $a(x) = f_X(x)/f_Y(x)$ in Lemma 7.3.1 we get the following theorem which generalizes Theorem 1.C.30 of Shaked and Shanthikumar (2007).

Theorem 7.4.1 *Let X and Y be two random variables having respective density functions f_X and f_Y such that $X \leqslant_{lr} Y$. For $p \in [0, 1]$, let us take W_p as defined above. Then, for $p \leqslant p'$, we have $W_{p'} \leqslant_{lr} W_p$.* □

The above theorem immediately gives the following corollary.

Corollary 7.4.1 *Let X and Y be two random variables having respective density functions f_X and f_Y and let W_p be as defined above. Then*

$$X \leq_{lr} Y \Rightarrow X \leq_{lr} W_p \leq_{lr} Y,$$

for $p \in [0, 1]$. □

Let $\bar{G}_\theta$ be the survival function of the Marshall and Olkin (2007) semi-parametric model given by

$$\bar{G}_\theta(x) = \frac{\theta \bar{F}(x)}{1 - \bar{\theta}\, \bar{F}(x)}$$

for some baseline survival function $\bar{F}$ where $\bar{\theta} = 1 - \theta$. Let X_θ be a random variable having survival function $\bar{G}_\theta(\cdot)$. Then Benduch-Fraszczak (2010) shows that

$$X_{\theta_1} \leq_{lr} X_{\theta_2}$$

whenever $\theta_1 \leq \theta_2$. It is easy to see that the M-O model is a proportional odds model.

The Skew Normal distribution defined by Azzalini (1985) was generalized by Gupta and Gupta (2004) as a random variable X_n having density function given by

$$f(x; n, \lambda) \propto \phi(x) \Phi^n(\lambda x), \quad x \in \mathbb{R}, n \in \mathbb{N}, \lambda \in \mathbb{R},$$

where ϕ and Φ are, respectively, the pdf and the cdf of $N(0, 1)$. This is known as Generalized Skew Normal (GSN) distribution. If a random variable X has the above density we write $X \sim GSN(x; n, \lambda)$. It is to be noted that a GSN distribution is unimodal and it belongs to the ILR class (and hence it is IFR, $DRHR$, and $DMRL$ as well). If all the iid components forming a k-out-of-n system have the GSN life distribution then the system life has ILR distribution.

A random variable X having density function f is said to be ILR (increasing likelihood ratio) if $X_t \geq_{lr} X_s$ whenever $t \leq s$, where $X_t = [X - t | X \geq t]$ is the residual life random variable (cf. Shaked and Shanthikumar, 2007). Many well-known distributions are ILR, which includes Exponential distribution, Weibull distribution with shape parameter greater than or equal to unity, Gamma distribution with shape parameter greater than or equal to unity, Normal distribution, Truncated Normal distribution (with support $\mathbb{R}^+$), and Laplace distribution among others. It is to be noted that, although many well-known distributions are ILR, all distributions do not belong to the ILR class, viz. Cauchy distribution is not ILR.

Miziula (2012) (see also Navarro and Pellerey, 2022) has shown that ILR class is closed under convocation of independent components, whereas it is neither closed under the formation of parallel systems (cf. Barlow and Proschan, 1975) (although it is closed under the formation of parallel system of iid components, Hazra et al., 2014) nor under the formation of series system (cf. Franco et al., 2003). It can be shown that if components are independent but not identically distributed then ILR

class is not closed under mixture (cf. Miziula, 2012). Navarro and Shaked (2010) have shown that, if the component lives are exchangeable random variables, then the ILR class is closed under the formation of series and parallel systems.

7.5 Mean Residual Life Order

Write $X_t = [X - t | X \geq t]$, the residual random variable, as defined in Sect. 2.4.2. This is the remaining life of a unit having lifetime X that has already survived for t units of time. In terms of the residual random variable, (7.3.4) can be written as $X_t \leq_{st} Y_t$, for all t. This tells that $X \leq_{fr} Y$ if and only if $X_t \leq_{st} Y_t$, for all t, which implies $m_X(t) = E(X_t) \leq E(Y_t) = m_Y(t)$ for all t. This comparison of mean (expectation) of residual life is known as mean residual life (MRL) order. This is formally defined as follows.

Definition 7.5.1 For two random variables X and Y having finite expectations, we say that X is smaller than Y in MRL order, written as $X \leq_{mr} Y$ if $m_X(t) \leq m_Y(t)$, for all t.

From the above discussion we see that $X \leq_{fr} Y \Rightarrow X \leq_{mr} Y$. Now an immediate question that arises is—out of usual stochastic order and mean residual life order, which one is stronger? It can be shown that none implies the other. However, Gupta and Kirmani (1987) have shown that if $m_Y(t)/E(Y) \leq m_X(t)/E(X)$ then $X \leq_{mr} Y \Rightarrow X \leq_{st} Y$. A generalization of this result is given in Hu et al. (2001).

It can be noted that

$$m_X(t) = \int_0^\infty \bar{F}_{X_t}(u) du$$
$$= \frac{\int_t^\infty \bar{F}(u) du}{\bar{F}(t)}$$
$$= \frac{\int_t^\infty x f(x) dx}{\bar{F}(t)} - t.$$

From (7.3.1) we have

$$m_X(t) = \int_0^\infty \bar{F}_{X_t}(u) du$$
$$= \int_0^\infty \frac{\bar{F}_X(u+t)}{\bar{F}_X(t)} du$$
$$= \int_0^\infty \frac{e^{-\int_0^{u+t} r_X(x) dx}}{e^{-\int_0^t r_X(x) dx}} du$$
$$= \int_0^\infty e^{-\int_t^{u+t} r_X(x) dx} du. \qquad (7.5.7)$$

Suppose X has Exponential distribution with constant hazard rate λ. Then (7.5.7) shows that $m_X(t) = 1/\lambda$. Differentiating (7.5.7) with respect to t we get

$$1 + \frac{d}{dt} m_X(t) = r_X(t) m_X(t)$$

which gives

$$d\left(\ln \bar{F}(t)\right) = -\frac{dt}{m_X(t)} - d\left(\ln m(t)\right).$$

Integrating both sides with respect to t we get

$$\ln \bar{F}_X(t) = -\int_0^t \frac{dx}{m_X(x)} - \ln\left(\frac{m(t)}{m(0)}\right) + c.$$

Taking $t = 0$ we get $c = 0$. Writing $m(0) = E(X) = \mu$, we get from the above expression

$$\ln \bar{F}_X(t) = -\int_0^t \frac{dx}{m_X(x)} - \ln\left(\frac{m_X(t)}{\mu}\right)$$

which gives

$$\bar{F}_X(t) = \frac{\mu}{m_X(t)} e^{-\int_0^t \frac{dx}{m_X(x)}}.$$

Let us consider the failure rate function, $r_e(\cdot)$, of a distribution as the reciprocal of the above-discussed MRL function. Then the resulting distribution is known as the equilibrium distribution of the original distribution. The equilibrium distribution arises as the limiting distribution of the forward recurrence time in a renewal process. That's why it is also known as stationary renewal excess distribution. See also Sect. 5.3. It is easy to see that the density function of the equilibrium distribution is given by

$$f_e(x) = \frac{\bar{F}_X(x)}{\mu}, \quad x > 0,$$

where $\mu = E(X)$. If someone is waiting for a bus where the inter-arrival times between successive arrivals of the buses are iid random variables having the same distribution as X then, in the long run, the waiting time of the person will have the density function $f_e(\cdot)$ (cf. Wolff, 1989).

Define $\bar{G}_X(x) = \int_x^\infty \bar{F}_X(u)\,du$ and $\bar{G}_Y(x) = \int_x^\infty \bar{F}_Y(u)\,du$. Suppose X and Y be such that $X \leqslant_{mrl} Y$. Take $a(x) = \bar{G}_X(x)/\bar{G}_Y(x)$ in Lemma 7.3.1. Then we immediately get the following.

Theorem 7.5.1 *Let X and Y be two random variables having respective distribution functions F_X and F_Y such that $X \leqslant_{mrl} Y$. For $p \in [0, 1]$, let W_p be as defined above. Then, for $p \leqslant p'$, we have $W_{p'} \leqslant_{mrl} W_p$.* □

The following corollary is immediate from the above theorem.

Corollary 7.5.1 *Let X and Y be two random variables having respective distribution functions F_X and F_Y and let W_p be as defined above. Then, for $p \in [0, 1]$,*

$$X \leq_{mrl} Y \Rightarrow X \leq_{mrl} W_p \leq_{mrl} Y.$$

7.6 Reversed Hazard Rate Order

Let X be a non-negative random variable denoting the life of certain system. Suppose the system has failed at or before time t. Then, for any positive ϵ, the conditional probability $P(t - \epsilon < X \leq t | X \leq t)$ denotes the probability that the system has failed in the interval $(t - \epsilon, t]$ and is approximately given by $\epsilon . \mu_X(t)$, where $\mu_X(t) = f(t)/F(t)$ over the set $\{t : F(t) > 0\}$. This function μ_X is known as reversed hazard rate (also known as retro-hazard rate or backward hazard rate or inactivity time) function.

7.6.1 Definition and Preliminaries

In the analysis of left-censored data, the reversed hazard rate function plays the same role as the hazard rate function plays in the analysis of right-censored data (cf. Andersen et al., 1993). Thus, we have $d \ln F_X(t) = \mu_X(t) dt$, which when integrated gives

$$- \ln F_X(t) = \int_t^\infty \mu_X(u) du + c.$$

Taking limit as $t \to \infty$, we get $c = 0$ so that

$$F_X(t) = e^{- \int_t^\infty \mu_X(u) du}. \qquad (7.6.8)$$

Thus, for a proper lifetime distribution, we have

$$\int_0^\infty \mu_X(u) du = \infty \qquad (7.6.9)$$

whereas, for all $t > 0$,

$$\int_t^\infty \mu_X(u) du < \infty. \qquad (7.6.10)$$

Equations (7.6.9) and (7.6.10) together give

$$\int_0^t \mu_X(u) du = \infty$$

for all $t > 0$. This immediately gives $\lim_{t \to 0} \mu_X(t) = \infty$. We have seen that $r_X(t) =$ constant gives Exponential distribution. However, from (7.6.8), it is clear that there does not exist any distribution over $\mathbb{R}^+$ such that $\mu_X(x)$ is constant. However, a separate proof can be obtained in Sengupta and Nanda (1997).

It can be noted that $\mu_X(x)/r_X(x) = \bar{F}_X(x)/F(x)$ which is always decreasing. Thus, if $r_X(x)$ is decreasing, i.e., X is DFR, $\mu_X(x)$ is decreasing. Thus, the DFR distributions are contained in the DRHR class of distributions.

Shaked and Shanthikumar (1994) have defined reversed hazard rate ordering as follows.

Definition 7.6.1 For two random variables X and Y, X is said to be smaller than Y in reversed hazard rate order, written as $X \leqslant_{rh} Y$, if $\mu_X(x) \leqslant \mu_Y(x)$ for all $x \in (\min\{l_X, l_Y\}, \infty)$.

In the above definition, we consider $a/0 = \infty$ for any $a > 0$. It is easy to see that

$$u_X \leqslant l_Y \Rightarrow X \leqslant_{rh} Y \quad \text{and} \quad X \leqslant_{rh} Y \Rightarrow (l_X \leqslant l_Y, u_X \leqslant u_Y).$$

If the density does not exist then the reversed hazard rate order can be defined as $X \leqslant_{rh} Y$ if and only if

$$\frac{F_Y(x)}{F_X(x)} \text{ is non} - \text{decreasing in } x \in (\min\{l_X, l_Y\}, \infty). \tag{7.6.11}$$

Clearly, this gives that, for all x,

$$\frac{F_X(x)}{F_Y(x)} \geqslant \lim_{x \to \infty} \frac{F_X(x)}{F_Y(x)} = 1.$$

Thus, we have

$$\mu_X(t) \leqslant \mu_Y(t) \Rightarrow \bar{F}_X(t) \leqslant \bar{F}_Y(t).$$

For discrete random variables X and Y having support $S_d = \{0, 1, 2, \ldots\}$, X is said to be smaller than Y in reversed hazard rate order, written as $X \leqslant^d_{rh} Y$, if

$$\frac{P(X = k)}{P(X \leqslant k)} \leqslant \frac{P(Y = k)}{P(Y \leqslant k)} \tag{7.6.12}$$

for all $k \in S_d$.

The reversed hazard rate of a distribution was introduced in Barlow et al. (1963) whereas the reversed hazard rate order was introduced by Keilson and Sumita (1982) where they have called this as 'X is uniformly smaller than Y in the negative direction'. For any $x \leqslant t$, (7.6.11) can equivalently be written as

$$\frac{F_X(x)}{F_X(t)} \geqslant \frac{F_Y(x)}{F_Y(t)}.$$

7.6 Reversed Hazard Rate Order

Or, equivalently
$$P(X \geqslant x | X \leqslant t) \leqslant P(Y \geqslant x | Y \leqslant t)$$

whenever $x \leqslant t$. This means that $[X | X \leqslant t] \leqslant [Y | Y \leqslant t]$ for all t. So, we get that

$$X \leqslant_{rh} Y \iff [X | X \leqslant t] \leqslant [Y | Y \leqslant t].$$

This shows that reversed hazard rate order is stronger than usual stochastic order. Suppose ϕ is a strictly decreasing continuous function on (l_X, u_X), then we have

$$P(\phi(X) \leqslant t) = P\left(X \geqslant \phi^{-1}(t)\right) = \bar{F}_X\left(\phi^{-1}(t)\right).$$

This immediately gives us the following one very important result. See Nanda and Shaked (2001) for more detailed discussion on this result.

Theorem 7.6.1 *For two random variables X and Y,*

(i) $X \leqslant_{hr} (resp. \leqslant_{rh}) Y \Rightarrow \phi(X) \geqslant_{rh} (resp. \geqslant_{hr}) \phi(Y)$ *for any continuous and strictly decreasing function ϕ;*
(ii) *if, for some continuous and strictly decreasing function ϕ, $\phi(X) \leqslant_{hr} (resp. \leqslant_{rh}) \phi(Y)$ then $X \geqslant_{rh} (resp. \geqslant_{hr}) Y$.* □

A special case of this result, when $\phi(X) = 1/X$ has been discussed in Sengupta et al. (1999) whereas this result is used in Nanda (2010) to prove that two random variables X_1 and X_2 have proportional hazards if and only if $\phi(X_1)$ and $\phi(X_2)$ have proportional reversed hazards.

Writing $a(x) = F_X(x)/F_Y(x)$ and proceeding similarly as in Theorem 7.3.1 we get the following theorem.

Theorem 7.6.2 *Let X and Y be two random variables having respective distribution functions F_X and F_Y such that $X \leqslant_{rh} Y$. For $p \in [0, 1]$, let W_p be as defined above. Then, for $p \leqslant p'$, we have $W_{p'} \leqslant_{rh} W_p$.* □

The following corollary is immediately obtained from Theorem 7.6.2.

Corollary 7.6.1 *Let X and Y be two random variables having respective distribution functions F_X and F_Y and let W_p be as defined earlier. Then, for $p \in [0, 1]$, $X \leqslant_{rh} Y \Rightarrow X \leqslant_{rh} W_p \leqslant_{rh} Y$.* □

Some of the properties of $DRHR$ class are discussed in Sect. 2.4.1. Below we state some more borrowed from Sengupta and Nanda (1999).

- Remaining life of a maintained unit with zero repair time, under either perfect repair or minimal repair (a repair which takes the condition of the system to that just prior to its failure is called minimal repair), tends to have a log-concave distribution.
- DFR class is a sub-class of $DRHR$ class.
- If X belongs to the IFR class and ϕ is a strictly decreasing function, then $\phi(X)$ belongs to the $DRHR$ class.

- The ILR and the DLR classes are sub-class of the $DRHR$ class.
- $DRHR$ class is the smallest class which is closed under limits of distributions and under the formation of parallel systems and contains all concave life distributions.
- A $DRHR$ distribution has at most one jump-discontinuity, which must be at the left-end point of its support.
- $DRHR$ class is not closed under mixture.

They have also given the sharp upper bound of the reliability of a $DRHR$ distribution with a specific mean whereas the sharp lower bound in this case is zero. The detailed bound is as follows.

Theorem 7.6.3 *Let F be a log-concave distribution with known mean μ. Then the sharp upper bound of the corresponding reliability function is given by*

$$\bar{F}(t) = \begin{cases} 1, & \text{if } t \leqslant \mu \\ 1 - \frac{1}{e}\left[\frac{t}{t-\mu}\right]^{\frac{t}{\mu}-1}, & \text{if } t > \mu \end{cases}.$$

A random variable X having distribution function F and RHR function μ_F is said to have decreasing reversed hazard rate average ($DRHA$) if $\int_0^x \mu_F(t)\, dt/x$ is decreasing in $x > 0$.

7.6.2 Component Redundancy Versus System Redundancy

A result similar to Theorem 7.3.2 holds for reversed hazard rate ordering as stated below (cf. Gupta and Nanda, 2001).

Theorem 7.6.4 *Suppose $X_1, X_2, \ldots, X_n$ denote the iid component lifetimes and $Y_1, Y_2, \ldots, Y_n$ denote the lifetimes of n iid matching spares. Further suppose τ denote the lifetime of the coherent system formed from the n components with h as the system reliability function. If $(1-p)h'(p)/(1-h(p))$ is increasing in p and $h(p) \leqslant p$ for all $p \in [0, 1]$, then*

$$\tau(X \vee Y) \geqslant_{rh} \tau(X) \vee \tau(Y)$$

where $h'(p) = dh(p)/dp$. □

One may wonder whether the given conditions of the above theorem are met in general. Actually, for any k-out-of-n system, $(1-p)h'(p)/(1-h(p))$ is increasing in p whereas the condition $h(p) \leqslant p$ is satisfied by any series system where each component is a k_j-out-of-n_j sub-system with at least one sub-system consisting of a single component.

In case of non-matching spares, Gupta and Nanda (2001) have shown that, for iid components and iid spares, the result holds if the system is formed out of 2 components. Boland and El-Neweihi (1995) have shown that this result cannot hold for hazard rate ordering in case of non-matching spares.

7.7 Mean Inactivity Time

Let X be a random variable having distribution function F. The mean of the inactivity time random variable corresponding to X, $\alpha_F(t) = E(_tX)$, is known as *mean inactivity time* (*MIT*) of X (see also Sect. 5.5). For two lifetime random variables X and Y, let us denote the corresponding inactivity time by $_tX$ and $_tY$, respectively. Then X is said to be smaller than Y in MIT order, written as $X \leqslant_{mit} Y$, if $E(_tX) \leqslant E(_tY)$, for all $t > 0$. Note that, for $t > 0$,

$$\alpha_F(t) = \frac{\int_0^t F(u)\,du}{F(t)} = t - \frac{\int_0^t xf(x)\,dx}{F(t)} \qquad (7.7.13)$$

so that

$$\frac{\int_0^t xf(x)\,dx}{F(t)} = t - \alpha_F(t).$$

It can be shown that $t - \alpha_F(t)$ is increasing in $t \geqslant 0$.

A random variable X having distribution function F is said to have increasing mean inactivity time (*IMIT*) if $\alpha_F(t)$ is increasing in $t > 0$. Sengupta and Nanda (1999) have shown that a random variable X having distribution function F will have $DRHR$ property if and only if, for all positive integers p and q and for all $x \geqslant 0$,

$$F^{p+q}(x) \geqslant F^p\left(\frac{q}{p}x\right).F^q\left(\frac{p}{q}x\right),$$

which has been shown to be equivalent to the fact that $_tX$ is increasing in t with respect to usual stochastic order. Note that $\mu_F(t)$ is decreasing in t if and only if, for all $x \geqslant 0$, $F(t-x)/F(t)$ is increasing in $t > 0$, which gives that $\int_0^t F(t-x)\,dx/F(t)$ is increasing in $t > 0$. This shows that $\alpha_F(t)$ is increasing in $t > 0$. So, $DRHR$ class is smaller than $IMIT$ class (cf. Nanda et al., 2003).

An immediate question that now arises is—whether there exists any decreasing mean inactivity life function. Note that $0 \leqslant \alpha_F(t) \leqslant t$, for all $t \geqslant 0$. Now taking limit as $t \to 0+$, we get $\lim_{t\to 0+} \alpha_F(t) = 0$. Now, if $\alpha_F(t)$ is decreasing in $t \geqslant 0$, then $\alpha_F(t) \leqslant 0$ for all $t \geqslant 0$, which is a contradiction because α_F cannot be negative, giving that $\alpha_F(t) \equiv 0$. This is not possible. Hence we conclude that there does not exist any non-negative random variable over the entire interval $(0, \infty)$ having decreasing α_F. Block et al. (1998) have shown that if a random variable has infinite support and the distribution has increasing reversed hazard rate over the support then the support must be of the form $(-\infty, a)$ with $a < \infty$.

Nanda et al. (2003) have defined the variance of the inactivity time as $\beta_F(t) = V_F(_tX)$, which immediately gives the coefficient of variation of inactivity time as $\gamma_F(t) = \sqrt{\beta_F(t)}/\alpha_F(t)$, for $t > 0$. It is easy to see that $\beta_F^2(t) = \mu_F(t)(\alpha_F^2(t) - \beta_F(t))$, for $t > 0$. It is shown that if $\alpha_F(t)$ is increasing then, for all $t > 0$, $\beta_F(t) \leqslant \alpha_F^2(t)$. They have defined the increasing variance inactivity time (*IVIT*) class as the one where the members F have β_F increasing. Then it is easy to see that $IMIT$ class is smaller than $IVIT$ class.

Let X and Y be two non-negative random variables having respective distribution functions F and G, and the MIT functions α_F and α_G. Nanda et al. (2003) have shown that $X \leqslant_{mit} Y$ if and only if, for all $t > 0$,

$$\frac{F(t)\alpha_F(t)}{G(t)\alpha_G(t)} = \frac{\int_0^t F(u)\,du}{\int_0^t G(u)\,du} \text{ is decreasing in } t > 0.$$

The following theorem is due to Kayid and Ahmad (2004).

Theorem 7.7.1 *Let X_1, X_2 and Y be three non-negative random variables with X_1 and X_2 being independent of another random variable Y. Suppose each of X_1, X_2 and Y has log-concave density. If $X_1 \leqslant_{mit} X_2$ then $X_1 + Y \leqslant_{mit} X_2 + Y$.* □

Since convolution of random variables with log-concave densities has log-concave density, from the above theorem we immediately get

Theorem 7.7.2 *If $\{X_i\}$ and $\{Y_i\}$ are two sequences of random variables all having log-concave densities such that $X_i \leqslant_{mit} Y_i$ for all i then we have*

$$\sum_{i=1}^n X_i \leqslant_{mit} \sum_{i=1}^n Y_i$$

for any $n = 1, 2, 3, \ldots$ □

It is well known that Exponential distribution belongs to the boundary of the most of the positive and negative ageing classes. However, the DRHR class and the IMIT class do not have their dual classes in the sense that there does not exist any distribution which has either increasing RHR or decreasing MIT over the entire positive half of the real line. So, these classes are not ageing classes in true sense. So, for testing whether a distribution belongs to any ageing class we generally test F is Exponential against the alternative that F belongs to one ageing class and not Exponential. However, in order to test whether F belongs to DRHR or IMIT class we cannot have null hypothesis analogous to the above. So, for testing of DRHR or of IMIT class we test $H_0 : F = F_0$ where F_0 is one given distribution against the alternative that F is RHR or IMIT and not F_0. This kind of testing for RHR class has been discussed in Nanda and Paul (2003), who have also considered testing for whether two variables are ordered in RHR order, whereas that for the IMIT class has been studied in Kayid and Ahmad (2004). A large sample test for the IMIT class has also been developed by Zhang and Cheng (2010) who have used a characterization of IMIT class in terms of *right spread* order. They have shown that X is IMIT if and only if $_tX \leqslant_{rs} (-X)$. Note that X (having cdf F) is said to be less than Y (having cdf G) in right spread order, written as $X \leqslant_{rs} Y$, if and only if

7.7 Mean Inactivity Time

$$\int_{F^{-1}(p)}^{\infty} \bar{F}(x)\,dx \leqslant \int_{G^{-1}(p)}^{\infty} \bar{G}(x)\,dx$$

for all $p \in (0, 1)$.

Based on MIT order, dependence concept between two random variables may be defined as follows.

Definition 7.7.1 Two non-negative random variables X and Y are said to be positive (resp. negative) MIT dependent, in short called PMIT (resp. NMIT) dependent, if

$$(X|Y = y_1) \leqslant_{mit} (resp. \geqslant_{mit})(X|Y = y_2),$$

for all $y_1 \leqslant y_2$. □

Zamani et al. (2017) have shown that if the conditional reversed hazard rate, $\mu(x|Y = y)$ (the reversed hazard rate of the conditional random variable $(X|Y = y)$, defined as $\mu(x|Y = y) = f_{X|Y=y}(x|y)/F_{X|Y=y}(x|y)$), is increasing (resp. decreasing) in y, then (X, Y) is PMIT (resp. NMIT) dependent.

They have given the following characterization of PMIT (resp. NMIT) dependence.

Theorem 7.7.3 *Two non-negative random variables X and Y are PMIT (resp. NMIT) dependent if and only if any one of the following equivalent conditions holds.*

(a) $\int_0^x \frac{F_{X|Y=y}(z|y)}{F_{X|Y=y}(x|y)}\,dz$ *is decreasing (resp. increasing) in y, for all $x \geqslant 0$.*

(b) $\frac{\int_0^x F_{X|Y=y_2}(z|y_2)dz}{\int_0^x F_{X|Y=y_1}(z|y_1)dz}$ *is increasing (resp. decreasing) in x, for all $y_1 \leqslant y_2$.*

(c) $\int_0^x F_{X|Y=y}(z|y)\,dz$ *is TP_2 (resp. RR_2) in $(x, y) \in (0, \infty)^2$.* □

The following result is due to Zamani et al. (2017).

Theorem 7.7.4 *(a) If (X, Y) are positively likelihood ratio dependent (that is, the joint density $f(x, y)$ is TP_2 in (x, y))[1] then they are PMIT dependent.*
(b) If (X, Y) are PMIT dependent then $E(X|Y = y)$ is increasing in y. □

Writing $F(t) = p$, for $p \in (0, 1)$, (7.7.13) can be rewritten as

$$\alpha_F(F^{-1}(p)) = \frac{1}{p}\int_0^{F^{-1}(p)} F(x)\,dx,$$

where F^{-1} is the left-continuous inverse of F. Based on this representation of inactivity function, Arriaza et al. (2017) have defined Quantile MIT order as follows.

Definition 7.7.2 For two non-negative random variables X and Y with respective distribution functions F and G, X is said to be less than Y in quantitle mean inactivity time order, written as $X \leqslant_{qmit} Y$, if

---

[1] See Definition 7.8.7 for TP_2 function.

$$\frac{\int_0^{G^{-1}(p)} G(x)\,dx}{\int_0^{F^{-1}(p)} F(x)\,dx} \text{ is increasing in } p \in (0,1).$$

They have studied some properties of quantile mean inactivity time.

The following result gives some bounds on the MIT function (see Khan et al., 2021; Asadi and Berred, 2012).

Theorem 7.7.5 *If a distribution has mean μ and $E(X^r) < \infty$ for some $r > 1$, then*

$$\max\left\{0, x - \left(\frac{E(X^r)}{F(x)}\right)^{1/r}\right\} \leqslant \alpha_F(x) \leqslant \min\left\{x, x - \mu + \left(\frac{E(X^r)}{F(x)}\right)^{1/r}\frac{\bar{F}(x)}{F(x)}\right\}$$

for all $x \geqslant 0$.

7.8 Generalized Orderings

Let F_X be the cdf of a non-negative random variable X. Write, for $x > 0$,

$$F_X^{(n+1)}(x) = \int_0^x F_X^{(n)}(t)\,dt$$

for $n \in \mathbb{N} = \{1, 2, \ldots\}$ with $F_X^{(1)} = F_X$. For another random variable Y having cdf F_Y, we similarly define $F_Y^{(n)}$. Fishburn (1980) (see also O'Brien, 1984) defines nth-order stochastic dominance as follows.

Definition 7.8.1 A non-negative random variable X is said to dominate another non-negative random variable Y, written as $X \leqslant_{(n)} Y$ if $F_X^{(n)}(x) \geqslant F_Y^{(n)}(x)$, for all $x \geqslant 0$.

It is easy to see that $X \leqslant_{(n)} Y \Rightarrow X \leqslant_{(n+1)} Y$. Further, note that the first-order ($n=1$) stochastic dominance is equivalent to the usual stochastic order. The first-order and the second-order ($n=2$) stochastic dominance were discussed in Hadar and Russel (1969) whereas the third-order ($n=3$) stochastic dominance was discussed in Whitmore (1970).

Note that the generalized orders defined above are based on the distribution function. Similar measures based on the survival functions are studied in Mukherjee and Chatterjee (1992) (see also Fagiuoli and Pellerey, 1993).

Let X be a non-negative random variable having distribution function F_X with the corresponding density function f_X. Define $\bar{T}_0(X, x) = f_X(x)$ and

$$\bar{T}_s(X, x) = \frac{\int_x^\infty \bar{T}_{s-1}(X, u)\,du}{\mu_{s-1}(X)}, \quad s \in \mathbb{N}, \tag{7.8.14}$$

7.8 Generalized Orderings

where

$$\mu_s(X) = \int_0^\infty \bar{T}_s(X, x)\, dx, \quad s \in \mathbb{N} \cup \{0\}.$$

Note that, for every $s \in \mathbb{N}$, $\bar{T}_s(X, \cdot)$ represents the survival function of some random variable. Also, $\mu_1(X) = E(X)$. Further, $\bar{T}_2(X, \cdot)$ is the survival function of the equilibrium distribution of X and $\bar{T}_s(X, \cdot)$ is the survival function of the equilibrium distribution of a distribution having survival function $\bar{T}_{s-1}(X, \cdot)$, for $s > 2$. Again, write

$$r_s(X, x) = \frac{\bar{T}_{s-1}(X, x)}{\int_x^\infty \bar{T}_{s-1}(X, u)\, du}, \quad s \in \mathbb{N}. \tag{7.8.15}$$

Note that $r_1(X, x) = r_X(x)$, the failure rate of X whereas $r_2(X, x) = 1/m_X(x)$, the reciprocal of the mean residual life at time x. A simple calculation shows that, for $s \in \mathbb{N}$,

$$\bar{T}_s(X, x) = e^{-\int_0^x r_s(X, u)\, du}. \tag{7.8.16}$$

The relationship between $\mu_s(X)$ and the ordinary raw moments as given in Nanda et al. (1996b) is

$$\mu_s(X) = \frac{E(X^s)}{sE\left(X^{s-1}\right)}, \quad s \in \mathbb{N}. \tag{7.8.17}$$

Below are some of the generalized orders.

Definition 7.8.2 Let X and Y be two non-negative random variables and let $s \in \mathbb{N} \cup \{0\}$. Then X is said to be smaller than Y in

(a) s-FR order, written as $X \leqslant_{s-FR} Y$, if

$$\frac{\bar{T}_s(X, x)}{\bar{T}_s(Y, x)} \quad \text{is decreasing in } x > 0,$$

which, for $s \in \mathbb{N}$, can equivalently be written as

$$r_s(X, x) \geqslant r_s(Y, x) \quad \text{for all } x > 0; \tag{7.8.18}$$

(b) s-ST order, written as $X \leqslant_{s-ST} Y$, if

$$\frac{\bar{T}_s(X, x)}{\bar{T}_s(Y, x)} \leqslant \frac{\bar{T}_s(X, 0)}{\bar{T}_s(Y, 0)} \quad \text{for all } x > 0,$$

which, for $s \in \mathbb{N}$, reduces to

$$\bar{T}_s(X, x) \leqslant \bar{T}_s(Y, x) \quad \text{for all } x > 0; \tag{7.8.19}$$

(c) *s-CX* order, written as $X \leqslant_{s-CX} Y$, if

$$\int_x^\infty \frac{\bar{T}_s(X,u)}{\bar{T}_s(X,0)} du \leqslant \int_x^\infty \frac{\bar{T}_s(Y,u)}{\bar{T}_s(Y,0)} du \quad \text{for all } x > 0,$$

which, for $s \in \mathbb{N}$, reduces to

$$\int_x^\infty \bar{T}_s(X,u) \, du \leqslant \int_x^\infty \bar{T}_s(Y,u) \, du \quad \text{for all } x > 0;$$

(d) *s-CV* order, written as $X \leqslant_{s-CV} Y$, if

$$\int_0^x \frac{\bar{T}_s(X,u)}{\bar{T}_s(X,0)} du \leqslant \int_0^x \frac{\bar{T}_s(Y,u)}{\bar{T}_s(Y,0)} du \quad \text{for all } x > 0,$$

which, for $s \in \mathbb{N}$, reduces to

$$\int_0^x \bar{T}_s(X,u) \, du \leqslant \int_0^x \bar{T}_s(Y,u) \, du \quad \text{for all } x > 0.$$

Note that

$X \leqslant_{0-FR} Y \Leftrightarrow X \leqslant_{lr} Y, \quad X \leqslant_{1-FR} Y \Leftrightarrow X \leqslant_{hr} Y, \quad X \leqslant_{2-FR} Y \Leftrightarrow X \leqslant_{mrl} Y$
$X \leqslant_{3-FR} Y \Leftrightarrow X \leqslant_{vrl} Y, \quad X \leqslant_{0-ST} Y \Leftrightarrow X \leqslant_{wlr} Y, \quad X \leqslant_{1-st} Y \Leftrightarrow X \leqslant_{st} Y$
$X \leqslant_{2-ST} Y \Leftrightarrow X \leqslant_{hmlr} Y, \quad X \leqslant_{1-CX} Y \Leftrightarrow X \leqslant_{icx} Y \quad X \leqslant_{1-CV} Y \Leftrightarrow X \leqslant_{icv} Y.$

Integrating both sides of (7.8.19) with respect to x from $(0, \infty)$, we get $\mu_s(X) \leqslant \mu_s(Y)$, for all $s \in \mathbb{N}$. So, we get

$$X \leqslant_{s-ST} Y \Rightarrow \mu_s(X) \leqslant \mu_s(Y), \quad \text{for } s \in \mathbb{N}.$$

Suppose $X \leqslant_{s-ST} Y$. Then, for all $x > 0$, we have $\bar{T}_s(X,x) \leqslant \bar{T}_s(Y,x)$, for $s \in \mathbb{N}$, which is equivalent to

$$\frac{1}{\mu_{s-1}(X)} \int_x^\infty \bar{T}_{s-1}(X,u) \, du \leqslant \frac{1}{\mu_{s-1}(Y)} \int_x^\infty \bar{T}_{s-1}(Y,u) \, du.$$

Dividing both sides by x and taking limit as $x \to 0+$, we get

$$\frac{1}{\mu_{s-1}(X)} \geqslant \frac{1}{\mu_{s-1}(Y)}.$$

Thus, we have the following.

7.8 Generalized Orderings

Remark 7.8.1 For $s \in \mathbb{N}$,

$$X \leqslant_{s\text{-}ST} Y \Rightarrow \mu_{s-1}(X) \leqslant \mu_{s-1}(Y).$$

Below we give a characterization result of $s\text{-}FR$ order (cf. Hu et al., 2001). Before that we state the following lemma due to Capéraà (1988).

Lemma 7.8.1 *Let X and Y be two non-negative random variables, and let α and β be two functions such that α/β and β are increasing and β is non-negative. Then $X \leqslant_{hr} Y$ if and only if*

$$\frac{\int_0^\infty \alpha(x) \, dF_X(x)}{\int_0^\infty \beta(x) \, dF_X(x)} \leqslant \frac{\int_0^\infty \alpha(x) \, dF_Y(x)}{\int_0^\infty \beta(x) \, dF_Y(x)}.$$

Theorem 7.8.1 *Let X and Y be two non-negative random variables, and let α and β be two functions such that α/β and β are increasing and β is non-negative. Then $X \leqslant_{s\text{-}FR} Y$ if and only if*

$$\frac{\int_0^\infty \alpha(x) \bar{T}_{s-1}(X, x) \, dx}{\int_0^\infty \beta(x) \bar{T}_{s-1}(X, x) \, dx} \leqslant \frac{\int_0^\infty \alpha(x) \bar{T}_{s-1}(Y, x) \, dx}{\int_0^\infty \beta(x) \bar{T}_{s-1}(Y, x) \, dx} \qquad (7.8.20)$$

for all $s \in \mathbb{N}$.

Proof Suppose $X \leqslant_{s\text{-}FR} Y$ for $s \in \mathbb{N}$. This means that

$$\frac{\bar{T}_s(X, x)}{\bar{T}_s(Y, x)} \text{ is decreasing in } x > 0.$$

Since $\bar{T}_s(X, \cdot)$ is a cdf, the corresponding density function is given by $h_{s-1}^X(\cdot) = \bar{T}_{s-1}(X, \cdot)/\mu_{s-1}(X)$. Similarly we define $h_{s-1}^Y(\cdot) = \bar{T}_{s-1}(Y, \cdot)/\mu_{s-1}(Y)$. Note that (7.8.20) can be rewritten as

$$\frac{\int_0^\infty \alpha(x) h_{s-1}^X(x) \, dx}{\int_0^\infty \beta(x) h_{s-1}^X(x) \, dx} \leqslant \frac{\int_0^\infty \alpha(x) h_{s-1}^Y(x) \, dx}{\int_0^\infty \beta(x) h_{s-1}^Y(x) \, dx}.$$

Now, the necessity part of the theorem follows from Lemma 7.8.1. To prove the sufficiency part, we define, for $t < t'$,

$$\alpha(x) = \begin{cases} 0, & \text{if } x \leqslant t' \\ 1, & \text{if } x > t' \end{cases} \quad \text{and} \quad \beta(x) = \begin{cases} 0, & \text{if } x \leqslant t \\ 1, & \text{if } x > t. \end{cases}$$

Now, (7.8.20) can be rewritten as

$$\frac{\int_{t'}^\infty \bar{T}_{s-1}(X, x) \, dx}{\int_{t'}^\infty \bar{T}_{s-1}(Y, x) \, dx} \leqslant \frac{\int_t^\infty \bar{T}_{s-1}(X, x) \, dx}{\int_t^\infty \bar{T}_{s-1}(Y, x) \, dx},$$

which gives

$$\frac{\bar{T}_s(X,x)}{\bar{T}_s(Y,x)} \text{ is decreasing in } x.$$

This means that $X \leqslant_{s-FR} Y$. Thus, the theorem is established. □

Taking $s = 1, 2, 3$ in the above theorem we get the following. It is to be mentioned here that Corollary 7.8.1(a) is given in Capéraà (1988) whereas (b) may be obtained in Joag-Dev et al. (1995). The variables A_X and A_Y mentioned below are the corresponding equilibrium distributions defined in Chap. 5.

Corollary 7.8.1 *Let X and Y be two non-negative random variables, and α and β be two functions such that α/β and β are increasing and β is non-negative. Then we have the following.*

(a) $X \leqslant_{hr} Y$ if and only if

$$\frac{\int_0^\infty \alpha(x)\,dF_X(x)}{\int_0^\infty \beta(x)\,dF_X(x)} \leqslant \frac{\int_0^\infty \alpha(x)\,dF_Y(x)}{\int_0^\infty \beta(x)\,dF_Y(x)}.$$

(b) $X \leqslant_{mrl} Y$ if and only if

$$\frac{\int_0^\infty \alpha(x)\,\bar{F}_X(x)\,dx}{\int_0^\infty \beta(x)\,\bar{F}_X(x)\,dx} \leqslant \frac{\int_0^\infty \alpha(x)\,\bar{F}_Y(x)\,dx}{\int_0^\infty \beta(x)\,\bar{F}_Y(x)\,dx}.$$

(c) $X \leqslant_{vrl} Y$ if and only if

$$\frac{\int_0^\infty \alpha(x)\,\bar{F}_{A_X}(x)\,dx}{\int_0^\infty \beta(x)\,\bar{F}_{A_X}(x)\,dx} \leqslant \frac{\int_0^\infty \alpha(x)\,\bar{F}_{A_Y}(x)\,dx}{\int_0^\infty \beta(x)\,\bar{F}_{A_Y}(x)\,dx}.$$

Out of the above-mentioned orders some are already discussed in Chap. 2, *viz. lr*, *hr*, *mrl*, and *st* orders. The *wlr* order mentioned above stands for *weak likelihood ratio* order defined in (cf. Singh, 1989).

In *mrl* order, the expected values $E(X_t)$ and $E(Y_t)$ of residual lives of two distributions are compared at every point t. In place of expected values, if we compare the harmonic means, by changing the point t, of two distributions, then what we get is called *harmonic mean residual life (hmrl)* order. To be more specific, given $m_X(t) = E(X_t)$, we define the harmonic mean residual life, by allowing t to vary, as

$$h_X(x) = \left(\frac{1}{x}\int_0^x \frac{dt}{m_X(t)}\right)^{-1}.$$

Let us define $h_Y(x)$ similarly. Thus, if a random variable X has *mrl* function m_X and another random variable Y has *mrl* function m_Y, then X is said to be smaller than Y in *hmrl* order, written as $X \leqslant_{hmrl} Y$, if $h_X(x) \leqslant h_Y(x)$, for all $x > 0$. It is easy to see that if $m_X(t) \leqslant m_Y(t)$ for all t then $h_X(t) \leqslant h_Y(t)$ for all t, which shows that

7.8 Generalized Orderings

$$X \leq_{mrl} Y \Rightarrow X \leq_{hmrl} Y.$$

Further, $X \leq_{hmrl} Y$ if and only if, for all $x > 0$,

$$\int_0^x \frac{dt}{m_X(t)} \geq \int_0^x \frac{dt}{m_Y(t)},$$

which can equivalently be written as

$$\int_0^x d\left(\ln \int_u^\infty \bar{F}_X(t)\, dt\right) \leq \int_0^x d\left(\ln \int_u^\infty \bar{F}_Y(t)\, dt\right).$$

This can be simplified to

$$\frac{\int_x^\infty \bar{F}_X(t)\, dt}{E(X)} \leq \frac{\int_x^\infty \bar{F}_Y(t)\, dt}{E(Y)}, \qquad (7.8.21)$$

for all $x > 0$. This is the comparison between the survival functions of the corresponding equilibrium distributions. Let us define

$$x_+ = \begin{cases} x, & \text{if } x > 0 \\ 0, & \text{otherwise} \end{cases} \quad \text{and} \quad x_- = \begin{cases} x, & \text{if } x < 0 \\ 0, & \text{otherwise}. \end{cases}$$

Clearly, $x_+ = -x_-$. So, we have

$$\int_x^\infty \bar{F}_X(u)\, du = \int_x^\infty \int_u^\infty f_X(t)\, dt\, du$$
$$= \int_x^\infty f_X(t) \int_x^t du\, dt$$
$$= \int_0^\infty (t - x)_+ f_X(t)\, dt$$
$$= E\left[(X - x)_+\right]. \qquad (7.8.22)$$

Similarly, proceeding as above, we get

$$\int_0^x \bar{F}_X(u)\, du = x - \int_0^x F_X(u)\, du$$
$$= x + \int_0^x (t - x) f_X(t)\, dt\, du$$
$$= x + \int_0^\infty (t - x)_- f_X(t)\, dt\, du$$
$$= x + E[(X - x)_-]. \qquad (7.8.23)$$

On using (7.8.21), this immediately tells that $X \leq_{hmrl} Y$ if and only if, for every $x > 0$,

$$\frac{E\left[(X-x)_+\right]}{E(X)} \leq \frac{E\left[(Y-x)_+\right]}{E(Y)}.$$

Note that $\lim_{x \to 0+} h_X(x) = E(X)$. So, taking limit as $x \to 0+$, we get that

$$X \leq_{hmrl} Y \Rightarrow E(X) \leq E(Y). \tag{7.8.24}$$

Let us now define the following.

Definition 7.8.3 For two non-negative random variables X and Y, X is said to be smaller than Y in increasing convex (resp. concave) order, written as $X \leq_{icx}$ (resp. $\leq_{icv}$) Y, if $E[\phi(X)] \leq E[\phi(Y)]$ for all increasing convex (resp. concave) functions ϕ. □

Since, for every choice of x, $(t - x)_+$ is an increasing convex function of t, we have

$$X \leq_{icx} Y \Rightarrow E\left[(X-x)_+\right] \leq E\left[(Y-x)_+\right]. \tag{7.8.25}$$

Further, since, for every choice of x, $(t - x)_-$ is an increasing concave function of t, we have

$$X \leq_{icv} Y \Rightarrow E\left[(X-x)_-\right] \leq E\left[(Y-x)_-\right]. \tag{7.8.26}$$

Again, since every increasing convex (resp. concave) function can be approximated by positive linear combinations of the functions of the form $(x - t)_+$ (resp. $(x - t)_-$) for various choices of t, the converse of (7.8.25) (resp. (7.8.26)) also holds. Thus, on using (7.8.22) (resp. (7.8.23)), we see that $X \leq_{icx} Y$ (resp. $X \leq_{icv} Y$) if and only if

$$\int_x^\infty \bar{F}_X(u)\,du \leq \int_x^\infty \bar{F}_Y(u)\,du \quad \left(\text{resp.} \int_0^x \bar{F}_X(u)\,du \leq \int_0^x \bar{F}_Y(u)\,du\right).$$

This, by considering (7.8.24), gives

$$X \leq_{hmrl} Y \Rightarrow X \leq_{icx} (\text{resp.} \leq_{icv}) Y.$$

We now define another generalized order given in Kaas et al. (1994).

Definition 7.8.4 For two random variables X and Y, X is said to be smaller than Y in stop loss order of degree $s \in \mathbb{N}$, written as $X \leq_{s-SL} Y$, if

(i) $E(X^k) \leq E(Y^k)$, for $k = 0, 1, 2, \ldots, s - 1$,
(ii) $E\left[\{(X-x)_+\}^s\right] \leq E\left[\{(Y-x)_+\}^s\right]$, for all $x \geq 0$.

It can be noted that 1-SL order is equivalent to icx order defined above.

Let $S \subseteq \mathbb{R}$. Suppose $\phi : S \to \mathbb{R}$ is a function. Note that ϕ is increasing on S if and only if $\phi(x_0) \leq \phi(x_1)$ for all $x_0 < x_1 \in S$, which can be rewritten as

7.8 Generalized Orderings

$$\begin{vmatrix} 1 & 1 \\ \phi(x_0) & \phi(x_1) \end{vmatrix} \geqslant 0.$$

Again, a function ϕ is said to be convex on S if and only if, for any $x_0 < x_1 < x_2 \in S$, we have

$$\frac{\phi(x_2) - \phi(x_0)}{x_2 - x_0} \geqslant \frac{\phi(x_1) - \phi(x_0)}{x_1 - x_0},$$

which can be rewritten as

$$\begin{vmatrix} 1 & 1 & 1 \\ x_0 & x_1 & x_2 \\ \phi(x_0) & \phi(x_1) & \phi(x_2) \end{vmatrix} \geqslant 0. \tag{7.8.27}$$

This representation now can easily be used to generalize the concept of convexity as follows.

Definition 7.8.5 A function $\phi : S \to \mathbb{R}$ with $S \subseteq \mathbb{R}$ is said to be s-convex if, for any $x_0 < x_1 < \cdots < x_s \in S$,

$$\begin{vmatrix} 1 & 1 & 1 & \cdots & 1 \\ x_0 & x_1 & x_2 & \cdots & x_s \\ x_0^2 & x_1^2 & x_2^2 & \cdots & x_s^2 \\ \vdots & \vdots & \vdots & \cdots & \vdots \\ x_0^{s-1} & x_1^{s-1} & x_2^{s-1} & \cdots & x_s^{s-1} \\ \phi(x_0) & \phi(x_1) & \phi(x_2) & \cdots & \phi(x_s) \end{vmatrix} \geqslant 0. \tag{7.8.28}$$

The function ϕ is defined by Bullen (1971) as s-concave if $-\phi$ is s-convex. It is said to be strictly s-convex if the inequality in (7.8.28) is strict. Since a function ϕ is concave if the inequality in (7.8.27) is reversed, Denuit et al. (1998) define ϕ to be s-concave if the inequality in (7.8.28) is reversed. According to the definition of Bullen, 1-convex (resp. 1-concave) functions are monotonic increasing (resp. decreasing) functions whereas both 1-convex and 1-concave functions are monotonic increasing functions as per the definition of Denuit et al. However, according to both the definitions, 2-convex (resp. 2-concave) functions are usual convex (resp. concave) functions. We mention here that, throughout our discussion, we shall consider the definition given by Denuit et al. It can be verified that, for a function ϕ which is s times differentiable, ϕ is s-convex if and only if $\phi^{(s)}(x) \geqslant 0$ for all $x \in S$, where $\phi^{(s)}$ denotes the s-th derivative of ϕ.

Definition 7.8.6 Let s be any natural number. For two non-negative random variables X and Y, X is said to be smaller than Y in s-convex order, written as $X \leqslant_{s-convex} Y$, if $E(\phi(X)) \leqslant E(\phi(Y))$ for all s-convex functions ϕ defined over the common support of X and Y. □

In the above definition, if we take ϕ as an s-concave function, then we say that X is smaller than Y in s-concave order and we write $X \leq_{s-concave} Y$. It is easy to verify that

$$X \leq_{1-convex} Y \equiv X \leq_{1-concave} Y \equiv X \leq_{st} Y,$$
$$X \leq_{2-convex} Y \equiv X \leq_{cx} Y$$
$$X \leq_{2-concave} Y \equiv X \leq_{cv} Y.$$

We must mention here that a non-negative random variable X is said to be smaller than another non-negative random variable Y in convex (resp. concave) order, written as $X \leq_{cx}$ (resp. $\leq_{cv}$)Y if and only if $E(\phi(X)) \leq E(\phi(Y))$ for any non-negative convex (resp. concave) function ϕ.

For any non-negative random variable Z, define

$$\bar{F}_Z^{(k)}(t) = \int_t^\infty \bar{F}_Z^{(k-1)}(u)\, du,$$

for $k = 2, 3, \ldots$ with $\bar{F}_Z^{(1)}(t) = \bar{F}_Z(t)$, the survival function of Z. Now, suppose, for $s = 1, 2, \ldots$, we have two non-negative random variables X and Y such that $X \leq_{s-ST} Y$. This means that, for $s = 1, 2, \ldots$, we have $\bar{T}_s(X, t) \leq \bar{T}_s(Y, t)$, for all $t \geq 0$. If the first $s - 1$ moments of X and Y are known to be equal, i.e., $E(X^k) = E(Y^k)$ for all $s = 1, 2, \ldots, s-1$ then, on using (7.8.17), $X \leq_{s-ST} Y$ can equivalently be written as $\bar{F}_X^{(s)}(t) \leq \bar{F}_Y^{(s)}(t)$, for all $t \geq 0$.

The following lemma may be obtained in Denuit et al. (1998).

Lemma 7.8.2 *For two non-negative random variables X and Y, $X \leq_{s-convex} Y$ if and only if the following two conditions are satisfied.*

(a) $E(X^k) = E(Y^k)$ for all $k = 0, 1, 2, \ldots, s-1$
(b) $\bar{F}_X^{(s)}(t) \leq \bar{F}_Y^{(s)}(t)$ for all $t > 0$. □

Remark 7.8.2 One may question on the feasibility of any practical cases where equality of first $(s - 1)$ moments is possible. A nice example as mentioned below has been considered by them. Suppose X and Y denote the waiting times of customers coming to two different shops which are located adjacent to each other and sale similar items with almost similar prices. If expected waiting time in the first shop is less than that in the second shop, people will prefer to go to the first shop and in order to survive in the business the second shop owner must take appropriate steps to ensure that the expected waiting time in his shop is not more than that in the other shop. This competitive situation will come to an equilibrium where the expected waiting time in both the shops are same. Once this is achieved usual stochastic order (which is same as 1-convex order) cannot be used since in this case X and Y will have identical distribution. Further, the expected waiting time being same, people

7.8 Generalized Orderings

may prefer to go to that shop where the variance of the waiting time distribution is less. A similar argument as above will ultimately lead to the equality in the two second-order moments. Note that

$$\int_0^\infty \int_x^\infty \bar{F}_X(u)\,du\,dx = \int_0^\infty u\bar{F}_X(u)\,du = \frac{E(X^2)}{2}.$$

So, if the second-order moments are same, we must have

$$\int_x^\infty \bar{F}_X(u)\,du = \int_x^\infty \bar{F}_Y(u)\,du,$$

which tells that X and Y have identical distribution. So, the cx order (or, equivalently 2-convex order) cannot be used for comparison. Thus, if first $(s-1)$ moments are identical, we have to go for s-convex order for comparison. □

Now we give a lemma which may be available in Denuit et al. (2000).

Lemma 7.8.3 *Suppose X and Y are two non-negative random variables such that $X \leq_{s-convex} Y$. If $E(\phi(X)) = E(\phi(Y))$ for some strictly s-convex function ϕ, then $X =_{st} Y$.* □

On using Lemma 7.8.2, we see that if we have $X \leq_{s-convex} Y$ and $E(X^k) = E(Y^k)$ for all $k = 0, 1, 2, \ldots, s-1$, then we have $\bar{F}_X^{(s)}(t) \leq \bar{F}_Y^{(s)}(t)$ for all $t \geq 0$. This means that $X \leq_{s-ST} Y$. Thus, on using Lemma 7.8.3, we have the following theorem which is given in Hu et al. (2001).

Theorem 7.8.2 *Let X and Y be two non-negative random variables such that $E(X^k) = E(Y^k)$ for all $k = 0, 1, 2, \ldots, s-1$ and $s \in \mathbb{N}$. If $X \leq_{s-ST} Y$ and $E(\phi(X)) = E(\phi(Y))$ for any function ϕ whose sth derivative is positive, then $X =_{st} Y$.* □

Note that every polynomial of degree at most $(s-1)$ is s-convex. Hence, if $X \leq_{s-convex} Y$ then we can easily see that $E(X^k) = E(Y^k)$ for all $k = 0, 1, 2, \ldots, s-1$. Thus, in order that two random variables are ordered in the s-convex order, they must have the first $s-1$ moments identical.

Now we give definition of totally positive function. For more on this topic, one may refer to Karlin (1968).

Definition 7.8.7 Suppose $\mathcal{X} \subseteq \mathbb{R}$ and $\mathcal{Y} \subseteq \mathbb{R}$ be two sets. Let K be a real-valued function (commonly known as kernel) defined on $\mathcal{X} \times \mathcal{Y}$. Then K is called totally positive of order 2, written as TP_2, if

$$\begin{vmatrix} K(x_1, y_1) & K(x_1, y_2) \\ K(x_2, y_1) & K(x_2, y_2) \end{vmatrix} \geq 0 \qquad (7.8.29)$$

for all $x_1, x_2 \in \mathcal{X}$, $y_1, y_2 \in \mathcal{Y}$ such that $x_1 < x_2$, $y_1 < y_2$. □

The function K will be called RR_2 (reverse regular of order 2) if the inequality in (7.8.29) is reversed.

Example 7.8.1 Consider the indicator function

$$K(x, y) = \begin{cases} 1, & \text{if } x \leq y \\ 0, & \text{otherwise} \end{cases}$$
$$= I_{[x,\infty)}(y), \text{ say}.$$

It is not difficult to check that K is a TP_2 function. □

The following theorem is similar to Theorem 11.2 of Karlin (1968, p. 324) where the parenthetical statement was not given. The proof of the parenthetical statement is similar.

Theorem 7.8.3 Suppose $\mathcal{X} \subseteq \mathbb{R}$ and $\mathcal{Y} \subseteq \mathbb{R}$ be two intervals. Let K be a nonnegative TP_2 (resp. RR_2) function defined on $\mathcal{X} \times \mathcal{Y}$ with $K(x, y) > 0$ for some (x, y). Suppose $f : \mathcal{X} \times \mathcal{Y} \to \mathbb{R}$ be such that

(a) for each x, $f(x, y)$ changes sign at most once, and if one sign change does occur it occurs from negative (resp. positive) to positive (resp. negative) values as y traverses from left to right;
(b) for each y, $f(x, y)$ is increasing (resp. decreasing) in x.

Define

$$g(x) = \int_{\mathcal{Y}} K(x, y) f(x, y) \, dy.$$

If $g(x)$ exists then it changes sign at most once and from negative (resp. positive) to positive (resp. negative) values as x traverses from left to right.

Proof Let $x_0 \in \mathcal{X}$ be such that $g(x_0) = 0$. Suppose $S \subseteq \mathcal{Y}$ be such that $K(x_0, y) > 0$ for all $y \in S$. Then there exists $y_0 \in S$ such that

$$f(x_0, y) \begin{cases} \leq 0, & \text{for all } y < y_0 \\ \geq 0, & \text{for all } y > y_0. \end{cases}$$

Now, since $g(x_0) = 0$, we have

$$g(x) K(x_0, y_0) = g(x) K(x_0, y_0) - g(x_0) K(x, y_0)$$
$$= \int_{\mathcal{Y}} K(x, y) f(x, y) K(x_0, y_0) \, dy - \int_{\mathcal{Y}} K(x_0, y) K(x, y_0) f(x_0, y) \, dy$$
$$= K(x_0, y_0) \int_{\mathcal{Y}} K(x, y) [f(x, y) - f(x_0, y)] \, dy$$
$$+ \int_{\mathcal{Y}} f(x_0, y) [K(x, y).K(x_0, y_0) - K(x_0, y).K(x, y_0)] \, dy$$
$$= I_1 + I_2, \text{ say,}$$

7.8 Generalized Orderings

where

$$I_1 = K(x_0, y_0) \int_y K(x, y) [f(x, y) - f(x_0, y)] \, dy$$

which is non-negative for all $x \geq x_0$, by (b) above, and since K is TP_2, we have

$$I_2 = \int_y f(x_0, y) [K(x, y).K(x_0, y_0) - K(x_0, y).K(x, y_0)] \, dy \geq 0,$$

for all $y \geq y_0$. Thus, the result is established. $\square$

Below we take $a/0 = \infty$ if $a > 0$, $= -\infty$ if $a < 0$.

Theorem 7.8.4 *Let $S \subseteq \mathbb{R}$, and let f and g be two real-valued continuous functions defined on S. Then f/g is increasing (resp. decreasing) on S if and only if, for all $c \in \mathbb{R}$, $f(x) - c g(x)$ changes sign at most once as x traverses S, and if one sign change does occur, it occurs from negative (resp. positive) to positive (resp. negative) values.*

Proof Let us prove the case when f/g is increasing. The proof for the decreasing case is similar.

Necessity: Let f/g be increasing. This means that the rate of increase in f is more than that in g. Let us fix an arbitrary c. For that c, if $f(x) - c g(x) \geq 0$ for all x then the result trivially holds. Suppose $f(x) - c g(x)$ changes sign. Since the rate of increase in f is more, if there exists some x_0 such that $f(x_0) - c g(x_0) > 0$, we must have $f(x) - c g(x) > 0$ for all $x > x_0$. Suppose $f(x_0) - c g(x_0) < 0$. Then since f increases faster than g, there must exist some $x_1 (> x_0)$ such that $f(x_1) - c g(x_1) = 0$ and this again ensures that $f(x) - c g(x) > 0$ for all $x > x_1$. This completes the proof of necessary part.

Sufficiency: Given that, for all $c \in \mathbb{R}$, $(f(x) - c g(x))$ gets at most one change in sign and if such a change does occur it occurs from negative to positive values. If possible, suppose $h_c(x) = f(x) - c g(x)$ is not increasing. Then there exists x_0 and x_1 with $x_0 < x_1$ such that h_c increases at x_0 and decreases at x_1. This means that there exists an $\epsilon > 0$ such that $x_0 + \epsilon < x_1$ with $h_c(x_0 + \epsilon) > h_c(x_0)$ and $h_c(x_1 + \epsilon) < h_c(x_1)$. Since h_c is continuous, we can always choose one c, say $c = c_0$, such that $h_{c_0}(x_0) < 0$, $h_{c_0}(x_0 + \epsilon) > 0$ and $h_{c_0}(x_1) < 0$. This violates the assumption of at most one change in sign. Hence the result follows by contradiction. $\square$

The following theorem (cf. Hu et al., 2001) shows that the lower order generalized FR order implies the higher one. This may be compared with Theorem 7.12.1.

Theorem 7.8.5 *Let X and Y be two non-negative random variables. Then, for $s \in \mathbb{N} \cup \{0\}$,*

$$X \leq_{s-FR} Y \Rightarrow X \leq_{(s+1)-FR} Y.$$

Proof Note that

$$\frac{\bar{T}_{s+1}(X, x)}{\bar{T}_{s+1}(X, x)} = \alpha . \frac{\int_x^\infty \bar{T}_s(X, u) \, du}{\int_x^\infty \bar{T}_s(Y, u) \, du}, \tag{7.8.30}$$

where $\alpha = \mu_s(Y)/\mu_s(X)$. It is given that

$$\frac{\bar{T}_s(X, u)}{\bar{T}_s(Y, u)} \text{ is decreasing in } u,$$

which, by Theorem 7.8.4, gives that, for any $c \in \mathbb{R}$, $\bar{T}_s(X, u) - c\bar{T}_s(Y, u)$ has at most one change of sign when u traverses from 0 to ∞, and if one such change does occur, it occurs from positive to negative. Note that

$$\int_x^\infty \bar{T}_s(X, u)\, du = \int_0^\infty I_{[x,\infty)}(u) \bar{T}_s(X, u)\, du.$$

Write

$$g(x) = \int_0^\infty I_{[x,\infty)}(u) \left[\bar{T}_s(X, u) - c\bar{T}_s(Y, u)\right] du.$$

From Example 7.8.1, we see that the indicator function $I_{[x,\infty)}(u)$ is TP_2 and hence, by Theorem 7.8.3, $g(x)$ has at most one change of sign when x traverses from 0 to ∞, and if one such change does occur, it occurs from positive to negative. Again,

$$g(x) = \int_x^\infty \bar{T}_s(X, u)\, du - c \int_x^\infty \bar{T}_s(Y, u)\, du,$$

which enjoys the sign change property, as mentioned above, for any $c \in \mathbb{R}$. Thus, by Theorem 7.8.4, we say that

$$\frac{\int_x^\infty \bar{T}_s(X, u)\, du}{\int_x^\infty \bar{T}_s(Y, u)\, du} \text{ is decreasing in } x.$$

Hence, the result follows from (7.8.30).

Corollary 7.8.2 *Taking $s = 0, 1, 2, 3$, we see that*

$$X \leqslant_{lr} Y \Rightarrow X \leqslant_{hr} Y \Rightarrow X \leqslant_{mrl} Y \Rightarrow X \leqslant_{vrl} Y.$$

A general question could be whether the converse of Theorem 7.8.5 holds. In the above corollary, we see that $X \leqslant_{lr} Y \Rightarrow X \leqslant_{hr} Y$. However, that the converse is not true is demonstrated by Boland et al. (1994) in their Example 3.2 where they have proved that if a series system and a parallel system are formed out of two independent components, then parallel system dominates the series system in hazard rate order. However, taking a specific distribution they have shown that they are not likelihood ratio order (although if such systems are formed out of *iid* components they are likelihood ratio ordered).

Although, in general, this is not true, this holds under certain condition as given below (cf. Hu et al., 2001).

7.8 Generalized Orderings

Theorem 7.8.6 *Let $s \in \mathbb{N}$. If $r_s(X, x)/r_s(Y, x)$ is decreasing in x, then*

$$X \leqslant_{s-FR} Y \Rightarrow X \leqslant_{(s-1)-FR} Y.$$

Proof Note that

$$\begin{aligned}
\bar{T}_{s-1}(X, x) &= r_s(X, x) \int_x^\infty \bar{T}_{s-1}(X, u)\, du \\
&= r_s(X, x) \mu_{s-1}(X) \bar{T}_s(X, x) \\
&= r_s(X, x) \mu_{s-1}(X) e^{-\int_0^x r_s(X, u)\, du}.
\end{aligned}$$

The first equality follows from (7.8.15), the second equality follows from (7.8.14) whereas the last equality follows from (7.8.16). So, $X \leqslant_{(s-1)-FR} Y$ if and only if

$$e^{-\int_0^x (r_s(X,u) - r_s(Y,u))\, du} \cdot \frac{r_s(X, x)}{r_s(Y, x)} \text{ is decreasing in } x,$$

Given that $X \leqslant_{s-FR} Y$. So, by (7.8.18), the first term of the above expression is decreasing in x whereas the second term (i.e., the ratio) in the above expression is decreasing by the given hypothesis. Hence the result follows. □

Taking $s = 1, 2$ in the above theorem, we get the following.

Corollary 7.8.3 *For two non-negative random variables X and Y,*

(a) if $r_X(x)/r_Y(x)$ is decreasing in x, then $X \leqslant_{hr} Y \Rightarrow X \leqslant_{lr} Y$;
(b) if $m_X(x)/m_Y(x)$ is increasing in x, then $X \leqslant_{mrl} Y \Rightarrow X \leqslant_{hr} Y$.

The result given in (*a*) above is given in Belzunce et al. (2001) whereas (*b*) is given in Gupta and Kirmani (1987).

The following theorem shows that, for any $s \in \mathbb{N}$, s-FR order is stronger that s-ST order.

Theorem 7.8.7 *Let X and Y be two non-negative random variables. Then, for $s \in \mathbb{N} \cup \{0\}$,*

$$X \leqslant_{s-FR} Y \Rightarrow X \leqslant_{s-ST} Y.$$

Proof Let $s \in \mathbb{N}$. Then, $\bar{T}_s(X, 0) = \bar{T}_s(Y, 0) = 1$. So, from the definition of s-FR order, it follows that, if $X \leqslant_{s-FR} Y$ then we have

$$\frac{\bar{T}_s(X, x)}{\bar{T}_s(Y, x)} \leqslant \frac{\bar{T}_s(X, 0)}{\bar{T}_s(Y, 0)} = 1$$

giving $X \leqslant_{s-ST} Y$. Now, take $s = 0$. Then we have

$$\frac{\bar{T}_0(X, x)}{\bar{T}_0(Y, x)} \leqslant \frac{\bar{T}_0(X, 0)}{\bar{T}_0(Y, 0)}$$

which means that
$$\frac{f_X(x)}{f_Y(x)} \le \frac{f_X(0)}{f_Y(0)}$$
so that $X \le_{wlr} Y$. Hence the result is established.

Corollary 7.8.4 *From the above result, we get the following.*

(a) $X \le_{lr} Y \Rightarrow X \le_{wlr} Y$.
(b) $X \le_{hr} Y \Rightarrow X \le_{st} Y$.
(c) $X \le_{mrl} Y \Rightarrow X \le_{hmrl} Y$.

Theorem 7.8.7 tells that s-FR order is stronger than s-ST order whereas Theorem 7.8.5 gives that s-FR order implies $(s+1)$-FR order. So, one might be interested to know whether there is any relationship between $(s+1)$-FR and s-ST orders. In general, there exists no such relationship. To see this let us consider the following counterexamples which may be obtained in Gupta and Kirmani (1987).

Counterexample 7.8.1 *Let us consider X and Y having respective distributions*

$$\bar{F}_X(x) = \begin{cases} 1, & \text{if } x < 0 \\ e^{-\sqrt{x}}, & \text{if } x \ge 0 \end{cases} \quad \text{and} \quad \bar{F}_Y(x) = \begin{cases} 1, & \text{if } x < 0 \\ e^{-x}, & \text{if } 0 \le x < 1 \\ e^{-\sqrt{x}}, & \text{if } x \ge 1. \end{cases}$$

It is easy to check that $\bar{F}_X(x) \le \bar{F}_Y(x)$ for all x. Thus, we have $X \le_{st} Y$. Further,

$$m_X(x) - m_Y(x) = \frac{\int_x^\infty \bar{F}_X(u)\,du}{\bar{F}_X(x)} - \frac{\int_x^\infty \bar{F}_Y(u)\,du}{\bar{F}_Y(x)}$$
$$= \begin{cases} 3e^{x-1} - 2\sqrt{x} - 1, & \text{if } 0 \le x \le 1 \\ 0, & \text{if } x > 1 \end{cases}$$
$$= h(x), \text{ say.}$$

Note that $h(x) > 0$ for all $x \in (0, p)$ and $h(x) < 0$ for all $x \in (p, 1)$ where p is seen from Fig. 7.2 as approximately 0.002851. So, $X \not\le_{mrl} Y$. □

Now, we give an example (borrowed from Gupta and Kirmani, 1987) which shows that mrl order does not imply usual stochastic order.

Counterexample 7.8.2 *Let X have the survival function*

$$\bar{F}_X(x) = \begin{cases} 1, & \text{if } x < 0 \\ e^{-x^2}, & \text{if } x \ge 0 \end{cases}$$

which has a linearly increasing failure rate given by $r_X(x) = 2x$. So, the distribution has increasing failure rate. Further,

$$E(X) = \int_0^\infty e^{-x^2}\,dx = \frac{\sqrt{\pi}}{2} < 1.$$

7.8 Generalized Orderings

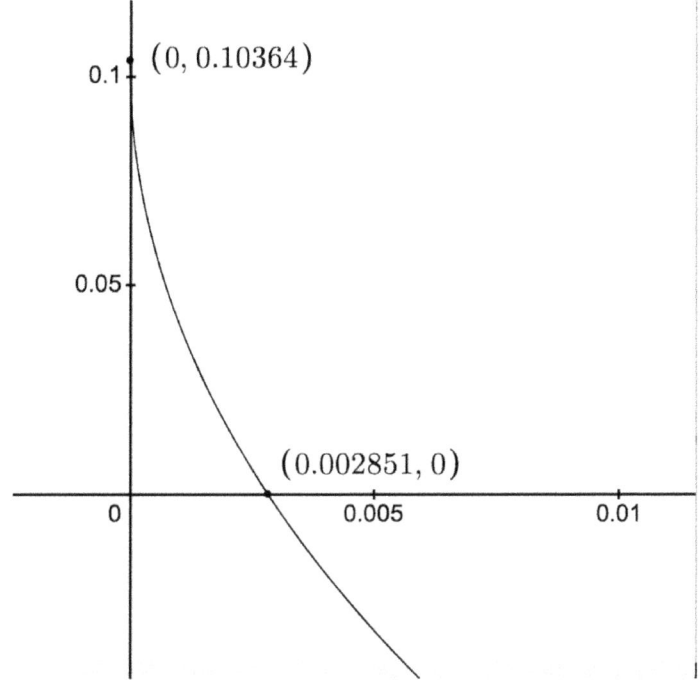

Fig. 7.2 Graph of $h(x)$ in Counterexample 7.8.1

We know that every IFR distribution is NBUE. Thus, we have $m_X(x) \leqslant E(X) < 1$. Now let Y be a standard Exponential random variable so that

$$\bar{F}_Y(x) = \begin{cases} 1, & \text{if } x < 0 \\ e^{-x}, & \text{if } x \geqslant 0 \end{cases}$$

which gives $m_Y(x) = 1$ for all x. Thus, we have $m_X(x) < 1 = m_Y(x)$ for all x. Hence $X \leqslant_{mrl} Y$. However,

$$\bar{F}_X(x) - \bar{F}_Y(x) = e^{-x^2} - e^{-x} \begin{cases} \geqslant 0, & \text{if } 0 \leqslant x \leqslant 1 \\ \leqslant 0, & \text{if } x \geqslant 1. \end{cases}$$

So, $X \not\leqslant_{st} Y$. □

Below we give a condition under which $(s+1)$-FR order implies s-FR order (cf. Hu et al., 2001). Before that we need the following lemma.

Lemma 7.8.4 *For $s \in \mathbb{N}$,*

$$\frac{\bar{T}_s(X, t)}{\bar{T}_s(X, 0)} = \frac{r_{s+1}(X, t)}{r_{s+1}(X, 0)} \cdot e^{-\int_0^t r_{s+1}(X, x)\, dx}.$$

Proof On using (7.8.15) we have

$$\frac{r_{s+1}(X,t)}{r_{s+1}(X,0)} = \frac{\bar{T}_s(X,t)}{\int_t^\infty \bar{T}_s(X,u)\,du} \cdot \frac{\mu_s(X)}{\bar{T}_s(X,0)}$$

which gives

$$\frac{\bar{T}_s(X,t)}{\bar{T}_s(X,0)} = \frac{r_{s+1}(X,t)}{r_{s+1}(X,0)} \frac{\int_t^\infty \bar{T}_s(X,u)\,du}{\mu_s(X)}$$

$$= \frac{r_{s+1}(X,t)}{r_{s+1}(X,0)} \cdot \bar{T}_{s+1}(X,t) = \frac{r_{s+1}(X,t)}{r_{s+1}(X,0)} \cdot e^{-\int_0^x r_{s+1}(X,x)\,dx}.$$

The second equality follows from (7.8.14) whereas the last equality follows from (7.8.16). □

We are now in a position to prove the following theorem (cf. Hu et al., 2001).

Theorem 7.8.8 *For any $s \in \mathbb{N}$, if $r_{s+1}(X,t)/r_{s+1}(Y,t) \leq \mu_s(Y)/\mu_s(X)$ for all $t \geq 0$, then*

$$X \leq_{(s+1)-FR} Y \implies X \leq_{s-ST} Y.$$

Proof On using Lemma 7.8.4 we have

$$\frac{\bar{T}_s(X,t)}{\bar{T}_s(X,0)} - \frac{\bar{T}_s(Y,t)}{\bar{T}_s(Y,0)} = \frac{r_{s+1}(X,t)}{r_{s+1}(X,0)} \cdot e^{-\int_0^x r_{s+1}(X,x)\,dx} - \frac{r_{s+1}(Y,t)}{r_{s+1}(Y,0)} \cdot e^{-\int_0^x r_{s+1}(Y,x)\,dx}$$

$$= r_{s+1}(X,t)\mu_s(X) \cdot e^{-\int_0^x r_{s+1}(X,x)\,dx} - r_{s+1}(Y,t)\mu_s(Y) \cdot e^{-\int_0^x r_{s+1}(Y,x)\,dx}$$

$$\leq r_{s+1}(Y,t)\mu_s(Y)\left[e^{-\int_0^x r_{s+1}(X,x)\,dx} - e^{-\int_0^x r_{s+1}(Y,x)\,dx}\right]$$

$$\leq 0.$$

The first inequality follows from the hypothesis. □

Taking $s = 1$ we get the following corollary which is given in Gupta and Kirmani (1987).

Corollary 7.8.5 *If $m_Y(t)/E(Y) \leq m_X(t)/E(X)$ for all $t \geq 0$, then*

$$X \leq_{mrl} Y \implies X \leq_{st} Y.$$

Following theorem shows that s-ST order is stronger than both s-CX and s-CV orders.

Theorem 7.8.9 *Let X and Y be two non-negative random variables. Then, for $s \in \mathbb{N}$, $X \leq_{s-ST} Y$ gives*

(a) $X \leq_{s-CX} Y$;
(b) $X \leq_{s-CV} Y$.

7.8 Generalized Orderings

Proof Given that

$$\frac{\bar{T}_s(X,t)}{\bar{T}_s(Y,t)} \leqslant \frac{\bar{T}_s(X,0)}{\bar{T}_s(Y,0)} \quad \text{for all } t \geqslant 0,$$

which can equivalently be written as

$$\frac{\bar{T}_s(X,t)}{\bar{T}_s(X,0)} \leqslant \frac{\bar{T}_s(Y,t)}{\bar{T}_s(Y,0)} \quad \text{for all } t \geqslant 0. \tag{7.8.31}$$

Integrating both sides of (7.8.31) with respect to t in $[x, \infty)$ we get

$$\int_x^\infty \frac{\bar{T}_s(X,t)}{\bar{T}_s(X,0)} \, dt \leqslant \int_x^\infty \frac{\bar{T}_s(Y,t)}{\bar{T}_s(Y,0)} \, dt \quad \text{for all } x \geqslant 0,$$

which proves (a). If we integrate (7.8.31) with respect to t in $[0, x]$, we get (b).

Corollary 7.8.6 *Taking $s = 1$ in the above theorem we get*

(i) $X \leqslant_{st} Y \Rightarrow X \leqslant_{icx} Y$;
(ii) $X \leqslant_{st} Y \Rightarrow X \leqslant_{icv} Y$. □

Now we show that $(s + 1)$-ST order implies s-CX order.

Theorem 7.8.10 *For $s \in \mathbb{N}$, $X \leqslant_{(s+1)-ST} Y \Rightarrow X \leqslant_{s-CX} Y$.*

Proof Given that, for all $x > 0$, $\bar{T}_{s+1}(X, x) \leqslant \bar{T}_{s+1}(Y, x)$, which gives

$$\frac{1}{\mu_s(X)} \int_x^\infty \bar{T}_s(X, u) \, du \leqslant \frac{1}{\mu_s(Y)} \int_x^\infty \bar{T}_s(Y, u) \, du. \tag{7.8.32}$$

On using Remark 7.8.1, (7.8.32) gives

$$\int_x^\infty \bar{T}_s(X, u) \, du \leqslant \int_x^\infty \bar{T}_s(Y, u) \, du.$$

This proves the result.

Corollary 7.8.7 *Taking $s = 1$ in the above theorem we get*

$$X \leqslant_{hmrl} Y \Rightarrow X \leqslant_{icx} Y.$$

Corollary 7.8.8 *From (7.8.32) it is clear that if $\mu_s(X) = \mu_s(Y)$, then we get*

$$X \leqslant_{(s+1)-ST} Y \iff X \leqslant_{s-CX} Y.$$

Remark 7.8.3 For $s = 1$, the above corollary tells that, for two random variables X and Y having same mean, the $hmrl$ order and the icx order are identical. □

The following diagram summarizes the above discussion.

```
0 − FR ⇔ lr      ⇒   1 − FR ⇔ hr        ⇒   2 − FR ⇒ mrl     ⇒   3 − FR ⇔ vrl ⇒   ...
   ⇓                    ⇓                      ⇓                     ⇓
0 − ST ⇔ wlr         1 − ST ⇔ st           2 − ST ⇔ hmrl         3 − ST             ...
   ⇓           ↘        ⇓           ↘        ⇓            ↘        ⇓         ↘
   ⇓         0 − CX     ⇓         1 − CX ⇔ icx   ⇓        2 − CX    ⇓        3 − CX ...
0 − CV              1 − CV ⇔ icv            2 − CV                3 − CV           ...
```

7.8.1 Closure Under Mixture of Distributions

For a random variable X, write $X_\theta = [X|\Theta = \theta]$ where Θ is a random variable having distribution G. Similarly, for another random variable Y, define $Y_\theta = [Y|\Theta = \theta]$. Then, on using mathematical induction on $s \in \mathbb{N} \cup \{0\}$, we have the following (see Nanda et al., 1996a where most of the results of this section may be available).

Lemma 7.8.5 *For $s \in \mathbb{N} \cup \{0\}$,*

(a) $\bar{T}_s(X, x) = \int_\mathbb{R} \lambda_{s-1}(\theta) \bar{T}_s(X_\theta, x) \, dG(\theta)$, *where*

$$\lambda_s(\theta) = \frac{\mu_0(X_\theta)\mu_1(X_\theta)\mu_2(X_\theta)\ldots\mu_s(X_\theta)}{\mu_0(X)\mu_1(X)\mu_2(X)\ldots\mu_s(X)} \qquad (7.8.33)$$

with $\lambda_{-1}(\theta) = 1$, and

(b) $\mu_s(X) = \int_\mathbb{R} \lambda_{s-1}(\theta) \mu_s(X_\theta) \, dG(\theta)$.

Proof We shall prove the result using mathematical induction. To prove (a), note that, for $s = 0$, the result is trivial. For $s = 1$, we have

$$\bar{T}_1(X, x) = \int_\mathbb{R} \bar{F}_\theta(x) \, dG(\theta) = \int_\mathbb{R} \lambda_0(\theta) \bar{T}_1(X_\theta, x) \, dG(\theta).$$

Let the result hold for some specific value of s. Then we have

$$\bar{T}_{s+1}(X, x) = \frac{1}{\mu_s(X)} \int_x^\infty \bar{T}_s(X, t) \, dt$$

$$= \frac{1}{\mu_s(X)} \int_x^\infty \left(\int_\mathbb{R} \lambda_{s-1}(\theta) \bar{T}_s(X_\theta, t) \, dG(\theta) \right) dt$$

$$= \int_\mathbb{R} \lambda_{s-1}(\theta) \left(\frac{1}{\mu_s(X)} \int_x^\infty \bar{T}_s(X_\theta, t) \, dt \right) dG(\theta)$$

$$= \int_\mathbb{R} \lambda_s(\theta) \left(\frac{1}{\mu_s(X_\theta)} \int_x^\infty \bar{T}_s(X_\theta, t) \, dt \right) dG(\theta)$$

$$= \int_\mathbb{R} \lambda_s(\theta) \bar{T}_{s+1}(X_\theta, x) \, dG(\theta).$$

7.8 Generalized Orderings

This proves (a). To prove (b), note that the result trivially holds for $s = 0$. Suppose (b) holds for some specific value of s. Then we have

$$\mu_{s+1}(X) = \int_0^\infty \bar{T}_{s+1}(X, x)\, dx$$

$$= \int_0^\infty \left(\int_\mathbb{R} \lambda_s(\theta) \bar{T}_{s+1}(X_\theta, x)\, dG(\theta) \right) dx$$

$$= \int_\mathbb{R} \lambda_s(\theta) \left(\int_0^\infty \bar{T}_{s+1}(X_\theta, x)\, dx \right) dG(\theta)$$

$$= \int_\mathbb{R} \lambda_s(\theta) \mu_{s+1}(X_\theta)\, dG(\theta).$$

Hence, (b) follows by using mathematical induction on s. $\square$

On using Lemma 7.8.5, we get the following result.

Theorem 7.8.11 *Let X_θ and Y_θ be as defined above. In (a), (c), and (d) below we take $s \in \mathbb{N}$ whereas (b) holds for $s \in \mathbb{N} \cup \{0\}$.*

(a) *Let $X_{\theta_1} \leq_{s-FR} Y_{\theta_2}$ for all θ_1 and θ_2 in the support of Θ. Then $X \leq_{s-FR} Y$.*
(b) *Let $X_{\theta_1} \leq_{s-ST} Y_{\theta_2}$ for all θ_1 and θ_2 in the support of Θ. Then $X \leq_{s-ST} Y$.*
(c) *Let $X_{\theta_1} \leq_{s-CX} Y_{\theta_2}$ for all θ_1 and θ_2 in the support of Θ. Then $X \leq_{s-CX} Y$.*
(d) *Let $X_{\theta_1} \leq_{s-CV} Y_{\theta_2}$ for all θ_1 and θ_2 in the support of Θ. Then $X \leq_{s-CV} Y$.*

Proof Given that

$$\frac{\bar{T}_s(X_{\theta_1}, x)}{\bar{T}_s(Y_{\theta_2}, x)} \text{ is decreasing in } x$$

for all θ_1 and θ_2. So, differentiating the above expression with respect to x, and using (7.8.14), we get

$$\frac{1}{\mu_{s-1}(Y_{\theta_2})} \bar{T}_s(X_{\theta_1}, x) \bar{T}_{s-1}(Y_{\theta_2}, x) \leq \frac{1}{\mu_{s-1}(X_{\theta_1})} \bar{T}_{s-1}(X_{\theta_1}, x) \bar{T}_s(Y_{\theta_2}, x), \quad (7.8.34)$$

for all $x \geq 0$, and all θ_1, θ_2 in the support of Θ. Now, define $\beta_s(\theta)$, for $s \in \mathbb{N} \cup \{0\}$, which is same as $\lambda_s(\theta)$ defined in (7.8.33) with X_θ replaced by Y_θ and X replaced by Y. Now, we multiply both sides of (7.8.34) by $\lambda_{s-1}(\theta_1) \cdot \beta_{s-1}(\theta_2)$ and then integrate with respect to $G(\theta_1)$ and $G(\theta_2)$ to get

$$\left(\int_\mathbb{R} \beta_{s-1}(\theta_2) \frac{\bar{T}_{s-1}(Y_{\theta_2}, x)}{\mu_{s-1}(Y_{\theta_2})} dG(\theta_2) \right) \left(\int_\mathbb{R} \lambda_{s-1}(\theta_1) \bar{T}_s(X_{\theta_1}, x)\, dG(\theta_1) \right)$$

$$- \left(\int_\mathbb{R} \lambda_{s-1}(\theta_1) \frac{\bar{T}_{s-1}(X_{\theta_1}, x)}{\mu_{s-1}(X_{\theta_1})} dG(\theta_1) \right) \left(\int_\mathbb{R} \beta_{s-1}(\theta_2) \bar{T}_s(Y_{\theta_2}, x)\, dG(\theta_2) \right) \leq 0$$

which can equivalently be written as

$$\left(\int_{\mathbb{R}} \beta_{s-1}(\theta) \frac{\bar{T}_{s-1}(Y_\theta, x)}{\mu_{s-1}(Y_\theta)} dG(\theta)\right) \left(\int_{\mathbb{R}} \lambda_{s-1}(\theta) \bar{T}_s(X_\theta, x) dG(\theta)\right)$$
$$- \left(\int_{\mathbb{R}} \lambda_{s-1}(\theta) \frac{\bar{T}_{s-1}(X_\theta, x)}{\mu_{s-1}(X_\theta)} dG(\theta)\right) \left(\int_{\mathbb{R}} \beta_{s-1}(\theta) \bar{T}_s(Y_\theta, x) dG(\theta)\right) \leqslant 0.$$

This can be rewritten as

$$\frac{d}{dx}\left(\frac{\int_{\mathbb{R}} \lambda_{s-1}(\theta) \bar{T}_s(X_\theta, x) dG(\theta)}{\int_{\mathbb{R}} \beta_{s-1}(\theta) \bar{T}_s(Y_\theta, x) dG(\theta)}\right) \leqslant 0$$

so that

$$\frac{\int_{\mathbb{R}} \lambda_{s-1}(\theta) \bar{T}_s(X_\theta, x) dG(\theta)}{\int_{\mathbb{R}} \beta_{s-1}(\theta) \bar{T}_s(Y_\theta, x) dG(\theta)} \text{ is decreasing in } x.$$

This, on using Lemma 7.8.5, gives

$$\frac{\bar{T}_s(X, x)}{\bar{T}_s(Y, x)} \text{ is decreasing in } x.$$

Hence, we have $X \leqslant_{s-FR} Y$. This proves (a). The proofs of the other parts follow similarly and hence we skip this.

Remark 7.8.4 Part (a) of the above theorem is proved in Nanda et al. (1996a). However, a similar result holds for $s = 0$ as well. This means the following.

Let $X_{\theta_1} \leqslant_{lr} Y_{\theta_2}$ for all θ_1 and θ_2 in the support of Θ. Then $X \leqslant_{lr} Y$.

To prove this, note from Lemma 7.8.5 that

$$f_X(x) = \int_{-\infty}^{\infty} f_{X_\theta}(x) dG(\theta). \tag{7.8.35}$$

Given that, for all θ_1, θ_2 from the support of Θ,

$$\frac{f_{X_{\theta_1}}(x)}{f_{X_{\theta_2}}(x)} \text{ is decreasing in } x,$$

which, after differentiating with respect to x, reduces to

$$f_{X_{\theta_2}}(x) \frac{d}{dx} f_{X_{\theta_1}}(x) - f_{X_{\theta_1}}(x) \frac{d}{dx} f_{X_{\theta_2}}(x) \leqslant 0.$$

7.8 Generalized Orderings

Let us first integrate the above expression with respect to θ_1 then with respect to θ_2 to get

$$\int_{-\infty}^{\infty} f_{X_{\theta_2}}(x)\,dG(\theta_2) \int_{-\infty}^{\infty} \frac{d}{dx} f_{X_{\theta_1}}(x)\,dG(\theta_1) - \int_{-\infty}^{\infty} f_{X_{\theta_1}}(x)\,dG(\theta_1) \int_{-\infty}^{\infty} \frac{d}{dx} f_{X_{\theta_2}}(x)\,dG(\theta_2) \leqslant 0$$

which can be rewritten as

$$\frac{d}{dx}\left[\frac{\int_{-\infty}^{\infty} f_{X_{\theta_1}}(x)\,dG(\theta_1)}{\int_{-\infty}^{\infty} f_{X_{\theta_2}}(x)\,dG(\theta_2)}\right] \leqslant 0.$$

This gives $X \leqslant_{lr} Y$.

Corollary 7.8.9 *The result given in (i) below is available in Shaked and Shanthikumar (1994), which is obtained from Theorem 7.8.11(a) by taking $s = 1$ whereas (ii) and (iii) are obtained by taking $s = 2$ and $s = 3$, respectively. Taking $s = 0, 1, 2$ in Theorem 7.8.11(b) we get (iv), (v), and (vi) below whereas (vii) and (viii) are obtained, respectively, from Theorem 7.8.11(c) and Theorem 7.8.11(d) by taking $s = 1$.*

(i) *Let $X_{\theta_1} \leqslant_{hr} Y_{\theta_2}$ for all θ_1 and θ_2 in the support of Θ. Then $X \leqslant_{hr} Y$.*
(ii) *Let $X_{\theta_1} \leqslant_{mrl} Y_{\theta_2}$ for all θ_1 and θ_2 in the support of Θ. Then $X \leqslant_{mrl} Y$.*
(iii) *Let $X_{\theta_1} \leqslant_{vrl} Y_{\theta_2}$ for all θ_1 and θ_2 in the support of Θ. Then $X \leqslant_{vrl} Y$.*
(iv) *Let $X_{\theta_1} \leqslant_{wlr} Y_{\theta_2}$ for all θ_1 and θ_2 in the support of Θ. Then $X \leqslant_{wlr} Y$.*
(v) *Let $X_{\theta_1} \leqslant_{st} Y_{\theta_2}$ for all θ_1 and θ_2 in the support of Θ. Then $X \leqslant_{st} Y$.*
(vi) *Let $X_{\theta_1} \leqslant_{hmrl} Y_{\theta_2}$ for all θ_1 and θ_2 in the support of Θ. Then $X \leqslant_{hmrl} Y$.*
(vii) *Let $X_{\theta_1} \leqslant_{icx} Y_{\theta_2}$ for all θ_1 and θ_2 in the support of Θ. Then $X \leqslant_{icx} Y$.*
(viii) *Let $X_{\theta_1} \leqslant_{icv} Y_{\theta_2}$ for all θ_1 and θ_2 in the support of Θ. Then $X \leqslant_{icv} Y$.*

Remark 7.8.5 Although the result given in (v) above is obtained by taking $s = 1$ in Theorem 7.8.11(b), it can be proved under less restrictive conditions as given below (cf. Shaked and Shanthikumar, 2007).

Let X, Y and Θ be random variables such that $[X|\Theta = \theta] \leqslant_{st} [Y|\Theta = \theta]$ for all θ in the support of Θ. Then $X \leqslant_{st} Y$. □

Remark 7.8.6 Theorem 7.8.11(a) holds for any real $s \in \mathbb{R}^+$. A proof may be obtained in Hu et al. (2004b). □

Consider a family of distributions $\{G_\theta, \theta \in \chi\}$, where $\chi \subseteq \mathbb{R}$. Let X_θ have the cdf G_θ and let Θ_i be a random variable having cdf F_i and support in χ. Write $Y_i =_{st} X(\Theta_i)$, for $i = 1, 2$.

Below we give a lemma which will be used to prove a mixture-type result.

Lemma 7.8.6 *For any* $s \in \mathbb{N} \cup \{0\}$,

(a) $\bar{T}_s(Y_i, y) = \dfrac{\int_X \prod_{j=1}^{s-1} \mu_j(X_\theta) \bar{T}_s(X_\theta, y) \, dF_i(\theta)}{\int_X \prod_{j=1}^{s-1} \mu_j(X_\theta) \, dF_i(\theta)}$;

(b) $\mu_s(Y_i) = \dfrac{\int_X \prod_{j=1}^{s} \mu_j(X_\theta) \, dF_i(\theta)}{\int_X \prod_{j=1}^{s-1} \mu_j(X_\theta) \, dF_i(\theta)}$,

where $\prod_{j=1}^{k} \mu_j(X_\theta) = 1$, *for* $k < 1$.

Proof We shall prove this result on using induction on s. Note that

$$\bar{F}_{Y_i}(y) = P(Y_i \geqslant y) = P(X(\Theta_i) \geqslant y)$$
$$= \int_X P(X_\theta \geqslant y) \, dF_i(\theta)$$
$$= \int_X \bar{G}_\theta(y) \, dF_i(\theta). \qquad (7.8.36)$$

Taking derivative of the expression with respect to y we get

$$f_{Y_i}(y) = \int_X g_\theta(y) \, dF_i(\theta),$$

where f_{Y_i} is the density function of Y_i and g_θ is the density of X_θ. This proves (a) for $s = 0, 1$. The proof of (b) for $s = 0$ is trivial. Integrating (7.8.36) with respect over y in $(0, \infty)$, we get

$$\mu_1(Y_i) = \int_0^\infty \int_X \bar{G}_\theta(y) \, dF_i(\theta) \, dy = \int_X \mu_1(X_\theta) \, dF_i(\theta).$$

This proves (b) for $s = 1$. Suppose both (a) and (b) hold for some specific s. Then we have

$$\bar{T}_{s+1}(Y_i, y) = \dfrac{\int_y^\infty \bar{T}_s(Y_i, x) \, dx}{\mu_s(Y_i)}$$
$$= \dfrac{\int_y^\infty \int_X \prod_{j=1}^{s-1} \mu_j(X_\theta) \bar{T}_s(X_\theta, x) \, dF_i(\theta) \, dx}{\int_0^\infty \int_X \prod_{j=1}^{s-1} \mu_j(X_\theta) \bar{T}_s(X_\theta, x) \, dF_i(\theta) \, dx}$$
$$= \dfrac{\int_X \prod_{j=1}^{s-1} \mu_j(X_\theta) \int_y^\infty \bar{T}_s(X_\theta, x) \, dx \, dF_i(\theta)}{\int_X \prod_{j=1}^{s-1} \mu_j(X_\theta) \int_0^\infty \bar{T}_s(X_\theta, x) \, dx \, dF_i(\theta)}$$
$$= \dfrac{\int_X \prod_{j=1}^{s} \mu_j(X_\theta) \bar{T}_{s+1}(X_\theta, y) \, dF_i(\theta)}{\int_X \prod_{j=1}^{s} \mu_j(X_\theta) \, dF_i(\theta)}. \qquad (7.8.37)$$

7.8 Generalized Orderings

Integrating (7.8.37) with respect to y over $(0, \infty)$, we get

$$\mu_{s+1}(Y_i) = \int_0^\infty \bar{T}_{s+1}(Y_i, y)\, dy$$

$$= \int_0^\infty \frac{\int_\chi \prod_{j=1}^s \mu_j(X_\theta) \bar{T}_{s+1}(X_\theta, y)\, dF_i(\theta)}{\int_\chi \prod_{j=1}^s \mu_j(X_\theta)\, dF_i(\theta)}\, dy$$

$$= \frac{\int_\chi \prod_{j=1}^{s+1} \mu_j(X_\theta)\, dF_i(\theta)}{\int_\chi \prod_{j=1}^s \mu_j(X_\theta)\, dF_i(\theta)}.$$

Thus, the result is proved by the method of mathematical induction. $\square$

A mixture-type result is also given in Hu et al. (2001) as mentioned below. However, before that we need a lemma which again needs basic composition formula (cf. Karlin (1968, p. 17), Barlow and Proschan (1981, p. 100)).

Lemma 7.8.7 Basic Composition Formula: Suppose $x \in \mathcal{X}$, $y \in \mathcal{Y}$ and $z \in \mathcal{Z}$, where $\mathcal{X}, \mathcal{Y}, \mathcal{Z} \subseteq \mathbb{R}$. Let $u : \mathcal{X} \times \mathcal{Y} \to \mathbb{R}$, $v : \mathcal{Y} \times \mathcal{Z} \to \mathbb{R}$ and $w : \mathcal{X} \times \mathcal{Z} \to \mathbb{R}$ be Borel measurable functions such that

$$w(x, z) = \int_\mathcal{Y} u(x, y) v(y, z)\, dy$$

converges absolutely. Then

$$\begin{vmatrix} w(x_1, z_1) & \ldots & w(x_1, z_n) \\ \vdots & \ldots & \vdots \\ w(x_n, z_1) & \ldots & w(x_n, z_n) \end{vmatrix}$$

$$= \int \ldots \int_{y_1 < \ldots < y_n} \begin{vmatrix} u(x_1, y_1) & \ldots & u(x_1, y_n) \\ \vdots & \ldots & \vdots \\ u(x_n, y_1) & \ldots & w(x_n, z_n) \end{vmatrix} \cdot \begin{vmatrix} v(y_1, z_1) & \ldots & v(y_1, z_n) \\ \vdots & \ldots & \vdots \\ v(y_n, z_1) & \ldots & v(y_n, z_n) \end{vmatrix} dy_1 \ldots dy_n$$

Lemma 7.8.8 Let $\psi(\theta, x)$ be any (non-negative) TP_2 function in $\theta \in \chi$ and $x \in \mathbb{R}$, and let $\bar{F}_i(\theta)$ be a differentiable survival function in θ and TP_2 in $i \in \{1, 2\}$ and $\theta \in \chi$. Assume that $\psi(\theta, x)$ is differentiable in θ for every x. Then

$$H_i(x) = \int_\chi \psi(\theta, x)\, dF_i(\theta) \text{ is } TP_2 \text{ in } x \in \mathbb{R} \text{ and } i \in \{1, 2\},$$

where $\bar{F}_i \equiv 1 - F_i$. Conversely, if $H_i(x)$ is TP_2 in $i \in \{1, 2\}$ and $x \in \mathbb{R}$ whenever $\bar{F}_i(\theta)$ is TP_2 in $i \in \{1, 2\}$ and $\theta \in \chi$, then $\psi(\theta, x)$ is TP_2 in $\theta \in \chi$ and $x \in \mathbb{R}$.

Proof If f_i is the density corresponding to the cdf F_i, for $i = 1, 2$, then, for $x_1 < x_2 \in \mathbb{R}$, we have

$$\begin{vmatrix} H_1(x_1) & H_1(x_2) \\ H_2(x_1) & H_2(x_2) \end{vmatrix} = \begin{vmatrix} \int_\chi \psi(\theta, x_1) dF_1(\theta) & \int_\chi \psi(\theta, x_1) dF_2(\theta) \\ \int_\chi \psi(\theta, x_2) dF_1(\theta) & \int_\chi \psi(\theta, x_2) dF_2(\theta) \end{vmatrix}$$

$$= \begin{vmatrix} \int_\chi \psi(\theta, x_1) f_1(\theta) d\theta & \int_\chi \psi(\theta, x_1) f_2(\theta) d\theta \\ \int_\chi \psi(\theta, x_2) f_1(\theta) d\theta & \int_\chi \psi(\theta, x_2) f_2(\theta) d\theta \end{vmatrix}$$

$$= \int\int_{\theta_1 < \theta_2} \begin{vmatrix} \psi(\theta_1, x_1) & \psi(\theta_1, x_2) \\ \psi(\theta_2, x_1) & \psi(\theta_2, x_2) \end{vmatrix} \cdot \begin{vmatrix} f_1(\theta_1) & f_2(\theta_1) \\ f_1(\theta_2) & f_2(\theta_2) \end{vmatrix} d\theta_1\, d\theta_2$$

$$= \int_\chi \left[\begin{vmatrix} \psi(\theta_1, x_1) & \psi(\theta_1, x_2) \\ \psi(\theta_2, x_1) & \psi(\theta_2, x_2) \end{vmatrix} \cdot \begin{vmatrix} f_1(\theta_1) & f_2(\theta_1) \\ -\bar{F}_1(\theta_2) & -\bar{F}_2(\theta_2) \end{vmatrix} \Big|_{\theta_2 = \theta_1}^{\infty} \right.$$

$$\left. + \int_{\theta_1}^\infty \begin{vmatrix} \psi(\theta_1, x_1) & \psi(\theta_1, x_2) \\ \frac{d}{d\theta_2}\psi(\theta_2, x_1) & \frac{\partial}{\partial\theta_2}\psi(\theta_2, x_2) \end{vmatrix} \cdot \begin{vmatrix} f_1(\theta_1) & f_2(\theta_1) \\ \bar{F}_1(\theta_2) & \bar{F}_2(\theta_2) \end{vmatrix} d\theta_2 \right] d\theta_1$$

$$= \int_{\theta_1 < \theta_2} \begin{vmatrix} \psi(\theta_1, x_1) & \psi(\theta_1, x_2) \\ \frac{\partial}{\partial\theta_2}\psi(\theta_2, x_1) & \frac{\partial}{\partial\theta_2}\psi(\theta_2, x_2) \end{vmatrix} \cdot \begin{vmatrix} f_1(\theta_1) & f_2(\theta_1) \\ \bar{F}_1(\theta_2) & \bar{F}_2(\theta_2) \end{vmatrix} d\theta_2\, d\theta_1 \quad (7.8.38)$$

. The third equality follows from basic composition formula. The second last equality is obtained by integration by parts, where the first term is zero because, for $\theta_2 = \theta_1$, the first determinant becomes zero whereas the second determinant becomes zero when $\theta_2 \to \infty$. The second determinant of the last equality is non-negative since this is equivalent to the fact that $\bar{F}_i(\theta)$ is TP_2. Further, $\psi(\theta, x)$ is given to be TP_2 in (θ, x). So, for $\theta_1 < \theta_2$, $x_1 < x_2$, we have

$$\frac{\psi(\theta_2, x_1)}{\psi(\theta_1, x_1)} \leq \frac{\psi(\theta_2, x_2)}{\psi(\theta_1, x_2)}, \quad (7.8.39)$$

which shows that

$$\frac{\psi(\theta_2, x_2)}{\psi(\theta_2, x_1)} \text{ is increasing in } \theta_2.$$

Differentiating the above expression with respect to θ_2, we get

$$\psi(\theta_2, x_1) \frac{\partial}{\partial \theta_2} \psi(\theta_2, x_2) \geq \psi(\theta_2, x_2) \frac{\partial}{\partial \theta_2} \psi(\theta_2, x_1)$$

which can be rewritten as

$$\frac{\frac{\partial}{\partial\theta_2}\psi(\theta_2, x_1)}{\psi(\theta_2, x_1)} \leq \frac{\frac{\partial}{\partial\theta_2}\psi(\theta_2, x_2)}{\psi(\theta_2, x_2)}. \quad (7.8.40)$$

7.8 Generalized Orderings

Multiplying (7.8.39) and (7.8.40) we get that $\psi(\theta, x)$ is TP_2 if and only if

$$\frac{\frac{\partial}{\partial \theta_2}\psi(\theta_2, x_1)}{\psi(\theta_1, x_1)} \leqslant \frac{\frac{\partial}{\partial \theta_2}\psi(\theta_2, x_2)}{\psi(\theta_1, x_2)}.$$

Hence, the first determinant of (7.8.38) is non-negative. So, we get that $H_i(x)$ is TP_2 is (i, x). This proves the theorem. □

We are now in a position to prove the following mixture-type result.

Theorem 7.8.12 *For any $s \in \mathbb{N}$, suppose*

$$X(\theta) \leqslant_{s-FR} X(\theta') \quad \text{whenever } \theta \leqslant \theta' \tag{7.8.41}$$

and

$$\Theta_1 \leqslant_{hr} \Theta_2. \tag{7.8.42}$$

Then $Y_1 \leqslant_{s-FR} Y_2$.

Proof Note from Lemma 7.8.6 that

$$\bar{T}_s(Y_i, y) = C_i^{-1} \int_{\chi} \prod_{j=1}^{s-1} \mu_j(X_\theta) \bar{T}_s(X_\theta, y) \, dF_i(\theta), \tag{7.8.43}$$

where $C_i = \int_{\chi} \prod_{j=1}^{s-1} \mu_j(X_\theta) \, dF_i(\theta)$. The statement in (7.8.41) tells that $\bar{T}_s(X_\theta, y)$ is TP_2 in θ and y. So, $\prod_{j=1}^{s-1} \mu_j(X_\theta)\bar{T}_s(X_\theta, y)$ is TP_2 in θ and y. Again, (7.8.42) tells that $\bar{F}_i(\theta)$ is TP_2 in $i \in \{1, 2\}$ and $\theta \in \chi$. Hence, by Lemma 7.8.8, we get that $\bar{T}_s(Y_i, y)$ is TP_2 in i and y. Hence, the result is established.

Corollary 7.8.10 *Let $\Theta_1 \leqslant_{hr} \Theta_2$. Then, from the above theorem, we get the following.*

(a) *If $X(\theta) \leqslant_{hr} X(\theta')$ for all $\theta \leqslant \theta'$, then $Y_1 \leqslant_{hr} Y_2$.*
(b) *If $X(\theta) \leqslant_{mrl} X(\theta')$ for all $\theta \leqslant \theta'$, then $Y_1 \leqslant_{mrl} Y_2$.*
(c) *If $X(\theta) \leqslant_{vrl} X(\theta')$ for all $\theta \leqslant \theta'$, then $Y_1 \leqslant_{vrl} Y_2$.*

Remark 7.8.7 Part (a) of the above corollary is given in Shaked and Wong (1995) whereas part (b) generalizes a result of Ahmed et al. (1996) where they have proved the result when Θ_1 and Θ_2 are likelihood ratio ordered.

7.8.2 Characterizations in Terms of Residual Lives

Let us write the residual lives corresponding to the random variables X and Y as $X_t = [X - t | X > t]$ and $Y_t = [Y - t | Y > t]$, respectively. Then we have the following lemma.

Lemma 7.8.9 *For* $s \in \mathbb{N} \cup \{0\}$,

(a) $\bar{T}_s(X_t, x) = \frac{\bar{T}_s(X, x+t)}{\bar{T}_s(X,t)}$;

(b) $\mu_s(X_t) = \frac{\bar{T}_{s+1}(X,t)\mu_s(X)}{\bar{T}_s(X,t)}$.

Proof We prove this by using induction on s. Note that, for $s = 0$, the result is trivial. For $s = 1$,

$$\bar{T}_1(X_t, x) = P(X - t \leqslant x | X > t) = \frac{\bar{F}_X(x+t)}{\bar{F}_X(t)} = \frac{\bar{T}_1(X, x+t)}{\bar{T}_1(X,t)}$$

and

$$\mu_1(X_t) = \frac{\int_0^\infty \bar{T}_1(X, x+t)\, dx}{\bar{T}_1(X,t)} = \frac{\int_t^\infty \bar{T}_1(X, u)\, du}{\bar{T}_1(X,t)} = \frac{\bar{T}_2(X,t)\mu_1(X)}{\bar{T}_1(X,t)}.$$

Let (a) and (b) hold for some specific value of s. Then

$$\bar{T}_{s+1}(X_t, x) = \frac{\int_x^\infty \bar{T}_s(X_t, u)\, du}{\mu_s(X_t)}$$

$$= \frac{\int_x^\infty \bar{T}_s(X, u+t)\, du}{\bar{T}_{s+1}(X,t)\mu_s(X)}$$

$$= \frac{\int_{x+t}^\infty \bar{T}_s(X, u)\, du}{\bar{T}_{s+1}(X,t)\mu_s(X)}$$

$$= \frac{\bar{T}_{s+1}(X, x+t)}{\bar{T}_{s+1}(X,t)}$$

and

$$\mu_{s+1}(X_t) = \int_0^\infty \bar{T}_{s+1}(X_t, x)\, dx = \frac{\int_t^\infty \bar{T}_{s+1}(X, u)\, du}{\bar{T}_{s+1}(X,t)} = \frac{\bar{T}_{s+2}(X,t)\mu_{s+1}(X)}{\bar{T}_{s+1}(X,t)}.$$

Hence, the result is established. □

Now, we need to prove the following lemma which characterizes s-FR order.

Lemma 7.8.10 *Let* $s \in \mathbb{N}$. *Then* $X \leqslant_{s-FR} Y$ *if and only if, for all* $x \leqslant z$,

$$\frac{\int_z^\infty \bar{T}_{s-1}(X, u)\, du}{\bar{T}_{s-1}(X, x)} \leqslant \frac{\int_z^\infty \bar{T}_{s-1}(Y, u)\, du}{\bar{T}_{s-1}(Y, x)}.$$

Proof Given that $X \leqslant_{s-FR} Y$. This means that

$$\frac{\bar{T}_s(X, x)}{\bar{T}_s(Y, x)} \text{ is decreasing in } x, \qquad (7.8.44)$$

7.8 Generalized Orderings

which give, for all $y \leqslant z$,

$$\frac{\bar{T}_s(X,z)}{\bar{T}_s(Y,z)} \leqslant \frac{\bar{T}_s(X,y)}{\bar{T}_s(Y,y)}. \tag{7.8.45}$$

Again, for $x < y$, (7.8.44) can be rewritten as

$$\bar{T}_s(X,y)\bar{T}_s(Y,x) \leqslant \bar{T}_s(X,x)\bar{T}_s(Y,y),$$

which can be rewritten as

$$\bar{T}_s(X,y)\left[\bar{T}_s(Y,y) + P(x < Y_{(s)} \leqslant y)\right] \leqslant \bar{T}_s(Y,y)\left[\bar{T}_s(X,y) + P(x < X_{(s)} \leqslant y)\right], \tag{7.8.46}$$

where $X_{(s)}$ and $Y_{(s)}$ are the random variables having respective survival functions $\bar{T}_s(X,\cdot)$ and $\bar{T}_s(Y,\cdot)$. Expression in (7.8.46) can be rewritten as

$$\frac{\bar{T}_s(X,y)}{\bar{T}_s(Y,y)} \leqslant \frac{P(x < X_{(s)} \leqslant y)}{P(x < Y_{(s)} \leqslant y)}. \tag{7.8.47}$$

For any random variable Z, we write $T_s(Z,\cdot) = 1 - \bar{T}_s(Z,\cdot)$. Then, from (7.8.45) and (7.8.47) we get, for all $x < y \leqslant z$,

$$\begin{aligned}\frac{\bar{T}_s(X,z)}{\bar{T}_s(Y,z)} &\leqslant \frac{P(x < X_{(s)} \leqslant y)}{P(x < Y_{(s)} \leqslant y)} \\ &= \frac{T_s(X,y) - T_s(X,x)}{T_s(Y,y) - T_s(Y,x)} \\ &= \frac{(T_s(X,y) - T_s(X,x))/(y-x)}{(T_s(Y,y) - T_s(Y,x))/(y-x)}.\end{aligned}$$

Note that the LHS of the above expression is independent of x and y whereas RHS is a function of x and y with the restriction that $x < y \leqslant z$. Since the above inequality holds for any x and y satisfying the restriction $x < y \leqslant z$, we immediately get

$$\begin{aligned}\frac{\bar{T}_s(X,z)}{\bar{T}_s(Y,z)} &\leqslant \lim_{y \to x} \frac{(T_s(X,y) - T_s(X,x))/(y-x)}{(T_s(Y,y) - T_s(Y,x))/(y-x)} \\ &= \frac{\lim_{y \to x}\left[(T_s(X,y) - T_s(X,x))/(y-x)\right]}{\lim_{y \to x}\left[(T_s(Y,y) - T_s(Y,x))/(y-x)\right]} \\ &= \frac{\frac{d}{dx}T_s(X,x)}{\frac{d}{dx}T_s(Y,x)} \\ &= \left(\frac{\bar{T}_{s-1}(X,x)}{\mu_{s-1}(X)}\right) \cdot \left(\frac{\mu_{s-1}(Y)}{\bar{T}_{s-1}(Y,x)}\right).\end{aligned}$$

The last equality follows from (7.8.14). Again, writing the LHS of the above expression using (7.8.14), we get after cancellation

$$\frac{\int_z^\infty \bar{T}_{s-1}(X,u)\,du}{\int_z^\infty \bar{T}_{s-1}(Y,u)\,du} \leqslant \frac{\bar{T}_{s-1}(X,x)}{\bar{T}_{s-1}(Y,x)}.$$

This gives the required result after rearranging the terms. □

On using the above lemma the following theorem can be established (cf. Hu et al., 2001).

Theorem 7.8.13 *For $s \in \mathbb{N}$,*

$$X \leqslant_{s-FR} Y \iff X_t \leqslant_{(s-1)-CX} Y_t$$

for all $t \geqslant 0$.

Proof Let us first prove the result for $s \geqslant 2$. Given that $X_t \leqslant_{(s-1)-CX} Y_t$, for $t \geqslant 0$. This means that, for all $x \geqslant 0$,

$$\int_x^\infty \bar{T}_{s-1}(X_t, u)\,du \leqslant \int_x^\infty \bar{T}_{s-1}(T_t, u)\,du,$$

which, on using Lemma 7.8.9, reduces to

$$\frac{\int_x^\infty \bar{T}_{s-1}(X, u+t)\,du}{\bar{T}_{s-1}(X,t)} \leqslant \frac{\int_x^\infty \bar{T}_{s-1}(Y, u+t)\,du}{\bar{T}_{s-1}(Y,t)}.$$

This can equivalently be written as

$$\frac{\int_{x+t}^\infty \bar{T}_{s-1}(X,u)\,du}{\bar{T}_{s-1}(X,t)} \leqslant \frac{\int_{x+t}^\infty \bar{T}_{s-1}(Y,u)\,du}{\bar{T}_{s-1}(Y,t)}.$$

Since this is true for all $x \geqslant 0$, the result follows for $s \geqslant 2$, on using Lemma 7.8.10. Now, note that the density functions of X_t and Y_t are given, respectively, by

$$f_{X_t}(x) = \frac{f_X(x+t)}{\bar{F}_X(t)} \quad \text{and} \quad f_{Y_t}(x) = \frac{f_Y(x+t)}{\bar{F}_Y(t)}.$$

Now, $X_t \leqslant_{0-CX} Y_t$ if and only if, for all $x \geqslant 0$,

$$\frac{\int_x^\infty f_{X_t}(u)\,du}{f_{X_t}(0)} \leqslant \frac{\int_x^\infty f_{Y_t}(u)\,du}{f_{Y_t}(0)}.$$

7.8 Generalized Orderings

which can be rewritten, on using Lemma 7.8.9, as

$$\frac{\int_x^\infty f_X(u+t)\,du}{f_X(t)} \leq \frac{\int_x^\infty f_Y(u+t)\,du}{f_Y(t)}.$$

Or, equivalently,

$$\frac{\int_{x+t}^\infty f_X(u)\,du}{f_X(t)} \leq \frac{\int_{x+t}^\infty f_Y(u)\,du}{f_Y(t)}.$$

On using Lemma 7.8.10 the above expression gives $X \leq_{s-FR} Y$. This proves the required result. □

Taking $s = 2$ in the above theorem we get the following corollary which may be obtained in Brown and Shanthikumar (1998).

Corollary 7.8.11 *For two random variables X and Y, $X \leq_{mrl} Y$ if and only if $X_t \leq_{icx} Y_t$ for all t.* □

The following theorem was proved in Mukherjee and Chatterjee (1992) for $s \in \mathbb{N}$.

Theorem 7.8.14 *For $s \in \mathbb{N} \cup \{0\}$,*

$$X \leq_{s-FR} Y \iff X_t \leq_{s-ST} Y_t$$

for all $t \geq 0$.

Proof Note that $X_t \leq_{s-ST} Y_t$ if and only if, for all $x \geq 0$,

$$\bar{T}_s(X_t, x) \leq \bar{T}_s(Y_t, x)$$

which, by Lemma 7.8.9, reduces to

$$\frac{\bar{T}_s(X, x+t)}{\bar{T}_s(X, t)} \leq \frac{\bar{T}_s(Y, x+t)}{\bar{T}_s(Y, t)}$$

for all $x \geq 0$. We can rewrite this as

$$\frac{\bar{T}_s(X, x+t)}{\bar{T}_s(Y, x+t)} \leq \frac{\bar{T}_s(X, t)}{\bar{T}_s(Y, t)}.$$

Since this holds for all $x, t \geq 0$, we can conclude from here that

$$\frac{\bar{T}_s(X, x)}{\bar{T}_s(Y, x)} \text{ is decreasing in } x \geq 0.$$

This means that $X \leq_{s-FR} Y$. This proves the result. □

Taking $s = 0, 1, 2$ in the above theorem we get the following corollary.

Corollary 7.8.12 (a) $X \leqslant_{lr} Y$ if and only if $X_t \leqslant_{wlr} Y_t$ for all t.
(b) $X \leqslant_{hr} Y$ if and only if $X_t \leqslant_{st} Y_t$ for all t.
(c) $X \leqslant_{mlr} Y$ if and only if $X_t \leqslant_{hmrl} Y_t$ for all t. □

Note, from Lemma 7.8.9, that, for $s \in \mathbb{N} \cup \{0\}$,

$$\frac{\bar{T}_s(X_t, x)}{\bar{T}_s(Y_t, x)} = C \cdot \frac{\bar{T}_s(X, t+x)}{\bar{T}_s(Y, t+x)},$$

where $C = \bar{T}_s(Y, t)/\bar{T}_s(X, t)$ is independent of x. Since the above expression holds for all $x, t \geqslant 0$, we have the following result.

Theorem 7.8.15 For $s \in \mathbb{N} \cup \{0\}$,

$$X \leqslant_{s-FR} Y \iff X_t \leqslant_{s-FR} Y_t$$

for all t. □

Corollary 7.8.13 Taking $s = 0, 1, 2, 3$ in the above theorem we get the following.

(a) $X \leqslant_{lr} Y$ if and only if $X_t \leqslant_{lr} Y_t$ for all t.
(b) $X \leqslant_{hr} Y$ if and only if $X_t \leqslant_{hr} Y_t$ for all t.
(c) $X \leqslant_{mrl} Y$ if and only if $X_t \leqslant_{mrl} Y_t$ for all t.
(d) $X \leqslant_{vrl} Y$ if and only if $X_t \leqslant_{vrl} Y_t$ for all t.

Remark 7.8.8 From Theorems 7.8.14 and 7.8.15 we see that the s-FR order and s-ST order between the residual lives are identical. We must mention here that Corollary 7.8.13 (b) is given in Nanda and Jain (1999) while studying some result on weighted distribution.

7.8.3 Characterizations in Terms of Equilibrium Distributions

We have discussed the equilibrium random variable of the original random variable X as one which has the survival function given by $\int_x^\infty \bar{F}_X(t)\,dt/E(X)$. Let us denote this random variable by A_X. Thus, we have

$$\bar{F}_{A_X}(x) = \frac{\int_x^\infty \bar{F}_X(t)\,dt}{E(X)}, \qquad (7.8.48)$$

which gives the corresponding density function as $f_{A_X}(x) = \bar{F}_X(x)/E(X)$.

7.8 Generalized Orderings

Below we give a lemma to be used in sequel.

Lemma 7.8.11 *For $s \in \mathbb{N}$,*

(a) $\bar{T}_s(A_X, x) = \bar{T}_{s+1}(X, x)$;
(b) $\mu_s(A_X) = \mu_{s+1}(X)$;
(c) $r_s(A_X, x) = r_{s+1}(X, x)$.

Proof From the above definition, we get that, for $s = 1$,

$$\bar{T}_1(A_X, x) = \frac{\int_x^\infty \bar{T}_1(X, u)\, du}{\mu_1(X)} = \bar{T}_2(X, x)$$

and

$$\mu_1(A_X) = \int_0^\infty \bar{T}_1(A_X, x)\, dx = \int_0^\infty \bar{T}_2(X, x)\, dx = \mu_2(X).$$

Let (a) and (b) above hold for some particular value of s. Then we have

$$\bar{T}_{s+1}(A_X, x) = \frac{\int_x^\infty \bar{T}_s(A_X, u)\, du}{\mu_s(A_X)}$$
$$= \frac{\int_x^\infty \bar{T}_{s+1}(X, u)\, du}{\mu_{s+1}(X)}$$
$$= \bar{T}_{s+2}(X, x)$$

and

$$\mu_{s+1}(A_X) = \int_0^\infty \bar{T}_{s+1}(A_X, x)\, dx = \int_0^\infty \bar{T}_{s+2}(X, x)\, dx = \mu_{s+2}(X).$$

Thus, (a) and (b) are proved by mathematical induction. To prove (c), note that

$$r_1(A_X, x) = \frac{\bar{T}_0(A_X, x)}{\bar{T}_1(A_X, x)} = \frac{\bar{T}_1(X, x)/E(X)}{\int_x^\infty \bar{T}_1(X, u)\, du / E(X)} = r_2(X, x).$$

The last equality follows from (7.8.15). Let the result hold for some specific value of s. Then we have, from (7.8.15),

$$r_{s+1}(A_X, x) = \frac{\bar{T}_s(A_X, x)}{\int_x^\infty \bar{T}_s(A_X, u)\, du}$$
$$= \frac{\bar{T}_{s+1}(X, x)}{\int_x^\infty \bar{T}_{s+1}(X, u)\, du}$$
$$= r_{s+2}(X, x).$$

Hence (c) follows by the method of mathematical induction. □

The following characterization result, which easily follows from Lemma 7.8.11, may be obtained in Hu et al. (2001).

Theorem 7.8.16 *For $s \in \mathbb{N}$,*

(a) $X \leq_{s-FR} Y \iff A_X \leq_{(s-1)-FR} A_Y$;
(b) $X \leq_{s-ST} Y \iff A_X \leq_{(s-1)-ST} A_Y$;
(c) $X \leq_{s-CX} Y \iff A_X \leq_{(s-1)-CX} A_Y$;
(d) $X \leq_{s-CV} Y \iff A_X \leq_{(s-1)-CV} A_Y$. □

Taking different values of s in the above theorem we get the following.

Corollary 7.8.14 (i) $X \leq_{hr} Y$ if and only if $A_X \leq_{lr} A_Y$.
(ii) $X \leq_{mrl} Y$ if and only if $A_X \leq_{hr} A_Y$.
(iii) $X \leq_{vrl} Y$ if and only if $A_X \leq_{mrl} A_Y$.
(iv) $X \leq_{st} Y$ if and only if $A_X \leq_{wlr} A_Y$.
(v) $X \leq_{hmrl} Y$ if and only if $A_X \leq_{st} A_Y$. □

Note that

$$E[((X-x)^+)^n] = \int_{-\infty}^{\infty} t^n \, dF_{(X-x)^+}(t)$$
$$= \int_0^{\infty} t^n f_X(x+t) \, dt$$
$$= \int_x^{\infty} (u-x)^n f_X(u) \, du. \qquad (7.8.49)$$

For two random variables X and Y, if we assume that $E(X) = E(Y)$ then we get the following theorem (cf. Kaas et al., 1994).

Theorem 7.8.17 *Let $E(X) = E(Y)$. Then*

$$X \leq_{s-SL} Y \iff A_X \leq_{(s-1)-SL} A_Y.$$

Proof Note that, on using (7.8.49), we get

$$E\left\{[(A_X - x)^+]^s\right\} = \int_x^{\infty} f_{A_X}(u)(u-x)^s \, du$$
$$= \frac{1}{E(X)} \int_x^{\infty} (u-x)^s \bar{F}_X(u) \, du$$
$$= \frac{1}{E(X)} \int_x^{\infty} (u-x)^s \int_u^{\infty} f_X(t) \, dt \, du$$
$$= \frac{1}{E(X)} \int_x^{\infty} \left(\int_x^t (u-x)^s \, du\right) f_X(t) \, dt$$
$$= \frac{1}{(s+1)E(X)} \int_x^{\infty} (t-x)^{s+1} f_X(t) \, dt.$$

7.8 Generalized Orderings

The second equality follows from (7.8.48). So,

$$E\left\{\left[(A_X - x)^+\right]^s\right\} \leq E\left\{\left[(A_Y - x)^+\right]^s\right\}$$

if and only if

$$\frac{1}{E(X)} \int_x^\infty (t - x)^{s+1} f_X(t)\, dt \leq \frac{1}{E(Y)} \int_x^\infty (t - x)^{s+1} f_Y(t)\, dt.$$

Now, if $E(X) = E(Y)$ then the above relation reduces to

$$\int_x^\infty (t - x)^{s+1} f_X(t)\, dt \leq \int_x^\infty (t - x)^{s+1} f_X(t)\, dt,$$

which is equivalent to

$$E\left\{\left[(X - x)^+\right]^{s+1}\right\} \leq E\left\{\left[(Y - x)^+\right]^{s+1}\right\}.$$

Further, the s-th moment of A_X is obtained as

$$\begin{aligned}
E\left(A_X^s\right) &= \int_0^\infty t^s f_{A_X}(t)\, dt \\
&= \frac{1}{E(X)} \int_0^\infty t^s \bar{F}_X(t)\, dt \\
&= \frac{1}{E(X)} \int_0^\infty t^s \left(\int_t^\infty f_X(u)\, du\right) dt \\
&= \frac{1}{E(X)} \int_0^\infty f_X(u) \int_0^u t^s\, dt\, du \\
&= \frac{1}{(s+1)E(X)} \int_0^\infty u^{s+1} f_X(u)\, du \\
&= \frac{E\left(X^{s+1}\right)}{(s+1)E(X)}.
\end{aligned}$$

Now, $A_X \leq_{(s-1)-SL} A_Y$ if and only if

(a) $E(A_X^k) \leq E(A_Y^k)$, for all $k = 1, 2, \ldots, (s-2)$,
(b) $E\left((A_X - x)^+\right)^{s-1} \leq E\left((A_Y - x)^+\right)^{s-1}$

which can equivalently be written as

(a') $\frac{E(X^{k+1})}{E(X)} \leq \frac{E(Y^{k+1})}{E(Y)}$, for all $k = 1, 2, \ldots, (s-2)$,
(b') $\frac{E((X-x)^+)^s}{E(X)} \leq \frac{E((Y-x)^+)^s}{E(X)}$.

If $E(X) = E(Y)$, then we have that $A_X \leqslant_{(s-1)-SL} A_Y$ if and only if

(i) $E(X^k) \leqslant E(Y^k)$, for all $k = 2, 3, \ldots, (s-1)$,
(ii) $E\left[(X-x)^+\right]^s \leqslant E\left[(Y-x)^+\right]^s$

which is same as $X \leqslant_{s-SL} Y$. □

Below we characterize the ordering between a given distribution and the corresponding equilibrium distribution.

Theorem 7.8.18 *Let $s \in \mathbb{N} \cup \{0\}$. Then $X \geqslant_{s-FR}$ (resp. $\leqslant_{s-FR}$) A_X if and only if X is $(s+1)$-IFR (resp. $(s+1)$-DFR).*

Proof For $s = 0$, the proof immediately follows by noting the fact that $f_{A_X}(x) = \bar{F}_X(x)/E(X)$. So, we take $s \in \mathbb{N}$. Then $X \geqslant_{s-FR} A_X$ if and only if

$$\frac{\bar{T}_s(X, x)}{\bar{T}_s(A_X, x)} \text{ is increasing in } x,$$

which is equivalent to saying that

$$\frac{\bar{T}_s(X, x)}{\bar{T}_{s+1}(X, x)} \text{ is increasing in } x.$$

This can further be written equivalently as

$$\frac{\bar{T}_s(X, x)}{\int_x^\infty \bar{T}_s(X, u)\, du} \text{ is increasing in } x,$$

which means that $r_{s+1}(X, x)$ is increasing in x. This means that X is $(s+1)$-IFR. The proof of the parenthetical statement is similar. □

Taking $s = 0, 1, 2$ in the above theorem, we immediately get the following corollary.

Corollary 7.8.15 (a) $X \geqslant_{lr}$ (resp. $\leqslant_{lr}$) A_X if and only if X is IFR (resp. (DFR).
(b) $X \geqslant_{hr}$ (resp. $\leqslant_{hr}$) A_X if and only if X is DMRL (resp. (IMRL).
(c) $X \geqslant_{mrl}$ (resp. $\leqslant_{mrl}$) A_X if and only if X is DVRL (resp. (IVRL).

Similar to the above theorem, below we give characterization of s-NBUFR class and its dual.

Theorem 7.8.19 *For $s \in \mathbb{N} \cup \{0\}$, X is $(s+1)$-NBUFR (resp. $(s+1)$-NWUFR) if and only if $X \geqslant_{s-ST}$ (resp. $\leqslant_{s-ST}$) A_X.*

Proof For $s = 0$, the proof immediately follows by noting the fact that $f_{A_X}(x) = \bar{F}_X(x)/E(X)$. So, we take $s \in \mathbb{N}$. Then $X \geqslant_{s-ST} A_X$ if and only if

7.8 Generalized Orderings

$$\bar{T}_s(X, x) \geqslant \bar{T}_{s+1}(X, x) = \frac{\int_x^\infty \bar{T}_x(X; u)\, du}{\mu_s(X)}$$

which immediately reduces to $r_{s+1}(X, x) \geqslant r_{s+1}(X, 0)$ giving that X is $(s + 1)$-$NBUFR$. The proof of the parenthetical statement is similar. □

Taking $s = 0, 1$ in the above theorem we get the following corollary.

Corollary 7.8.16 (a) $X \geqslant_{wlr}$ (resp. $\leqslant_{wlr}$) A_X if and only if X is $NBUFR$ (resp. $NWUFR$).
(b) $X \geqslant_{st}$ (resp. $\leqslant_{st}$) A_X if and only if X is $NBUE$ (resp. $NWUE$).

7.8.4 Characterizations in Terms of Laplace Transform

Let X be a non-negative random variable having finite moments of all orders. Define, for $\lambda > 0$,

$$\phi_X(\lambda) = \int_0^\infty e^{-\lambda x}\, dF_X(x),$$

the Laplace transform corresponding to X, and denote

$$\bar{\alpha}_\lambda^X(n) = \lambda^n \left[\frac{(-1)^{n-1}}{(n-1)!} \frac{d^{n-1}}{d\lambda^{n-1}} \left(\frac{1 - \phi_X(\lambda)}{\lambda} \right) \right], \quad n \in \mathbb{N} \qquad (7.8.50)$$

with $\bar{\alpha}_\lambda^X(0) = 1$. We shall show that $\bar{\alpha}_\lambda^X(n)$ is the survival function of some random variable. However, to do show, we need the following theorem which gives another representation of $\bar{\alpha}_\lambda^X(n)$.

Theorem 7.8.20 *For $n \in \mathbb{N}$,*

$$\bar{\alpha}_\lambda^X(n) = \lambda \int_0^\infty \frac{(\lambda x)^{n-1}}{(n-1)!} e^{-\lambda x} \bar{F}_X(x)\, dx.$$

Proof We shall prove this by induction on n. Take $n = 1$. Let Z be an Exponentially distributed random variable having mean $1/\lambda$. If Z is independent of X, then we have from (7.8.50)

$$\bar{\alpha}_\lambda^X(1) = 1 - \phi_X(\lambda) = \int_0^\infty \left(1 - e^{-\lambda x}\right) dF_X(x)$$
$$= P(X \geqslant Z)$$
$$= \int_0^\infty \bar{F}_X(x)\, dF_Z(x)$$
$$= \lambda \int_0^\infty e^{-\lambda x} \bar{F}_X(x)\, dx.$$

Let the result be true for some fixed $n = k$. Then, taking $n = k+1$, we have

$$\bar{\alpha}_\lambda^X(k+1) = \lambda^{k+1} \frac{(-1)^k}{k!} \frac{d^k}{d\lambda^k}\left[\frac{1-\phi_X(\lambda)}{\lambda}\right]$$

$$= -\frac{\lambda^{k+1}}{k} \cdot \frac{d}{d\lambda}\left[\frac{(-1)^{k-1}}{(k-1)!} \frac{d^{k-1}}{d\lambda^{k-1}}\left[\frac{1-\phi_X(\lambda)}{\lambda}\right]\right]$$

$$= -\frac{\lambda^{k+1}}{k} \frac{d}{d\lambda}\left[\frac{\lambda \int_0^\infty (\lambda x)^{k-1} e^{-\lambda x} \bar{F}_X(x)\, dx}{(k-1)!\lambda^k}\right]$$

$$= -\frac{\lambda^{k+1}}{k} \frac{d}{d\lambda}\left[\frac{\int_0^\infty x^{k-1} e^{-\lambda x} \bar{F}_X(x)\, dx}{(k-1)!}\right]$$

$$= \lambda \int_0^\infty \frac{(\lambda x)^k}{k!} e^{-\lambda x} \bar{F}_X(x)\, dx.$$

This establishes the result. □

Suppose $Z_1, Z_2, \ldots Z_n$ are the iid Exponentially distributed random variables having mean $1/\lambda$. Then $\zeta = \sum_{i=1}^n Z_i$ will have the density

$$f_\zeta(x) = \lambda e^{-\lambda x} \frac{(\lambda x)^{n-1}}{(n-1)!}, \quad x > 0, \lambda > 0.$$

If X is independent of Z_i's then, from Theorem 7.8.20, we have

$$P(X \geq \zeta) = \int_0^\infty P(X \geq x) f_\zeta(x)\, dx = \bar{\alpha}_\lambda^X(n). \tag{7.8.51}$$

Similarly, for a non-negative random variable Y having cdf G, we define $\bar{\alpha}_\lambda^Y(n)$. This shows that, for every $n \in \mathbb{N}$, $\bar{\alpha}_\lambda^X$ and $\bar{\alpha}_\lambda^Y$ are the discrete survival functions. Denote the corresponding discrete random variables by N_λ^X and N_λ^Y, respectively.

Define, for any $g : \mathbb{R}^+ \to \mathbb{R}$,

$$\Gamma_\lambda g(n) = \lambda \int_0^\infty \frac{(\lambda x)^{n-1}}{(n-1)!} e^{-\lambda x} g(x)\, dx, \tag{7.8.52}$$

provided the integral is finite.

The following result is due to Billingsley (1968).

Lemma 7.8.12 *Suppose $A \subseteq \mathbb{R}$. Let $f : A \to \mathbb{R}$ be any bounded, measurable function and let S be the set of discontinuity points of f. Suppose $\{G_n\}$ is a sequence of distribution functions which converges weakly to another distribution function G. Then*

7.8 Generalized Orderings

$$\lim_{n\to\infty} \int_A f(x)\,dG_n(x) = \int_A f(x)\,dG(x),$$

provided $P(S) = 0$.

Lemma 7.8.13 *Let $x > 0$ be a continuity point of F. Let $\lambda = \lambda(n, x)$ satisfy $\lim_{n\to\infty} (n/\lambda) = x$. Write*

$$dG_n(t) = \frac{\lambda^{n+1}}{n!} e^{-\lambda t} t^n\,dt.$$

Then G_n converges weakly to G, where G is the distribution degenerate at x.

Proof Let X_i, $i = 1, 2, \ldots, n$, be *iid* Exponentially distributed random variables with mean $1/\lambda$. Define $X = X_1 + X_2 + \cdots + X_n$. Then, by the strong law of large numbers, as $n \to \infty$,

$$\frac{1}{n}(X_1 + X_2 + \cdots + X_n) \xrightarrow{a.s.} \frac{1}{\lambda}.$$

This gives, as $n \to \infty$,

$$\frac{\lambda x}{n}(X_1 + X_2 + \cdots + X_n) \xrightarrow{a.s.} x.$$

Thus, by the hypothesis, we have, as $n \to \infty$,

$$X_1 + X_2 + \cdots + X_n \xrightarrow{a.s.} x.$$

The above expression gives

$$X \xrightarrow{L} x, \quad \text{as } n \to \infty$$

since x is a continuity point of F. $\square$

The following lemma which is a generalization of Lemma 2.3 of Block and Savits (1980) may be obtained in Nanda (1995).

Lemma 7.8.14 *Let $x > 0$ be a continuity point of F. Let $\lambda = \lambda(n, x)$ satisfy $\lim_{n\to\infty} (n/\lambda) = x$. Then, for any non-negative integer s,*

$$\lim_{n\to\infty} \Gamma_\lambda \bar{T}_s(X, n) = \bar{T}_s(X, x).$$

Proof By definition,

$$\Gamma_\lambda \bar{T}_s(X, n) = \int_0^\infty \bar{T}_s(X, t)\,dG_{n-1}(t),$$

where G_n is as given in Lemma 7.8.13. Thus, G_n converges weakly to a distribution degenerate at x. So, by Lemma 7.8.12, we get the required result. □

Now, we want to discuss some results related to shock models. However, before that we need to discuss the following.

For any random variable Z with support $\mathbb{N} \cup \{0\}$, we define

$$\bar{T}_s(Z, n) = \sum_{k=n+1}^{\infty} \frac{\bar{T}_{s-1}(Z, k)}{\mu_{s-1}(Z)} \tag{7.8.53}$$

for $s \in \mathbb{N}$ with $\bar{T}_0(Z, \cdot)$ as the mass function of Z and

$$\mu_s(Z) = \sum_{k=1}^{\infty} \bar{T}_s(Z, k) \tag{7.8.54}$$

for $s \in \mathbb{N}$ with $\mu_0(Z) = 1$.

Below we give the definitions of generalized orderings for discrete random variables.

Definition 7.8.8 Let X and Y be two non-negative discrete random variables. Then X is said to be smaller than Y in

(a) s-FR order, written as $X \leqslant_{s-FR} Y$, if

$$\frac{\bar{T}_s(X, k)}{\bar{T}_s(Y, k)} \text{ is decreasing in } k \in \mathbb{N},$$

for any $s \in \mathbb{N} \cup \{0\}$;

(b) s-ST order, written as $X \leqslant_{s-ST} Y$, if

$$\bar{T}_s(X, k) \leqslant \bar{T}_s(Y, k),$$

for all $k \in \mathbb{N} \cup \{0\}$, and for all $s \in \mathbb{N}$;

(c) s-CX order, written as $X \leqslant_{s-CX} Y$, if

$$\sum_{k=n}^{\infty} \bar{T}_s(X, k) \leqslant \sum_{k=n}^{\infty} \bar{T}_s(Y, k)$$

for all $n \in \mathbb{N} \cup \{0\}$ and $s \in \mathbb{N}$;

(d) s-CV order, written as $X \leqslant_{s-CV} Y$, if

$$\sum_{k=0}^{n} \bar{T}_s(X, k) \leqslant \sum_{k=0}^{n} \bar{T}_s(Y, k)$$

for all $n \in \mathbb{N} \cup \{0\}$ and $s \in \mathbb{N}$. □

7.8 Generalized Orderings

The following lemma will be used in sequel. The proof is due to Nanda (1998).

Lemma 7.8.15 *For any $s \in \mathbb{N}$,*

(i) $\sum_{k=n+1}^{\infty} \Gamma_\lambda \bar{T}_s(Z, k) = \lambda \mu_s(Z) \Gamma_\lambda \bar{T}_{s+1}(Z, n);$
(ii) $\sum_{k=0}^{\infty} \Gamma_\lambda \bar{T}_s(Z, k) = 1 + \mu_s(Z).$

Proof Note, after interchanging the sum and the integral, that

$$\sum_{k=n+1}^{\infty} \Gamma_\lambda \bar{T}_s(Z, k) = \lambda \int_0^\infty e^{-\lambda x} \sum_{k=n+1}^{\infty} \frac{(\lambda x)^{k-1}}{(k-1)!} \bar{T}_s(Z, x) \, dx$$

$$= \lambda \int_0^\infty e^{-\lambda x} \sum_{k=n}^{\infty} \frac{(\lambda x)^k}{k!} \bar{T}_s(Z, x) \, dx$$

$$= \lambda \int_0^\infty \frac{1}{\Gamma(n)} \left[\int_0^{\lambda x} e^{-t} t^{n-1} \, dt \right] \bar{T}_s(Z, x) \, dx$$

$$= \lambda \int_0^\infty \frac{1}{\Gamma(n)} e^{-t} t^{n-1} \left[\int_{t/\lambda}^\infty \bar{T}_s(Z, x) \, dx \right] dt$$

$$= \lambda \mu_s(Z) \int_0^\infty \frac{1}{\Gamma(n)} e^{-t} t^{n-1} \bar{T}_s\left(Z, \frac{t}{\lambda}\right) dt$$

$$= \lambda^2 \mu_s(Z) \int_0^\infty \frac{1}{\Gamma(n)} e^{-t} (\lambda t)^{n-1} \bar{T}_s(Z, t) \, dt$$

$$= \lambda \mu_s(Z) \Gamma_\lambda \bar{T}_{s+1}(Z, n).$$

The first equality is obtained by interchanging the sum and the integral whereas the fourth equality is obtained by interchanging the order of integration. This proves (i). To prove (ii), note that

$$\sum_{k=0}^{\infty} \Gamma_\lambda \bar{T}_s(Z, k) = \Gamma_\lambda \bar{T}_s(Z, 0) + \sum_{k=1}^{\infty} \Gamma_\lambda \bar{T}_s(Z, k)$$

$$= 1 + \lambda \mu_s(Z) \Gamma_\lambda \bar{T}_{s+1}(Z, 0)$$

$$= 1 + \lambda \mu_s(Z).$$

Hence the result follows. □

Suppose that a device is subject to a sequence of shocks which occur randomly as the events in a Poisson process having (constant) intensity $\lambda > 0$. Suppose each shock causes some random amount of damage to the system, which is accumulated additively. The device fails as soon as the accumulated damage exceeds a threshold (which is again random). Consider two such systems having lifetimes X and Y with two different failure thresholds denoted, respectively, by N_λ^X and N_λ^Y. The following theorems say that if the thresholds are ordered in some order, the discrete lifetime of the systems (experienced by a sequence of Exponentially distributed shocks) are also ordered in the same sense and conversely. But before that we need the following lemma that may be required to prove the main results of this section (cf. Nanda, 1998).

Lemma 7.8.16 *For $s \in \mathbb{N}$,*

$$\bar{T}_s\left(N_\lambda^Z, n\right) = \Gamma_\lambda \bar{T}_s(Z, n)$$

for all $n \in \mathbb{N} \cup \{0\}$.

Proof We prove the result by induction on s. For $s = 1$, we have

$$\bar{T}_1\left(N_\lambda^Z, n\right) = \lambda \int_0^\infty \frac{(\lambda x)^{n-1}}{(n-1)!} e^{-\lambda x} \bar{T}_1(Z, x)\, dx$$
$$= \Gamma_\lambda \bar{T}_1(Z, n).$$

Let the result be true for some $s = k$. Then we have from (7.8.53) and (7.8.54)

$$\bar{T}_{k+1}\left(N_\lambda^Z, n\right) = \frac{1}{\mu_k\left(N_\lambda^Z\right)} \sum_{i=n+1}^\infty \bar{T}_k\left(N_\lambda^Z, i\right)$$
$$= \frac{\sum_{i=n+1}^\infty \bar{T}_k\left(N_\lambda^Z, i\right)}{\sum_{i=1}^\infty \bar{T}_k\left(N_\lambda^Z, i\right)}$$
$$= \frac{\sum_{i=n+1}^\infty \Gamma_\lambda \bar{T}_k(Z, i)}{\sum_{i=1}^\infty \Gamma_\lambda \bar{T}_k(Z, i)}$$
$$= \Gamma_\lambda \bar{T}_{k+1}(Z, n).$$

The last equality follows from Lemma 7.8.15(i) whereas the second last equality follows from the hypothesis along with the fact that $\Gamma_\lambda \bar{T}_{s+1}(Z, 0) = 1$. □

The following result may be obtained in Nanda (1995).

Theorem 7.8.21 *For any $s \in \mathbb{N} \cup \{0\}$,*

$$X \leqslant_{s-FR} Y \text{ if and only if } N_\lambda^X \leqslant_{s-FR} N_\lambda^Y, \text{ for all } \lambda > 0.$$

Proof Given that $X \leqslant_{s-FR} Y$. So,

$$\frac{\bar{T}_s(X, x)}{\bar{T}_s(Y, x)} \text{ is decreasing in } x,$$

7.8 Generalized Orderings

which, by Theorem 7.8.4, gives that, for any $c \in \mathbb{R}$, $\bar{T}_s(X, x) - c.\bar{T}_s(Y, x)$ has at most one change of sign and if such a change does occur it occurs from positive to negative values as x traverses from left to right. Further, it is easy to verify that

$$K(n, x) = e^{-\lambda x} \frac{(\lambda x)^{n-1}}{(n-1)!} \text{ is } TP_2 \text{ in } (n, x). \tag{7.8.55}$$

So, by Theorem 7.8.3, we see that

$$\Gamma_\lambda \bar{T}_s(X, n) - c.\Gamma_\lambda \bar{T}_s(Y, n) = \int_0^\infty e^{-\lambda x} \frac{(\lambda x)^{n-1}}{(n-1)!} \bar{T}_s(X, x) \, dx - c. \int_0^\infty e^{-\lambda x} \frac{(\lambda x)^{n-1}}{(n-1)!} \bar{T}_s(Y, x) \, dx$$

$$= \int_0^\infty e^{-\lambda x} \frac{(\lambda x)^{n-1}}{(n-1)!} \left(\bar{T}_s(X, x) - c.\bar{T}_s(Y, x) \right) dx$$

has at most one change of sign and if such a change does occur it occurs from positive to negative values as x traverses from left to right. Note here that condition (b) of Theorem 7.8.3 is trivially satisfied because $\bar{T}_s(X, x) - c.\bar{T}_s(Y, x)$ is constant with respect to n. Thus, by Theorem 7.8.4, we can conclude that

$$\frac{\Gamma_\lambda \bar{T}_s(X, n)}{\Gamma_\lambda \bar{T}_s(Y, n)} \text{ is decreasing in } n,$$

which, by Lemma 7.8.16, gives that

$$N_\lambda^X \leq_{s-FR} N_\lambda^Y.$$

Conversely, write

$$C_s(\lambda, n) = \frac{\Gamma_\lambda \bar{T}_s(X, n)}{\Gamma_\lambda \bar{T}_s(Y, n)}.$$

Given that

$$C_s(\lambda, n) \text{ is decreasing in } n.$$

Differentiating the above expression with respect to λ, write

$$\frac{d}{d\lambda} [C_s(\lambda, n)] = \frac{N}{D}, \text{ say,}$$

where

$$N = \Gamma_\lambda \bar{T}_s(Y, n) . \frac{d}{d\lambda} \left(\Gamma_\lambda \bar{T}_s(X, n) \right) - \Gamma_\lambda \bar{T}_s(X, n) . \frac{d}{d\lambda} \left(\Gamma_\lambda \bar{T}_s(Y, n) \right).$$

Note that

$$\frac{d}{d\lambda}\left(\Gamma_\lambda \bar{T}_s(X,n)\right) = \int_0^\infty \frac{x^{n-1}}{(n-1)!} \frac{d}{d\lambda}\left(e^{-\lambda x}\lambda^n\right) \bar{T}_s(X,x)\,dx$$

$$= \int_0^\infty \frac{x^{n-1}}{(n-1)!} \left(n\lambda^{n-1}e^{-\lambda x} - x\lambda^n e^{-\lambda x}\right) \bar{T}_s(X,x)\,dx$$

$$= \frac{n}{\lambda}.\lambda \int_0^\infty e^{-\lambda x} \frac{(\lambda x)^{n-1}}{(n-1)!} \bar{T}_s(X,x)\,dx - \frac{n}{\lambda}.\lambda \int_0^\infty e^{-\lambda x} \frac{(\lambda x)^n}{n!} \bar{T}_s(y,x)\,dx$$

$$= \frac{n}{\lambda}\left[\Gamma_\lambda \bar{T}_x(X,n) - \Gamma_\lambda \bar{T}_x(X,n+1)\right].$$

So, after cancellation of terms we have

$$N = \frac{n}{\lambda}\left[\Gamma_\lambda \bar{T}_s(X,n).\Gamma_\lambda \bar{T}_x(Y,n+1) - \Gamma_\lambda \bar{T}_s(Y,n).\Gamma_\lambda \bar{T}_x(X,n+1)\right] \geqslant 0$$

since $C_s(\lambda,n)$ is given to be decreasing in n. Considering N as a function of λ, we see that $C_s(\lambda,n)$ is increasing in λ. So, writing $\lambda = n/x$, we have that

$$C_s\left(\frac{n}{x},n\right) \text{ is decreasing in } x.$$

Let $0 < x_1 < x_2$ be two continuity points of both $\bar{T}_s(X,\cdot)$ and $\bar{T}_s(Y,\cdot)$. Then, for all $n \in \mathbb{N}$, we have

$$C_s\left(\frac{n}{x_1},n\right) \geqslant C_s\left(\frac{n}{x_2},n\right).$$

By taking limit as $n \to \infty$ on both sides of the above expression, and using Lemma 7.8.14, we get

$$\frac{\bar{T}_s(X,x_1)}{\bar{T}_s(Y,x_1)} \geqslant \frac{\bar{T}_s(X,x_2)}{\bar{T}_s(Y,x_2)},$$

for all $x_1 < x_2$. Thus, we have that

$$\frac{\bar{T}_s(X,x)}{\bar{T}_s(Y,x)} \text{ is decreasing in } x \in \mathbb{R}.$$

This proves the required result.

Corollary 7.8.17 *Taking $s = 0, 1, 2, 3$ in the above theorem, we have the following. However, (a) and (b) below are given in Kebir (1994).*

(a) $X \leqslant_{lr} Y$ if and only if $N_\lambda^X \leqslant_{lr} N_\lambda^Y$;
(b) $X \leqslant_{hr} Y$ if and only if $N_\lambda^X \leqslant_{hr} N_\lambda^Y$;
(c) $X \leqslant_{mrl} Y$ if and only if $N_\lambda^X \leqslant_{mrl} N_\lambda^Y$;
(d) $X \leqslant_{vrl} Y$ if and only if $N_\lambda^X \leqslant_{vrl} N_\lambda^Y$. □

The following result is due to Nanda (1995).

7.8 Generalized Orderings

Theorem 7.8.22 *For any $s \in \mathbb{N}$,*

$$X \leqslant_{s-ST} Y \text{ if and only if } N_\lambda^X \leqslant_{s-ST} N_\lambda^Y, \text{ for all } \lambda > 0.$$

Proof Given that $\bar{T}_s(X, x) \leqslant \bar{T}_s(Y, x)$, for all $x \geqslant 0$. This gives that

$$\lambda e^{-\lambda x} \frac{(\lambda x)^{n-1}}{(n-1)!} \bar{T}_s(X, x) \leqslant \lambda e^{-\lambda x} \frac{(\lambda x)^{n-1}}{(n-1)!} \bar{T}_s(Y, x),$$

for all $x \geqslant 0$ and for all $n \in \mathbb{N}$. Integrating both sides of the above expression with respect to x we get

$$\int_0^\infty \lambda e^{-\lambda x} \frac{(\lambda x)^{n-1}}{(n-1)!} \bar{T}_s(X, x)\, dx \leqslant \int_0^\infty \lambda e^{-\lambda x} \frac{(\lambda x)^{n-1}}{(n-1)!} \bar{T}_s(Y, x)\, dx$$

for all $n \in \mathbb{N}$. This means that, for all $n \in \mathbb{N}$,

$$\Gamma_\lambda \bar{T}_s(X, n) \leqslant \Gamma_\lambda \bar{T}_s(Y, n).$$

This, by Lemma 7.8.16, gives

$$N_\lambda^X \leqslant_{s-ST} N_\lambda^Y.$$

Conversely, given that

$$\Gamma_\lambda \bar{T}_s(X, n) \leqslant \Gamma_\lambda \bar{T}_s(Y, n),$$

for all $n \in \mathbb{N}$, and all $\lambda > 0$. Suppose $x > 0$ is a continuity point both of $\bar{T}_s(X, \cdot)$ and $\bar{T}_s(Y, \cdot)$. Then, for each $n \in \mathbb{N}$,

$$\Gamma_{(n/x)} \bar{T}_s(X, n) \leqslant \Gamma_{(n/x)} \bar{T}_s(Y, n).$$

Now, by taking limit as $n \to \infty$ on both sides of the above expression, we get from Lemma 7.8.14,

$$\bar{T}_s(X, x) \leqslant \bar{T}_s(Y, x),$$

for all x, where $\bar{T}_s(X, \cdot)$ and $\bar{T}_s(Y, \cdot)$ are continuous. This gives $X \leqslant_{s-ST} Y$. □

Taking $s = 1, 2$ in the above theorem we get the following corollary of which (*a*) is given in Kebir (1994).

Corollary 7.8.18 (a) $X \leqslant_{st} Y$ *if and only if* $N_\lambda^X \leqslant_{st} N_\lambda^Y$, *for all* $\lambda > 0$;
(b) $X \leqslant_{hmrl} Y$ *if and only if* $N_\lambda^X \leqslant_{hmrl} N_\lambda^Y$, *for all* $\lambda > 0$. □

The following theorem may be obtained in Hu et al. (2001).

Theorem 7.8.23 *For any $s \in \mathbb{N}$, $X \leqslant_{s-CX} Y$ if and only if $N_\lambda^X \leqslant_{s-CX} N_\lambda^Y$, for all $\lambda > 0$.*

Proof Note that, for all $\lambda > 0$, $N_\lambda^X \leq_{s-CX} N_\lambda^Y$ if and only if, for all $n \in \mathbb{N} \cup \{0\}$,

$$\sum_{k=n}^{\infty} \bar{T}_s\left(N_\lambda^X, k\right) \leq \sum_{k=n}^{\infty} \bar{T}_s\left(N_\lambda^Y, k\right),$$

which, by Lemma 7.8.16, gives, for all $n \in \mathbb{N} \cup \{0\}$,

$$\sum_{k=n}^{\infty} \Gamma_\lambda \bar{T}_s(X, k) \leq \sum_{k=n}^{\infty} \Gamma_\lambda \bar{T}_s(Y, k),$$

which again, with the help of Lemma 7.8.15, can be rewritten as

$$\mu_s(X) \Gamma_\lambda \bar{T}_{s+1}(X, n-1) \leq \mu_s(Y) \Gamma_\lambda \bar{T}_{s+1}(X, n-1). \tag{7.8.56}$$

On using (7.8.52), this can equivalently be written as

$$\mu_s(X) \int_0^\infty e^{-\lambda x} (\lambda x)^{n-2} \bar{T}_{s+1}(X, x)\,dx \leq \mu_s(Y) \int_0^\infty e^{-\lambda x} (\lambda x)^{n-2} \bar{T}_{s+1}(Y, x)\,dx \tag{7.8.57}$$

for all $n \geq 2$.

Necessity: Given that $X \leq_{s-CX} Y$. Then, from (7.8.14), we get

$$\mu_s(X) \bar{T}_{s+1}(X, x) \leq \mu_s(Y) \bar{T}_{s+1}(Y, x)$$

for all $x > 0$. Multiplying both sides of the above inequality by $e^{-\lambda x}(\lambda x)^{n-2}$ and integrating with respect to x in $(0, \infty)$, we get

$$\mu_s(X) \int_0^\infty e^{-\lambda x} (\lambda x)^{n-2} \bar{T}_{s+1}(X, x)\,dx \leq \mu_s(Y) \int_0^\infty e^{-\lambda x} (\lambda x)^{n-2} \bar{T}_{s+1}(Y, x)\,dx$$

which, by using (7.8.57), gives $N_\lambda^X \leq_{s-CX} N_\lambda^Y$.

Sufficiency: Given that $N_\lambda^X \leq_{s-CX} N_\lambda^Y$, which by (7.8.56), is equivalent to

$$\mu_s(X) \Gamma_\lambda \bar{T}_{s+1}(X, n-1) \leq \mu_s(Y) \Gamma_\lambda \bar{T}_{s+1}(X, n-1)$$

for all $n \geq 2$. Now, taking limit as $n \to \infty$, we get from Lemma 7.8.14,

$$\mu_s(X) \bar{T}_{s+1}(X, x) \leq \mu_s(Y) \bar{T}_{s+1}(X, x).$$

On using (7.8.14), this gives $X \leq_{s-CX} Y$. □

Taking $s = 1$ in the above theorem, we get

Corollary 7.8.19 $X \leq_{icx} Y$ if and only if $N_\lambda^X \leq_{icx} N_\lambda^Y$.

7.8 Generalized Orderings

In a similar fashion as above, we can prove the following theorem which is available in Hu et al. (2001).

Theorem 7.8.24 *Let $s \in \mathbb{N}$. Then, for two non-negative random variables X and Y,*

$$X \leqslant_{s-CV} Y \text{ if and only if } N_\lambda^X \leqslant_{s-CV} N_\lambda^Y,$$

for all $\lambda > 0$. □

For $s = 1$, the above theorem gives that $X \leq_{icv} Y$ if and only if $N_\lambda^X \leq_{icv} N_\lambda^Y$, for all $\lambda > 0$.

7.8.5 Dispersion-Type Stochastic Orders

Let X and Y be two random variables with respective distribution functions F_X and F_Y, and survival functions $\bar{F}_X$ and $\bar{F}_Y$. Suppose two distributions are such that, for all $0 \leqslant \alpha \leqslant \beta \leqslant 1$,

$$F_X^{-1}(\beta) - F_X^{-1}(\alpha) \leqslant F_Y^{-1}(\beta) - F_Y^{-1}(\alpha) \tag{7.8.58}$$

which can equivalently be written as

$$\bar{F}_X^{-1}(\alpha) - \bar{F}_X^{-1}(\beta) \leqslant \bar{F}_Y^{-1}(\alpha) - \bar{F}_Y^{-1}(\beta). \tag{7.8.59}$$

This means that the difference between any two quantiles of the distribution F_X is less than that of another distribution F_Y. The properties of the distributions F_X and F_Y satisfying (7.8.58) were studied by Saunders and Moran (1978) and the property mentioned in (7.8.58) was later called dispersive order by Lewis and Thompson (1981). We define the same formally as follows. For different properties of dispersive order one may also see Shaked (1982). Most of the results of this section may be obtained in Nanda and Kundu (2009) who have proved some more general results.

Definition 7.8.9 X is said to be less than Y in dispersive order (written as $X \leq_{disp} Y$) if either (7.8.58) or (7.8.59) holds. ◊

Belzunce et al. (2003) have introduced dispersion-type variability orders and proved some interesting results.

Let us define the dispersion-type generalized order as follows.

Definition 7.8.10 Let X and Y be two random variables with distribution functions F_X and F_Y, respectively. Then, for $s \in \mathbb{N} \cup \{0\}$, X is said to be smaller than Y in s-$\star$ order of the dispersion type (written as $X \leq_{disp-(s-\star)} Y$) if, for all $p \in (0, 1)$,

$$[X - F_X^{-1}(p) | X > F_X^{-1}(p)] \leq_{s-\star} [Y - F_Y^{-1}(p) | Y > F_Y^{-1}(p)], \tag{7.8.60}$$

where $\star = FR, ST, CV, CX$. □

Let us discuss the following lemma which will be used in sequel. The proof is omitted.

Lemma 7.8.17 *Let s be any non-negative integer. If $X_{F^{-1}(p)}$, $p \in (0, 1)$, denotes the random variable $[X - F^{-1}(p) | X > F^{-1}(p)]$, then*

$$\bar{T}_s\left(X_{F^{-1}(p)}, x\right) = \frac{\bar{T}_s(X, x + F^{-1}(p))}{1 - p},$$

for all $x \geq 0$. □

The following theorem shows that $disp$-(s-ST) order is stronger than s-ST order.

Theorem 7.8.25 *For any $s \in \mathbb{N} \cup \{0\}$, $X \leq_{disp-(s-ST)} Y$ implies $X \leq_{s-ST} Y$.*

Proof Note that $X \leq_{disp-(s-ST)} Y$ if and only if, for all $x \geq 0$ and $p \in (0, 1)$,

$$\frac{\bar{T}_s\left(X_{F_X^{-1}(p)}, x\right)}{\bar{T}_s\left(X_{F_X^{-1}(p)}, 0\right)} \leq \frac{\bar{T}_s\left(X_{F_Y^{-1}(p)}, x\right)}{\bar{T}_s\left(X_{F_Y^{-1}(p)}, 0\right)},$$

which, by Lemma 7.8.17, reduces to

$$\frac{\bar{T}_s\left(X, x + F_X^{-1}(p)\right)}{\bar{T}_s\left(X, F_X^{-1}(p)\right)} \leq \frac{\bar{T}_s\left(Y, x + F_Y^{-1}(p)\right)}{\bar{T}_s\left(Y, F_Y^{-1}(p)\right)}.$$

The result now follows by taking $p \to 0+$. □

Taking $s = 0, 1, 2$ in the above theorem, we get the following corollary.

Corollary 7.8.20 *For any non-negative random variables X and Y, $X \leq_{disp-\star} Y \Rightarrow X \leq_\star Y$, where $\star$ stands for wlr, st, and hmrl.* □

The relationship between $disp$-(s-CV) order and s-CV order is given next.

Theorem 7.8.26 *For any $s \in \mathbb{N} \cup \{0\}$, if $X \leq_{disp-(s-CV)} Y$, then $X \leq_{s-CV} Y$.*

Proof For any $s \in \mathbb{N} \cup \{0\}$, $X \leq_{disp-(s-CV)} Y$ if and only if,

$$\frac{\int_0^x \bar{T}_s\left(X_{F_X^{-1}(p)}, t\right) dt}{\bar{T}_s\left(X_{F_X^{-1}(p)}, 0\right)} \leq \frac{\int_0^x \bar{T}_s\left(Y_{F_Y^{-1}(p)}, t\right) dt}{\bar{T}_s\left(Y_{F_Y^{-1}(p)}, 0\right)},$$

which, by Lemma 7.8.17, is equivalent to the fact that

$$\frac{\int_0^x \bar{T}_s\left(X, t + F_X^{-1}(p)\right) dt}{\bar{T}_s\left(X, F_X^{-1}(p)\right)} \leq \frac{\int_0^x \bar{T}_s\left(Y, t + F_Y^{-1}(p)\right) dt}{\bar{T}_s\left(Y, F_Y^{-1}(p)\right)}.$$

Taking limit as $p \to 0+$, the above inequality gives the required result. □

7.8 Generalized Orderings

Taking $s = 0$ in the above theorem we see that $X \leq_{disp-icv} Y \Rightarrow X \leq_{icv} Y$. In a similar fashion the following theorem may be proved.

Theorem 7.8.27 *For any $s \in \mathbb{N} \cup \{0\}$, if $X \leq_{disp-(s-CX)} Y$, then $X \leq_{s-CX} Y$.* □

This immediately gives that $X \leq_{disp-icx} Y \Rightarrow X \leq_{icx} Y$.

The following theorem shows that $disp$-$(s$-FR$)$ order implies s-FR order.

Theorem 7.8.28 *For any $s \in \mathbb{N} \cup \{0\}$, $X \leq_{disp-(s-FR)} Y$ implies $X \leq_{s-FR} Y$.*

Proof Note that $X \leq_{disp-(s-FR)} Y$ if and only if

$$\frac{\bar{T}_s\left(X_{F_X^{-1}(p)}, x\right)}{\bar{T}_s\left(Y_{F_Y^{-1}(p)}, x\right)} \text{ is decreasing in } x \geq 0,$$

for all $p \in (0, 1)$, which, by Lemma 7.8.17, is equivalent to the fact that

$$\frac{\bar{T}_s\left(X, x + F_X^{-1}(p)\right)}{\bar{T}_s\left(Y, x + F_Y^{-1}(p)\right)} \text{ is decreasing in } x \geq 0,$$

for all $p \in (0, 1)$. Taking limit as $p \to 0+$, we immediately get the required result. □

Taking $s = 0, 1, 2, 3$ in the above theorem we get the following.

For non-negative random variables X and Y, $X \leq_{disp-\star} Y \Rightarrow X \leq_{\star} Y$, where $\star$ stands for lr, hr, mrl, and vrl.

Although the converse of Theorem 7.8.28 does not hold in general, it is shown in the following theorem that the converse does hold under certain condition(s).

Theorem 7.8.29 *If X or Y is s-DFR, $X \leq_{s-FR} Y$ and $X \leq_{st} Y$, then $X \leq_{disp-(s-FR)} Y$, for $s \in \mathbb{N} \cup \{0\}$.*

An immediate consequence of Theorem 7.8.29 is given in the following corollary which may be obtained in Belzunce et al. (2003).

Corollary 7.8.21 *(i) If X or Y is DLR and $X \leq_{lr} Y$, then $X \leq_{disp-lr} Y$.*
(ii) If X or Y is DFR and $X \leq_{hr} Y$, then $X \leq_{disp-hr} Y$.
(iii) If X or Y is IMRL, $X \leq_{st} Y$ and $X \leq_{mrl} Y$, then $X \leq_{disp-mrl} Y$.
(iv) If X or Y is IVRL, $X \leq_{st} Y$ and $X \leq_{vrl} Y$, then $X \leq_{disp-vrl} Y$.
□

The ordering between two random variables in terms of ordering between their residual lives is given below.

Theorem 7.8.30 *Let X and Y be two non-negative random variables and $s \in \mathbb{N} \cup \{0\}$. If*

(i) $X \leq_{st} Y$
(ii) $X \leq_{disp-(s-FR)} Y$
(iii) X or Y is s-IFR,

then $X_t \leq_{disp-(s-FR)} Y_t$, for all $t \geq 0$. □

We get the following by taking $s = 0, 1, 2, 3$ in the above theorem. Here (i)–(iii) may be obtained in Belzunce et al. (2003).

Corollary 7.8.22 *Let X and Y be two non-negative random variables. If*

(i) *X or Y is ILR and $X \leq_{disp-lr} Y$, then $X_t \leq_{disp-lr} Y_t$ for all $t \geq 0$;*
(ii) *X or Y is IFR and $X \leq_{disp-hr} Y$, then $X_t \leq_{disp-hr} Y_t$ for all $t \geq 0$;*
(iii) *X or Y is $DMRL$, $X \leq_{st} Y$ and $X \leq_{disp-mrl} Y$ then $X_t \leq_{disp-mrl} Y_t$ for all $t \geq 0$;*
(iv) *X or Y is $DVRL$, $X \leq_{st} Y$ and $X \leq_{disp-vrl} Y$ then $X_t \leq_{disp-vrl} Y_t$ for all $t \geq 0$.* □

The following theorem is easy to prove.

Theorem 7.8.31 *Let X and Y be two non-negative random variables. Then, for $s \in \mathbb{N} \cup \{0\}$,*

(i) $X \leq_{disp-(s-FR)} Y$ *implies* $X \leq_{disp-(s-ST)} Y$;
(ii) $X \leq_{disp-(s-FR)} Y$ *implies* $X \leq_{disp-((s+1)-FR)} Y$;
(iii) $X \leq_{disp-(s-ST)} Y$ *implies* $X \leq_{disp-(s-CV)} Y$.

7.8.6 Some Characterizations of Generalized Dispersion-Type Orders

Here some characterizations of generalized dispersion-type orders are discussed. The following theorem gives characterization of $disp$-$(s$-FR$)$ order between the random variables which are known to be ordered by usual stochastic order.

Theorem 7.8.32 *Let X and Y be two non-negative random variables such that $X \leq_{st} Y$. Then, for $s \in \mathbb{N} \cup \{0\}$, $X \leq_{disp-(s-FR)} Y$ if and only if $X \leq_{s-FR} Y_t$, for all $t \geq 0$.*

Proof Note that $X \leq_{disp-(s-FR)} Y$ if and only if

$$\frac{\bar{T}_s\left(X_{F_X^{-1}(p)}, x\right)}{\bar{T}_s\left(Y_{F_Y^{-1}(p)}, x\right)} \text{ is decreasing in } x \geq 0,$$

7.8 Generalized Orderings 221

for all $p \in (0, 1)$. By Lemma 7.8.17, this can equivalently be written as

$$\frac{\bar{T}_s\left(X, x + F_X^{-1}(p)\right)}{\bar{T}_s\left(Y, x + F_Y^{-1}(p)\right)} \text{ is decreasing in } x \geqslant 0,$$

for all $p \in (0, 1)$. Now, since $X \leq_{st} Y$, replacing $x + F_X^{-1}(p)$ by x and $x + F_Y^{-1}(p)$ by $x + t$, the above statement can be restated as for all $t \geq 0$,

$$\frac{\bar{T}_s(X, x)}{\bar{T}_s(Y_t, x)} \text{ is decreasing in } x.$$

This proves the result. □

Taking $s = 0, 1, 2, 3$ in the above theorem, we have the following.

Corollary 7.8.23 *Let X and Y be two non-negative random variables. Then*

(i) $X \leq_{disp-lr} Y$ *if and only if* $X \leq_{lr} Y_t$, *for all* $t \geq 0$.
(ii) $X \leq_{disp-hr} Y$ *if and only if* $X \leq_{hr} Y_t$, *for all* $t \geq 0$.
  Further, if $X \leq_{st} Y$, *then,*
(iii) $X \leq_{disp-mrl} Y$ *if and only if* $X \leq_{mrl} Y_t$, *for all* $t \geq 0$.
(iv) $X \leq_{disp-vrl} Y$ *if and only if* $X \leq_{vrl} Y_t$, *for all* $t \geq 0$. ◇

In the following theorem, characterization of $disp$-(s-FR) order between a random variable X and a standard Exponential random variable is given. Before that we need the following lemma.

Lemma 7.8.18 *Let X be a non-negative random variable with standard Exponential distribution. Then, for $s \in \mathbb{N} \cup \{0\}$ and for all $x \geq 0$, $\bar{T}_s(X, x) = e^{-x}$.*

Theorem 7.8.33 *Let X be a non-negative random variable with distribution function F and Y be a standard Exponential random variable. Then, for $s \in \mathbb{N} \cup \{0\}$, $X \leq_{disp-(s-FR)} Y$ if and only if $X_t \leq_{s-FR} Y$, for all $t \geq 0$.*

Proof If $X \leq_{disp-(s-FR)} Y$, we have

$$\frac{\bar{T}_s\left(X_{F_X^{-1}(p)}, x\right)}{\bar{T}_s\left(Y_{F_Y^{-1}(p)}, x\right)} \text{ is decreasing in } x,$$

for all $p \in (0, 1)$. On using Lemmas 7.8.17 and 7.8.18, the above statement reduces to

$$\frac{\bar{T}_s\left(X, x + F_X^{-1}(p)\right)}{e^{-x}(1-p)} \text{ is decreasing in } x,$$

for all $p \in (0, 1)$. Writing $F_X^{-1}(p) = t$, the above statement can be restated as

$$\frac{\bar{T}_s(X, x+t)}{\bar{T}_s^Y(x)} \text{ is decreasing in } x,$$

for all $t \geq 0$. This proves the theorem. □

By taking $s = 0, 1, 2, 3$ in the above theorem we get the following, of which first two are given in Belzunce et al. (2003).

Corollary 7.8.24 *Let X be a non-negative random variable with distribution function F and Y be a standard exponential random variable. Then*

(i) $X \leq_{disp-lr} Y$ if and only if $X_t \leq_{lr} Y$, for all $t \geq 0$.
(ii) $X \leq_{disp-hr} Y$ if and only if $X_t \leq_{hr} Y$, for all $t \geq 0$.
(iii) $X \leq_{disp-mrl} Y$ if and only if $X_t \leq_{mrl} Y$, for all $t \geq 0$.
(iv) $X \leq_{disp-vrl} Y$ if and only if $X_t \leq_{vrl} Y$, for all $t \geq 0$. □

Characterization of s-IFR class and its dual by means of dispersion-type stochastic orders, using the residual lifetime, is given below. Here $s \in \mathbb{N} \cup \{0\}$.

Theorem 7.8.34 *The following conditions are equivalent.*

(a) X is s-IFR(s-DFR).
(b) $X_l \geq_{disp-(s-FR)} (\leq_{disp-(s-FR)}) X_t$ whenever $l < t$.
(c) $X \geq_{disp-(s-FR)} (\leq_{disp-(s-FR)}) X_t$ for all $t \geq 0$.

The following corollary is immediate where (i)-(iii) may be obtained in Hu et al. (2004).

Corollary 7.8.25 *(i) For a random variable X, the following conditions are equivalent.*

  (a) X is ILR (DLR).
  (b) $X_l \geq_{disp-lr} (\leq_{disp-lr}) X_t$ whenever $l < t$.
  (c) $X \geq_{disp-lr} (\leq_{disp-lr}) X_t$ for all $t \geq 0$.

(ii) For a random variable X, the following conditions are equivalent.

  (a) X is IFR (DFR).
  (b) $X_l \geq_{disp-hr} (\leq_{disp-hr}) X_t$ whenever $l < t$.
  (c) $X \geq_{disp-hr} (\leq_{disp-hr}) X_t$ for all $t \geq 0$.

(iii) For a random variable X, the following conditions are equivalent.

  (a) X is $DMRL$ ($IMRL$).
  (b) $X_l \geq_{disp-mrl} (\leq_{disp-mrl}) X_t$ whenever $l < t$.
  (c) $X \geq_{disp-mrl} (\leq_{disp-mrl}) X_t$ for all $t \geq 0$.

7.8 Generalized Orderings

(iv) For a random variable X, the following conditions are equivalent.

(a) X is $DVRL$ ($IMRL$).
(b) $X_l \geq_{disp-vrl} (\leq_{disp-vrl}) X_t$ whenever $l < t$.
(c) $X \geq_{disp-vrl} (\leq_{disp-vrl}) X_t$ for all $t \geq 0$. □

Another equivalence of s-IFR class and its dual is given next.

Theorem 7.8.35 *Let X be a random variable with distribution function F. Then, for $s \in \mathbb{N} \cup \{0\}$, the following conditions are equivalent.*

(a) X is s-IFR(s-DFR).
(b) $X_l \geq_{disp-(s-ST)} (\leq_{disp-(s-ST)}) X_t$, whenever $l < t$.
(c) $X \geq_{disp-(s-ST)} (\leq_{disp-(s-ST)}) X_t$, for all $t \geq 0$. □

By taking $s = 0, 1, 2, 3$ in the above theorem, we get the following corollary.

Corollary 7.8.26 *(i) For a random variable X, the following conditions are equivalent:*

(a) X is ILR (DLR).
(b) $X_l \geq_{disp-wlr} (\leq_{disp-wlr}) X_t$, whenever $l < t$.
(c) $X \geq_{disp-wlr} (\leq_{disp-wlr}) X_t$ for all $t \geq 0$.

(ii) For a random variable X, the following conditions are equivalent:

(a) X is IFR (DFR).
(b) $X_l \geq_{disp-st} (\leq_{disp-st}) X_t$, whenever $l < t$.
(c) $X \geq_{disp-st} (\leq_{disp-st}) X_t$ for all $t \geq 0$.

(iii) For a random variable X, the following conditions are equivalent:

(a) X is $DMRL$ ($IMRL$).
(b) $X_l \geq_{disp-hmrl} (\leq_{disp-hmrl}) X_t$, whenever $l < t$.
(c) $X \geq_{disp-hmrl} (\leq_{disp-hmrl}) X_t$ for all $t \geq 0$. □

Another similar characterization result is given below.

Theorem 7.8.36 *Let X be a random variable with distribution function F. Then, for $s \in \mathbb{N}$, the following conditions are equivalent:*

(a) X is $(s+1)$-IFR $((s+1)$-$DFR)$.
(b) $X_l \geq_{disp-(s-CX)} (\leq_{disp-(s-CX)}) X_t$ whenever $l < t$.
(c) $X \geq_{disp-(s-CX)} (\leq_{disp-(s-CX)}) X_t$ for all t.

The following corollary follows by taking $s = 1$ in the above theorem.

Corollary 7.8.27 *For a random variable X, the following statements are equivalent:*

(a) *X is DMRL (IMRL).*
(b) *$X_l \geq_{disp-icx} (\leq_{disp-icx}) X_t$ whenever $l < t$.*
(c) *$X \geq_{disp-icx} (\leq_{disp-icx}) X_t$, for all $t \geq 0$.*

7.9 Generalized Ageing Properties

Let X be a non-negative random variable. The following definition may be obtained in Fagiuoli and Pellerey (1993).

Definition 7.9.1 For $s \in \mathbb{N} \cup \{0\}$, X is said to be

(a) $s\text{-}IFR$ (resp. $s\text{-}DFR$) if

$$r_s(X, x) \text{ is increasing (resp. decreasing) in } x > 0; \qquad (7.9.61)$$

(b) $s\text{-}IFRA$ (resp. $s\text{-}DFRA$) if $\frac{1}{x}\int_0^x r_s(X, u)\, du$ is increasing (resp. decreasing) in $x > 0$;
(c) $s\text{-}NBU$ (resp. $s\text{-}NWU$) if $\bar{T}_s(X, x+t).\bar{T}_s(X, 0) \leq (resp. \geq) \bar{T}_s(X, x).\bar{T}_s(X, t)$, for all $x, t > 0$;
(d) $s\text{-}NBUFR$ (resp. $s\text{-}NWUFR$) if

$$r_s(X, x) \geq (resp. \leq) r_s(X, 0), \text{ for all } x \geq 0; \qquad (7.9.62)$$

(e) $s\text{-}NBUCX$ (resp. $s\text{-}NWUCX$) if

$$\bar{T}_s(X, 0)\int_x^\infty \bar{T}_s(X, t+u)\, du \leq (resp. \geq) \bar{T}_s(X, t)\int_x^\infty \bar{T}_s(X, u)\, du. \qquad (7.9.63)$$

In the above definition we assume that $\bar{T}_{-1}(X, x) = \frac{d}{dx} f_X(x)$. Note that, on using (7.8.15), for every $s = 2, 3, \ldots$, (7.9.62) can equivalently be written as $\bar{T}_s(X, x) \leq \bar{T}_{s-1}(X, x)$, for all $x \geq 0$. Thus, for $s = 2, 3, \ldots$, a non-negative random variable X is $s\text{-}NBUFR$ (resp. $s\text{-}NWUFR$) if and only if

$$\bar{T}_s(X, x) \leq (resp. \geq) \bar{T}_{s-1}(X, x). \qquad (7.9.64)$$

The following equivalences can easily be noted:

7.9 Generalized Ageing Properties

0-$IFR \Leftrightarrow ILR$,   1-$IFR \Leftrightarrow IFR$,   2-$IFR \Leftrightarrow DMRL$
3-$IFR \Leftrightarrow DVRL$,   1-$IFRA \Leftrightarrow IFRA$,   1-$NBU \Leftrightarrow NBU$
1-$NBUFR \Leftrightarrow NBUFR$, 2-$NBUFR \Leftrightarrow NBUE$, 0-$DFR \Leftrightarrow DLR$
1-$DFR \Leftrightarrow DFR$,   2-$DFR \Leftrightarrow IMRL$,   3-$DFR \Leftrightarrow IVRL$
1-$DFRA \Leftrightarrow DFRA$,   1-$NWU \Leftrightarrow NWU$,   1-$NWUFR \Leftrightarrow NWUFR$
2-$NWUFR \Leftrightarrow NWUE$

On using (7.8.15), (7.9.61) can be rewritten as, for all $x, t > 0$,

$$\frac{\bar{T}_{s-1}(X, x)}{\int_x^\infty \bar{T}_{s-1}(X, u)\, du} \leqslant (resp. \geqslant) \frac{\bar{T}_{s-1}(X, x+t)}{\int_{x+t}^\infty \bar{T}_{s-1}(X, u)\, du}. \qquad (7.9.65)$$

On using (7.8.14), the above expression can equivalently be written as

$$\frac{\bar{T}_{s-1}(X, x)}{\bar{T}_s(X, x)} \leqslant (resp. \geqslant) \frac{\bar{T}_{s-1}(X, x+t)}{\bar{T}_s(X, x+t)},$$

for all $x, t > 0$. This means that

$$\frac{\bar{T}_s(X, x)}{\bar{T}_{s-1}(X, x)} \text{ is decreasing (resp. increasing) in } x. \qquad (7.9.66)$$

This is equivalent to the fact that, for all $h > 0$,

$$\bar{T}_s(X, x+h) \cdot \bar{T}_{s-1}(X, x) \leqslant (resp. \geqslant) \bar{T}_s(X, x) \bar{T}_{s-1}(X, x+h).$$

Let us rewrite this as

$$\bar{T}_s(X, x) \frac{d}{dx} \bar{T}_s(X, x+h) \leqslant (resp. \geqslant) \bar{T}_s(X, x+h) \frac{d}{dx} \bar{T}_s(X, x),$$

which is equivalent to

$$\frac{\bar{T}_s(X, x+h)}{\bar{T}_s(X, x)} \text{ decreasing (resp. increasing) in } x. \qquad (7.9.67)$$

Further, on using (7.8.14), (7.9.66) gives

$$\frac{\frac{d}{dx}\bar{T}_s(X, x)}{\bar{T}_s(X, x)} \text{ is decreasing (increasing) in } x,$$

which means that

$$\bar{T}_s(X, x) \text{ is } \log - \text{concave} \ (\log - \text{convex}) \text{ in } x. \qquad (7.9.68)$$

Based on the above discussion we have the following.

Theorem 7.9.1 *The following conditions are equivalent:*

(a) X is s-IFR (resp. s-DFR).
(b) $r_s(X, x)$ is increasing (resp. decreasing) in $x > 0$.
(c) $\dfrac{\bar{T}_{s-1}(X,x)}{\int_x^\infty \bar{T}_{s-1}(X,u)\,du} \leqslant (resp. \geqslant) \dfrac{\bar{T}_{s-1}(X,x+t)}{\int_{x+t}^\infty \bar{T}_{s-1}(X,u)\,du}$.
(d) $\dfrac{\bar{T}_s(X,x)}{\bar{T}_{s-1}(X,x)}$ is decreasing (resp. increasing) in x.
(e) $\dfrac{\bar{T}_s(X,x+h)}{\bar{T}_s(X,x)}$ is decreasing (resp. increasing) in x, for all $h \geqslant 0$. □

On using (7.3.1), Definition 7.9.1(b) can be rewritten as

$$\frac{1}{x}\ln\left(\frac{1}{\bar{F}(x)}\right) \text{ is increasing (resp. decreasing) in } x,$$

which is equivalent to

$$\left(\bar{F}(x)\right)^{1/x} \text{ is decreasing (resp. increasing) in } x. \qquad (7.9.69)$$

Below we give corresponding discrete counterpart of Definition 7.9.1.

Definition 7.9.2 For $s = 0, 1, 2, \ldots$, a non-negative discrete random variable X is said to be

(a) s-IFR (resp. s-DFR) if

$$\frac{\bar{T}_s(X, n)}{\bar{T}_{s-1}(X, n)} \text{ is decreasing (resp. increasing) in } n \in \mathbb{N}; \qquad (7.9.70)$$

(b) s-$IFRA$ (resp. s-$DFRA$) if

$$\left(\bar{T}_s(X, n)\right)^{1/n} \text{ is decreasing (resp. increasing) in } n \in \mathbb{N};$$

(c) s-NBU (resp. s-NWU) if

$$\bar{T}_s(X, m+n)\bar{T}_s(X, 0) \leqslant \bar{T}_s(X, m).\bar{T}_s(X, n),$$

for all $m, n \in \mathbb{N} \cup \{0\}$. □

Below we give some results related to s-IFR order, which may be obtained in Hu et al. (2001). Before that we need the following lemma.

Lemma 7.9.1 *For $s \in \mathbb{N}$, and $a(\neq 0) \in \mathbb{R}$,*

(a) $\bar{T}_s(aX, x) = \bar{T}_s\left(X, \frac{x}{a}\right)$;
(b) $\mu_s(aX) = a\mu_s(X)$;
(c) $r_s(aX, x) = \frac{1}{a}r_s\left(X, \frac{x}{a}\right)$.

7.9 Generalized Ageing Properties

Proof We prove this using induction on s simultaneously for (a) and (b). Note that, for $s = 1$, (a) and (b) follow easily. Suppose both hold for some specific value of s. Then we have

$$\bar{T}_{s+1}(aX, x) = \frac{\int_x^\infty \bar{T}_s(aX, u)\, du}{\mu_s(aX)}$$

$$= \frac{\int_x^\infty \bar{T}_s\left(X, \frac{u}{a}\right) du}{a\mu_s(X)}$$

$$= \frac{\int_{x/a}^\infty \bar{T}_s(X, u)\, du}{\mu_s(X)}$$

$$= \bar{T}_{s+1}\left(X, \frac{x}{a}\right).$$

The first and the last equality follow from (7.8.14) whereas the second equality follows from the hypotheses. Thus, (a) and (b) follow from the method of mathematical induction. To prove (c), note that

$$r_s(aX, x) = \frac{\bar{T}_{s-1}(aX, x)}{\int_x^\infty \bar{T}_{s-1}(aX, u)\, du}$$

$$= \frac{\bar{T}_{s-1}\left(X, \frac{x}{a}\right)}{\int_x^\infty \bar{T}_{s-1}\left(X, \frac{u}{a}\right) du}$$

$$= \frac{\bar{T}_{s-1}\left(X, \frac{x}{a}\right)}{a \int_{x/a}^\infty \bar{T}_{s-1}(X, u)\, du}$$

$$= \frac{1}{a} \cdot r_s\left(X, \frac{x}{a}\right).$$

The first and the last equalities follow from (7.8.15) whereas the second equality follows from (a) above. Hence the result is established. □

In relation to an insurance policy, let the risk a direct insurer face is denoted by X and let ϕ denote the corresponding reinsurance contract. Then, for $a \in (0, 1)$, $\phi(X) = aX$ represents a reinsurance agreement, known as quota-share treaty. The following theorem shows that if the risk has an s-IFR distribution then the quota-share treaty is dominated by the risk in s-FR order (cf. Hu et al., 2001).

Theorem 7.9.2 *Let X be a non-negative s-IFR random variable. Then, for any $0 < a \leq 1$ and $s \in \mathbb{N}$, $aX \leq_{s-FR} X$.*

Proof Note that $aX \leq_{s-FR} X$ if and only if

$$\frac{\bar{T}_s(aX, x)}{\bar{T}_s(X, x)} \text{ is decreasing in } x,$$

which, on using Lemma 7.9.1(a), reduces to

$$\frac{\bar{T}_s\left(X, \frac{x}{a}\right)}{\bar{T}_s(X, x)} \text{ is decreasing in } x.$$

Since $a \in (0, 1)$, on using Lemma 7.9.1(e), we get that X is s-IFR. □

Taking $s = 1, 2, 3$ in the above theorem we get the following corollary, of which (i) is given in Kochar (1979) whereas (ii) may be obtained in Shaked and Shanthikumar (2007).

Corollary 7.9.1 *Let X be a non-negative random variable. Take $0 < a < 1$.*

(i) *If X is IFR then $aX \leq_{hr} X$.*
(ii) *If X is $DMRL$ then $aX \leq_{mrl} X$.*
(iii) *If X is $DVRL$ then $aX \leq_{vrl} X$.* □

The following lemma is stated in Hu et al. (2001). We give a proof of this which is similar to the one given in Pellerey et al. (2000) where the result is proved for $s = 1$.

Lemma 7.9.2 *For any $s \in \mathbb{N}$, if $r_s(X, x)$ is log-concave then it is increasing.*

Proof Let us fix an $s \in \mathbb{N}$. We prove this result assuming that $r_s(X, \cdot)$ is differentiable. If possible, suppose $r_s(X, x)$ is log-concave but it is not increasing. Then there exists some $x_0 \in (0, \infty)$ where it is decreasing. Write

$$\left.\frac{d}{dx} r_s(X, x)\right|_{x=x_0} = \alpha.$$

Thus, we have $\alpha < 0$. So, we have

$$\left.\frac{d}{dx} \ln r_s(X, x)\right|_{x=x_0} = \frac{\alpha}{r_s(X, x_0)} < 0,$$

since $r_s(X, x_0) < \infty$. This is because, otherwise, the support of the distribution of X will be $[0, x_0]$ and in this case the assumption that $r_s(X, x)$ is decreasing at x_0 does not make sense. The log-concavity of $r_s(X, x)$ gives that, for all $x \geq x_0$, $\ln r_s(X, x)$ must be strictly decreasing, otherwise, if there exists x_1 ($> x_0$) such that $r_s(X, x)$ is decreasing at x_0 and increasing at x_1, then $\ln r_s(X, x)$ becomes convex in $[x_0, x_1]$ which is a contradiction to the fact that $r_s(X, x)$ is log-concave. Hence,

$$\frac{d}{dx} \ln r_s(X, x) \leq \frac{\alpha}{r_s(X, x_0)}, \text{ for all } x \geq x_0,$$

which is equivalent to the fact that

$$\frac{\frac{d}{dx} r_s(X, x)}{r_s(X, x)} \leq \frac{\alpha}{r_s(X, x_0)}, \text{ for all } x \geq x_0.$$

7.9 Generalized Ageing Properties

This can be rewritten as

$$r_s(X, x) \leqslant \frac{r_s(X, x_0)}{\alpha} \left[\frac{d}{dx} r_s(X, x) \right], \quad \text{for all } x \geqslant x_0,$$

since $\alpha < 0$. Integrating both sides of the above expression we get

$$\int_{x_0}^{\infty} r_s(X, x)\, dx \leqslant \frac{r_s(X, x_0)}{\alpha} \int_{x_0}^{\infty} dr_s(X, x)$$

$$= \frac{r_s(X, x_0)}{\alpha} \left[\lim_{x \to \infty} r_s(X, x) - r_s(X, x_0) \right]. \quad (7.9.71)$$

Since $r_s(X, x_0)$ exists and is decreasing on $[x_0, \infty)$, it follows that $\lim_{x \to \infty} r_s(X, x) < \infty$. So, (7.9.71) gives $\int_{x_0}^{\infty} r_s(X, x)\, dx < \infty$, which is a contradiction (because, for $s \in \mathbb{N}$, $r_s(X, \cdot)$ is the failure rate of some random variable, which we have denoted earlier by $X_{(s)}$, and hence $\lim_{x \to \infty} r_s(X, x)$ must be infinity). Hence $r_s(X, x)$ must be increasing in $\mathbb{R}^+$. Now, since the result is true for any arbitrary $s \in \mathbb{N}$, the result is established. □

We have seen that if a random variable X is s-IFR then it is $(s-1)$-IFR. However, the converse is not always true. Below we give a condition under which the converse is true (cf. Hu et al., 2001).

Theorem 7.9.3 *Let $s \in \mathbb{N}$. If $r_s(X, x)$ is log-concave, then it is $(s-1)$-IFR.*

Proof Note that

$$\bar{T}_{s-1}(X, x) = r_s(X, x) \int_x^{\infty} \bar{T}_{s-1}(X, u)\, du$$

$$= r_s(X, x) \mu_{s-1}(X)\, \bar{T}_s(X, x)$$

$$= r_s(X, x) \mu_{s-1}(X)\, e^{-\int_0^x r_s(X, u)\, du}.$$

The first equality follows from (7.8.15), the second equality follows from (7.8.14) whereas the last equality follows from (7.8.16). Taking log on both sides of the above expression we get

$$\ln \bar{T}_{s-1}(X, x) = \ln r_s(X, x) + \ln \mu_{s-1}(X) - \int_0^x r_s(X, u)\, du.$$

Note that the first term in the right-hand side of the above expression is concave by hypothesis whereas the last term $\left[-\int_0^x r_s(X, u)\, du \right]$ can easily be shown to be concave. Thus, we get that the left-hand side of the above expression is concave. Now, on using (7.9.68), we get that X is $(s-1)$-IFR. □

Taking $s = 1, 2, 3$ in the above theorem, we see that if X has log-concave failure rate (resp. MRL, VRL) then it is ILR (resp. IFR, $DMRL$).

Remark 7.9.1 Let us consider two random variables both having some positive ageing property. Then one might be interested to know which one is ageing faster. This kind of study is meaningful because if one group of components is known to have less rate of ageing compared to another group of components then while designing a system the design engineers will be able to construct a system using the former group of components in order to get better reliable system. This is considered in detail by Hazra and Nanda (2016) and Nanda et al. (2017) where the generalized ageing classes have been studied using generalized orderings.

7.10 Residual Life at Random Time

So far we have discussed residual lifetime random variable X_t where t is fixed. If t is replaced by some random variable Y independent of X, then $X_Y = [X - Y | X > Y]$, will be called residual life at random time ($RLRT$). Here some results on generalized orderings and ageing classes of residual lifetime random variables are discussed.

Let X and Y have distribution functions F and G, respectively. Then the survival function of X_Y can be written as

$$\overline{F}_G(t) = \frac{\int_0^\infty \overline{F}(t+y)dG(y)}{\int_0^\infty \overline{F}(y)dG(y)}, \quad \text{for all } t \geq 0.$$

Suppose two systems A and B have respective independent lifetimes X and Y. Then X_Y gives the residual life of A when B fails. For a $GI/G/1$ queue, let T represent inter-arrival time, W the waiting time, and S the service time. Then $I = [T - (W + S) | T > W + S]$ represents the idle time which can be considered as $RLRT$ (cf. Marshall, 1968). See also Belzunce et al. (2003), Stoyan (1983), and Yue and Cao (2000) for discussion on $RLRT$. Similar to X_Y, the inactivity time at the random time ($ITRT$) is defined in Li and Zuo (2004) as $X_{(Y)} = (Y - X | X \leq Y)$ and its properties have been studied.

The following lemmas will be used in proving the upcoming theorems.

Lemma 7.10.1 *If X and Y be two independent random variables with respective distribution functions F and G, then*

$$\bar{T}_s(X_Y, x) = \frac{\int_0^\infty \bar{T}_s(X, x+y)dG(y)}{\int_0^\infty \bar{F}(y)dG(y)}.$$

Lemma 7.10.2 *(i) Let $\psi(\theta, x)$ be an RR_2 [resp. a TP_2] function (and not necessarily a survival function) in $\theta \in \mathcal{X} \subseteq \mathbb{R}$ and $x \in \mathcal{R} \subseteq \mathbb{R}$, and $F_i(\theta)$ be TP_2 in $i \in \{1, 2\}$ and $\theta \in \mathcal{X}$, where $F_i(\theta)$ is a distribution function in θ for each i. Assume that $\psi(\theta, x)$ is decreasing in θ for every x. Then*

$$\phi_i(x) = \int_\mathcal{X} \psi(\theta, x) dF_i(\theta)$$

7.10 Residual Life at Random Time

is RR$_2$ [resp. TP$_2$] in $x \in \mathcal{R}$ and $i \in \{1, 2\}$. Conversely, if $\phi_i(x)$ is RR$_2$ [resp. TP$_2$] in $x \in \mathcal{R}$ and $i \in \{1, 2\}$ whenever $F_i(\theta)$ is TP$_2$ in $i \in \{1, 2\}$ and $\theta \in \mathcal{X}$, then $\psi(\theta, x)$ is RR$_2$ [resp. TP$_2$] in $\theta \in \mathcal{X}$ and $x \in \mathcal{R}$.

(ii) Let $\psi(\theta, x)$ be any RR$_2$ [resp. a TP$_2$] function (and not necessarily a distribution function) in $\theta \in \mathcal{X}$ and $x \in \mathcal{R}$ and $\overline{F}_i(\theta)$ be a survival function in θ and be TP_2 in $i \in \{1, 2\}$ and $\theta \in \mathcal{X}$. Assume that $\psi(\theta, x)$ is increasing in θ for every x. Then

$$H_i(x) = \int_{\mathcal{X}} \psi(\theta, x) dF_i(\theta)$$

is RR$_2$ [resp. TP$_2$] in $x \in \mathcal{R}$ and $i \in \{1, 2\}$, where $\overline{F}_i \equiv 1 - F_i$. Conversely, if $H_i(x)$ is TP_2 in $i \in \{1, 2\}$ and $x \in \mathcal{R}$ whenever $\overline{F}_i(\theta)$ is TP_2 in $i \in \{1, 2\}$ and $\theta \in \mathcal{X}$, then $\psi(\theta, x)$ is RR$_2$ [resp. TP$_2$] in $\theta \in \mathcal{X}$ and $x \in \mathcal{R}$. ◇

The following theorem characterizes the s-IFR class and its dual (cf. Nanda and Kundu, 2009).

Theorem 7.10.1 *Let three non-negative random variables X, Y_1, and Y_2, where X is independent of Y_1 and Y_2 be such that $Y_1 \leq_{rh} Y_2$. Then, X is s-IFR (s-DFR), if and only if $X_{Y_1} \geq_{s-FR} (\leq_{s-FR}) X_{Y_2}$, where rh denotes reversed hazard rate order and $s \in \mathbb{N} \cup \{0\}$.*

Proof Here the proof for s-IFR is given. The proof for s-DFR is similar and hence omitted. Let G_1 and G_2 be distribution functions of Y_1 and Y_2, respectively. As $Y_1 \leq_{rh} Y_2$, $G_i(y)$ is TP_2 in $y \in [0, \infty)$ and $i \in \{1, 2\}$. Now, as X is s-IFR,

$$\frac{\overline{T}_s(X, x + y_1)}{\overline{T}_s(X, x + y_2)} \text{ is increasing in } x,$$

for all $y_1 \leq y_2$. Thus, by Lemma 7.10.2 (i),

$$\frac{\phi_1(x)}{\phi_2(x)} \text{ is increasing in } x,$$

where $\phi_i(x) = \int_0^\infty \overline{T}_s(X, x + y) dG_i(y)$, $i = 1, 2$. This gives

$$\frac{\int_0^\infty \overline{T}_s(X, x + y) dG_1(y)}{\int_0^\infty \overline{T}_s(X, x + y) dG_2(y)} \text{ is increasing in } x,$$

which, by Lemma 7.10.1, gives $X_{Y_1} \geq_{s-FR} X_{Y_2}$.

Conversely, if $X_{Y_1} \geq_{s-FR} X_{Y_2}$, the result follows from Lemma 7.10.2(i) and the fact that $G_i(y)$ is TP_2 in $y \in [0, \infty)$ and $i \in \{1, 2\}$. □

By taking $s = 0, 1, 2, 3$ in the above theorem we have the following corollary where (iii) may be obtained in Li and Zuo (2004).

Corollary 7.10.1 *(i) For Y_1 and Y_2, two non-negative random variables independent of X, let $Y_1 \leq_{rh} Y_2$. Then, $X_{Y_1} \geq_{lr} (\leq_{lr}) X_{Y_2}$ if and only if X be ILR (DLR).*

(ii) For Y_1 and Y_2, two non-negative random variables independent of X, let $Y_1 \leq_{rh} Y_2$. Then, $X_{Y_1} \geq_{hr} (\leq_{hr}) X_{Y_2}$ if and only if X be IFR (DFR).

(iii) For Y_1 and Y_2, two non-negative random variables independent of X, let $Y_1 \leq_{rh} Y_2$. Then, $X_{Y_1} \geq_{mrl} (\leq_{mrl}) X_{Y_2}$ if and only if X be $DMRL$ $(IMRL)$.

(iv) For Y_1 and Y_2, two non-negative random variables independent of X, let $Y_1 \leq_{rh} Y_2$. Then, $X_{Y_1} \geq_{vrl} (\leq_{vrl}) X_{Y_2}$ if and only if X be $DVRL$ $(IVRL)$.

The theorem given below gives a condition for the residual life random variable to belong to s-IFR class.

Theorem 7.10.2 *Let X and Y be two independent random variables and Y be $DRHR$. Then, for $s \in \mathbb{N} \cup \{0\}$, X is s-IFR if and only if X_Y is s-IFR.*

Proof X is s-IFR if and only if

$$\frac{\bar{T}_s(X-t, x)}{\bar{T}_s(X, x)} \text{ is decreasing in } x. \tag{7.10.72}$$

Now, by Lemma 7.10.1, X_Y is s-IFR if and only if

$$\frac{\int_0^\infty \bar{T}_s(X, x+t+y) dG(y)}{\int_0^\infty \bar{T}_s(X, x+y) dG(y)} \text{ is decreasing in } x,$$

for all $t \geq 0$. Or, equivalently,

$$\frac{\int_x^\infty \bar{T}_s(X-t, z) dG(z-x)}{\int_x^\infty \bar{T}_s(X, z) dG(z-x)} \text{ is decreasing in } x, \tag{7.10.73}$$

for all $t \geq 0$. Let us fix t and write $X - t$ as X_1 and X as X_2. Then (7.10.72) can be restated as

$\bar{T}_s(X_i, z)$ is TP_2 in $i \in \{1, 2\}$ and $z \in [0, \infty)$.

Further, as Y is $DRHR$, $G(z-x)$ is TP_2 in $z, x \in [0, \infty)$. Thus, by Lemma 7.10.2 (i), (7.10.72) and (7.10.73) are equivalent, provided Y is $DRHR$. □

By taking different values of s, the following corollary follows from the above theorem.

Corollary 7.10.2 *Let X and Y be two independent random variables and let Y be $DRHR$. Then X_Y is ILR (resp. IFR, $DMRL$, $DVRL$) if and only if X is ILR (resp. IFR, $DMRL$, $DVRL$).* □

Below is a characterization result of s-FR order in terms of $RLRT$.

7.10 Residual Life at Random Time

Theorem 7.10.3 *Let X and Y be two non-negative random variables and let Z be a DRHR random variable which is independent of X and Y. Then, for any $s \in \mathbb{N} \cup \{0\}$, $X \leq_{s-FR} Y$ if and only if $X_Z \leq_{s-FR} Y_Z$.*

Proof If $X \leq_{s-FR} Y$, then

$$\bar{T}_s(X_i, z) \text{ is } TP_2 \text{ in } i \in \{1, 2\} \text{ and } z \in [0, \infty),$$

where X_1 stands for X and X_2 stands for Y. Now, if $G(z)$ is the distribution function of Z, then, as Z is $DRHR$, we have

$$G(z - x) \text{ is } TP_2 \text{ in } z, x \in [0, \infty).$$

So, by Lemma 7.10.2(i),

$$\int_0^\infty \bar{T}_s(X_i, z) dG(z - x) \text{ is } TP_2 \text{ in } i \in \{1, 2\} \text{ and } x \in [0, \infty). \qquad (7.10.74)$$

This is equivalent to

$$\frac{\int_0^\infty \bar{T}_s(X_2, z) dG(z - x)}{\int_0^\infty \bar{T}_s(X_1, z) dG(z - x)} = \frac{\int_0^\infty \bar{T}_s(Y, x + z) dG(z)}{\int_0^\infty \bar{T}_s(X, x + z) dG(z)}$$

is decreasing in $x \geq 0$. Thus, from Lemma 7.10.1, the above statement can be restated as

$$\frac{\bar{T}_s(Y_Z, x)}{\bar{T}_s(X_Z, x)} \text{ is decreasing in } x \geq 0,$$

which proves the result. $\square$

The following corollary follows from the above theorem by taking $s = 0, 1, 2, 3$.

Corollary 7.10.3 *Let X and Y be two non-negative random variables and Z be a DRHR random variable which is independent of X and Y. Then $X \leq_\star Y$ if and only if $X_Z \leq_\star Y_Z$, where $\star$ may be lr, hr, mrl, vrl.*

7.10.1 Characterization in Terms of Laplace Transform

We start this section by giving a characterization result of s-IFR class in terms of Laplace transform (cf. Nanda, 2000). However, before that we need to state the following lemma (cf. Barlow and Proschan (1981), p. 112 and p. 187).

Lemma 7.10.3 *Define* $\lambda_r = E(X^r)/\Gamma(r+1)$.

(a) *If X is IFRA (resp. DFRA) then $\lambda_r^{1/r}$ is decreasing (resp. increasing) in $r \geq 0$.*
(b) *If X is NBU (resp. NWU) then, for all $m, n \geq 0$, $\lambda_{m+n} \leq (\text{resp.} \geq) \lambda_m . \lambda_n$.*

Theorem 7.10.4 *For $s \in \mathbb{N}$, a non-negative continuous random variable X is*

(a) *s-IFR (resp. s-DFR) if and only if, for all $\lambda > 0$, N_λ^X is s-IFR (resp. s-DFR);*
(b) *s-IFRA (resp. s-DFRA) if and only if, for all $\lambda > 0$, N_λ^X is s-IFRA (resp. s-DFRA);*
(c) *s-NBU (resp. s-NWU) if and only if, for all $\lambda > 0$, N_λ^X is s-NBU (resp. s-NWU),*

where N_λ^X is as defined before Theorem 7.8.21.

Proof We prove the result for positive ageing classes and the parenthetical statements follow similarly. From (7.9.66) we see that X is s-IFR if and only if

$$\frac{\bar{T}_s(X, x)}{\bar{T}_{s-1}(X, x)} \text{ is decreasing in } x.$$

So, by Theorem 7.8.4, we can say that, for any real c, $\bar{T}_s(X, x) - c.\bar{T}_{s-1}(X, x)$ changes sign at most once and if the change of sign does occur it occurs from positive to negative as x traverses from left to right. Since $K(n, x)$ given in (7.8.55) is TP_2 in (n, x), on using Theorem 7.8.3, we have that

$$\int_0^\infty e^{-\lambda x} \frac{(\lambda x)^{n-1}}{(n-1)!} \bar{T}_s(X, x) \, dx - c. \int_0^\infty e^{-\lambda x} \frac{(\lambda x)^{n-1}}{(n-1)!} \bar{T}_{s-1}(X, x) \, dx$$

changes sign at most once and if the change of sign does occur it occurs from positive to negative as n traverses from left to right, which means that $\Gamma_\lambda \bar{T}_s(X, n) - c.\Gamma_\lambda \bar{T}_{s-1}(X, n)$ changes sign at most once and if the change of sign does occur it occurs from positive to negative as n traverses from left to right. On using Theorem 7.8.4, this can equivalently be written as

$$\frac{\Gamma_\lambda \bar{T}_s(X, n)}{\Gamma_\lambda \bar{T}_{s-1}(X, n)} \text{ is decreasing in } n \in \mathbb{N},$$

for all $\lambda > 0$. On using Lemma 7.8.16, the above expression can be rewritten as

$$\frac{\bar{T}_s(N_\lambda^X, n)}{\bar{T}_{s-1}(N_\lambda^X, n)} \text{ is decreasing in } n \in \mathbb{N},$$

which, by (7.9.70), tells that N_λ^X is s-IFR. This proves the necessity part of the theorem. Conversely, given that

7.10 Residual Life at Random Time

$$\frac{\Gamma_\lambda \bar{T}_s(X, n)}{\Gamma_\lambda \bar{T}_{s-1}(X, n)} \text{ is decreasing in } n \in \mathbb{N},$$

which gives that, for any $k \in \mathbb{N}$,

$$\Gamma_\lambda \bar{T}_s(X, n)\Gamma_\lambda \bar{T}_{s-1}(X, n+k) \geq \Gamma_\lambda \bar{T}_s(X, n+k)\Gamma_\lambda \bar{T}_{s-1}(X, n). \quad (7.10.75)$$

Let $x\ (>0)$ and $x+y\ (>0)$ be two points of continuity both of $\bar{T}_{s-1}$ and $\bar{T}_s$. Write $\lambda = n/x$ and $k = [ny/x]$, where $[x]$ is the largest integer contained in x. Then $\lim_{n\to\infty} (n+k)/\lambda = x+y$. Thus, by using Lemma 7.8.13, we get that G_{n+k-1} converges weakly to a distribution G which is degenerate at $x+y$, where $G_n(t)$ is as given in Lemma 7.8.13. Hence, by Lemma 7.8.14, we have

$$\lim_{n\to\infty} \Gamma_\lambda \bar{T}_s(X, n+k) = \bar{T}_s(X, x+y).$$

By taking limit as $n \to \infty$, (7.10.75) gives

$$\bar{T}_s(X, x)\bar{T}_{s-1}(X, x+y) \geq \bar{T}_s(X, x+y)\bar{T}_{s-1}(X, x),$$

which means that

$$\frac{\bar{T}_s(X, x)}{\bar{T}_{s-1}(X, x)} \geq \frac{\bar{T}_s(X, x+y)}{\bar{T}_{s-1}(X, x+y)}.$$

Since the discontinuity points of a survival function are at most countable, the above expression gives that

$$\frac{\bar{T}_s(X, x)}{\bar{T}_{s-1}(X, x)} \text{ is decreasing in } x.$$

On using (7.9.66), the above expression gives that X is s-IFR. This proves (a). To prove (b), define, for $n \in \mathbb{N}$,

$$\bar{a}_\lambda^{(s)}(n) = \frac{1}{\lambda^{n+1}} \Gamma_\lambda \bar{T}_s(X, n+1).$$

Then we have

$$\bar{a}_\lambda^{(s)}(n) = \frac{1}{n!} \int_0^\infty x^n e^{-\lambda x} \bar{T}_s(X, x)\, dx$$

$$= \frac{1}{n!} \int_0^\infty x^n \bar{T}_s(Y, x)\, dx$$

$$= \frac{E\left(X_{(s)}^{n+1}\right)}{(n+1)!},$$

where $\bar{T}_s(Y, x) = e^{-\lambda x}\bar{T}_s(X, x)$ and the first equality follows from (7.8.52). Since, for every $s \in \mathbb{N}$, $\bar{T}_s(Y, \cdot)$ is a survival function, let us write the corresponding random variable as $X_{(s)}$. Clearly, X is s-$IFRA$ if and only if $X_{(s)}$ is $IFRA$. Thus, on using Lemma 7.10.3, we say that

$$\left(\bar{a}_\lambda^{(s)}(n)\right)^{1/(n+1)} \text{ is decreasing in } n.$$

Equation (7.8.52) immediately tells that

$$\left(\Gamma_\lambda \bar{T}_s(X, n)\right)^{1/n} \text{ is decreasing in } n.$$

Thus, by Lemma 7.8.16, we get that

$$\left(\bar{T}_s\left(N_\lambda^X, n\right)\right)^{1/n} \text{ is decreasing in } n.$$

So, N_λ^X is s-$IFRA$. This proves the necessity part of (b). To prove the sufficiency, it is given that, for all $n, k \in \mathbb{N}$ and for all $\lambda > 0$,

$$\left(\Gamma_\lambda \bar{T}_s(X, n)\right)^{1/n} \geqslant \left(\Gamma_\lambda \bar{T}_s(X, n+k)\right)^{1/(n+k)},$$

which can equivalently be written as

$$\Gamma_\lambda \bar{T}_s(X, n) \geqslant \left(\Gamma_\lambda \bar{T}_s(X, n+k)\right)^{n/(n+k)}.$$

If $x\ (>0)$ and $x + y\ (>0)$ are two continuity points of $\bar{T}_s(X, \cdot)$, then using a similar argument as in (a) above, we get

$$\bar{T}_s(X, x) \geqslant \left(\bar{T}_s(X, x+y)\right)^{x/(x+y)}.$$

This is true for all continuity points $x, x+y$ of $\bar{T}_s(X, \cdot)$. Since, for all $s \in \mathbb{N}$, $\bar{T}_s(X, \cdot)$ is a survival function and the discontinuity points of a survival function are countable, the above expression gives that

$$\left(\bar{T}_s(X, x)\right)^{1/x} \text{ is decreasing in } x.$$

This means that X is s-$IFRA$. This proves (b). To prove (c) note, from Definition 7.9.1, that, for all $x, t > 0$,

$$\bar{T}_s(X, x+t) \leqslant \bar{T}_s(X, x).\bar{T}_s(X, t). \tag{7.10.76}$$

7.10 Residual Life at Random Time

Note that, for all $s \in \mathbb{N}$, $\bar{T}_s(X, \cdot)$ is a survival function, which gives that $e^{-\lambda x}\bar{T}_s(X, x)$ represents a survival function in x. Let us denote this by $\bar{F}_{Y_s}(x)$. Thus, (7.10.76) can be rewritten as

$$\bar{F}_{Y_s}(x+t) \leqslant \bar{F}_{Y_s}(x)\bar{F}_{Y_s}(t).$$

On using Lemma 7.10.3(b), the above relation gives, for all $m, n \geqslant 0$, $\lambda_{m+n} \leqslant \lambda_m.\lambda_n$, which is equivalent to

$$\int_0^\infty \frac{u^{m+n-1}}{(m+n-1)!}e^{-\lambda x}\bar{T}_s(X, u)\,du \leqslant \left(\int_0^\infty \frac{u^{m-1}}{m!}e^{-\lambda x}\bar{T}_s(X, u)\,du\right) \cdot \left(\int_0^\infty \frac{u^{n-1}}{n!}e^{-\lambda x}\bar{T}_s(X, u)\,du\right).$$

Multiplying both sides of the above expression by λ^{m+n} we get, from (7.8.52),

$$\Gamma_\lambda \bar{T}_s(X, m+n) \leqslant \Gamma_\lambda \bar{T}_s(X, m).\Gamma_\lambda \bar{T}_s(X, n).$$

Now, the necessity part follows from Lemma 7.8.16. To prove the sufficiency, it is given that, for all $\lambda > 0$ and for all $m, n \geqslant 0$,

$$\Gamma_\lambda \bar{T}_s(X, m+n) \leqslant \Gamma_\lambda \bar{T}_s(X, m).\Gamma_\lambda \bar{T}_s(X, n).$$

Let x and y be two points of continuity of the function $\bar{T}_s(X, \cdot)$. Then using the same argument as in (a) above and taking limit as $n \to \infty$, we get that, for all $x, y > 0$,

$$\bar{T}_s(X, x+y) \leqslant \bar{T}_s(X, x).\bar{T}_s(X, y).$$

Thus, the theorem is established. □

Taking different values of s in the above theorem we get the following.

Corollary 7.10.4 *Let X be a continuous random variable. Then*

(a) *X is IFR if and only if N_λ^X is IFR;*
(b) *X is $DMRL$ if and only if N_λ^X is $DMRL$;*
(c) *X is $DVRL$ if and only if N_λ^X is $DVRL$;*
(d) *X is $IFRA$ if and only if N_λ^X is $IFRA$;*
(e) *X is NBU if and only if N_λ^X is NBU;*
(f) *X is DFR if and only if N_λ^X is DFR;*
(g) *X is $IMRL$ if and only if N_λ^X is $IMRL$;*
(g) *X is $IVRL$ if and only if N_λ^X is $IVRL$;*
(h) *X is $DFRA$ if and only if N_λ^X is $DFRA$;*
(i) *X is NWU if and only if N_λ^X is NWU.* □

The following theorem characterizes s-$NBUFR$ class of life distributions, which may be obtained in Hu et al. (2001).

Theorem 7.10.5 *For $s = 2, 3, \ldots$, a non-negative random variable X is s-$NBUFR$ (resp. s-$NWUFR$) if and only if N_λ^X is s-$NBUFR$ (resp. s-$NWUFR$), for every $\lambda > 0$.*

Proof Since $\bar{T}_s(X, 0) = 1$ for all $s \geq 1$, on using (7.8.15), we have that X is s-$NBUFR$ if and only if

$$\bar{T}_s(X, x) \leq \bar{T}_{s-1}(X, x) \text{ for all } x \geq 0.$$

Multiplying both sides of the above expression by $\lambda e^{-\lambda x}(\lambda x)^{n-1}/(n-1)!$ and integrating with respect to x we get, by (7.8.52),

$$\Gamma_\lambda \bar{T}_s(X, n) \leq \Gamma_\lambda \bar{T}_{s-1}(X, n),$$

for all $\lambda > 0$. With the help of Lemma 7.8.16, this gives

$$\bar{T}_s\left(N_\lambda^X, n\right) \leq \bar{T}_{s-1}\left(N_\lambda^X, n\right),$$

for all $n \in \mathbb{N}$. On using (7.9.64), we immediately get that N_λ^X is s-$NBUFR$.

To prove the sufficiency part, it is given that $\Gamma_\lambda \bar{T}_s(X, n) \leq \Gamma_\lambda \bar{T}_{s-1}(X, n)$, for all $\lambda > 0$. Let x be a continuity point both of $\bar{T}_s(X, \cdot)$ and $\bar{T}_{s-1}(X, \cdot)$. Then, taking limit as $n \to \infty$ and on using a similar argument as in the proof of Theorem 7.10.4, we get $\bar{T}_s(X, x) \leq \bar{T}_{s-1}(X, x)$. Since x is arbitrary and the set of continuity points of a survival function is countable, on using (7.8.15) we get that N_λ^X is s-$NBUFR$. The proof of the parenthetical statement is similar and is omitted. □

Taking $s = 2$ in the above theorem we get that a non-negative random variable X is $NBUFR$ (resp. $NWUFR$) if and only if N_λ^X is $NBUFR$ (resp. $NWUFR$), for every $\lambda > 0$.

Suppose, for any $\lambda > 0$, X_t is the residual random variable. The following theorem proves a result similar to the above result for s-$NBUCX$ order using X_t.

Theorem 7.10.6 *For $s \in \mathbb{N}$, X is s-$NBUCX$ (resp. s-$NWUCX$) if and only if*

$$\bar{T}_{s+1}\left(N_\lambda^{X_t}, n\right) \leq \bar{T}_s\left(N_\lambda^X, n\right),$$

for all $n \in \mathbb{N} \cup \{0\}$.

Proof Given that, for all $x, t \geq 0$,

$$\int_t^\infty \bar{T}_s(X, t+u)\,du \leq \bar{T}_s(X, x) \int_t^\infty \bar{T}_s(X, u)\,du$$

which, by using (7.8.14), reduces to

$$\bar{T}_{s+1}(X, x+t) \leq \bar{T}_s(X, x)\bar{T}_{s+1}(X, x+1).$$

This can be rewritten as

$$\frac{\bar{T}_{s+1}(X, x+t)}{\bar{T}_{s+1}(X, t)} \leq \bar{T}_s(X, x),$$

which, by using Lemma 7.8.9, can be expressed as

$$\bar{T}_{s+1}(X_t, x) \leqslant \bar{T}_s(X, x).$$

Now, multiplying both sides of this expression by $\lambda e^{-\lambda x}(\lambda x)^{n-1}/(n-1)!$ and integrating with respect to x in $(0, \infty)$, we get, with the help of (7.8.52),

$$\Gamma_\lambda \bar{T}_{s+1}(X_t, n) \leqslant \Gamma_\lambda \bar{T}_s(X, n),$$

for all $n \in \mathbb{N} \cup \{0\}$. On using Lemma 7.8.16, this reduces to

$$\bar{T}_{s+1}\left(N_\lambda^{X_t}, n\right) \leqslant \bar{T}_s\left(N_\lambda^X, n\right),$$

for all $n \in \mathbb{N} \cup \{0\}$. This proves the necessity part of the theorem. To prove the sufficiency, it is given that, for all $\lambda > 0$, and $n \in \mathbb{N} \cup \{0\}$,

$$\Gamma_\lambda \bar{T}_{s+1}(X_t, n) \leqslant \Gamma_\lambda \bar{T}_s(X, n).$$

If x is a continuity point both of $\bar{T}_{s+1}(X, \cdot)$ and $\bar{T}_{s-1}(X, \cdot)$. Then, taking limit as $n \to \infty$ and on using a similar argument as in the proof of Theorem 7.10.4, we get

$$\bar{T}_{s+1}(X_t, x) \leqslant \bar{T}_s(X, x)$$

which, by Lemma 7.8.9, on using (7.8.14), reduces to

$$\int_{x+t}^\infty \bar{T}_s(X, u)\, d \leqslant \int_t^\infty \bar{T}_s(X, u)\, du . \bar{T}_s(X, x).$$

This gives that X is s-$NBUCX$. This proves the sufficiency part. Now, the proof of the parenthetical statement follows on a similar line.

7.11 Some Results in Terms of Excess Lifetime

Let $\{X_n\}$, $n \in \mathbb{N}$, be a sequence of iid non-negative random variables with a common distribution function F such that $F(0) = 0$. Suppose shocks are coming one after another which may affect the working of a system. Define $S_n = \sum_{i=1}^n X_i$, the time of the nth arrival with $S_0 = 0$. Let $N(t) = \sup\{n : S_n \leqslant t\}$ represent the number of arrivals that occur till time t. Then $\{N(t) : t \geqslant 0\}$ is called a renewal process. Define $\gamma(t)$ to be the excess lifetime at time t, i.e., $\gamma(t) = S_{N(t)+1} - t$. The detailed proof may be obtained in Hu et al. (2004a).

Theorem 7.11.1 *Take $0 \leq t_1 < t_2$ and any $s \in \mathbb{N}$.*

(a) *If $\gamma(t_1) \geq_{s-FR}$ (resp. $\leq_{s-FR}$) $\gamma(t_2)$ then X is s-IFR (resp. s-DFR).*
(b) *If $\gamma(t_1) \geq_{s-ST}$ (resp. $\leq_{s-ST}$) $\gamma(t_2)$ then X is s-NBU (resp. s-NWU).*
(c) *If $\gamma(t_1) \geq_{s-CX}$ (resp. $\leq_{s-CX}$) $\gamma(t_2)$ then X is s-NBUCX (resp. s-NWUCX).*
(d) *If $\gamma(t_1) \geq_{s-CV}$ (resp. $\leq_{s-CV}$) $\gamma(t_2)$ then X is s-NBUCV (resp. s-NWUCV).* □

Taking different values of s, we get the following corollary.

Corollary 7.11.1 *For $0 \leq s < t$,*

(a) *If $\gamma(s) \geq_{FR}$ (resp. $\leq_{FR}$) $\gamma(t)$ then X is IFR (resp. DFR).*
(b) *If $\gamma(s) \geq_{mrl}$ (resp. $\leq_{mrl}$) $\gamma(t)$ then X is DMRL (resp. IMRL).*
(c) *If $\gamma(s) \geq_{vrl}$ (resp. $\leq_{vrl}$) $\gamma(t)$ then X is DVRL (resp. IVRL).*
(d) *If $\gamma(t_1) \geq_{ST}$ (resp. $\leq_{ST}$) $\gamma(t_2)$ then X is NBU (resp. NWU).*
(e) *If $\gamma(t_1) \geq_{icx}$ (resp. $\leq_{icx}$) $\gamma(t_2)$ then X is NBUC (resp. NWUC).*
(f) *If $\gamma(t_1) \geq_{icv}$ (resp. $\leq_{icv}$) $\gamma(t_2)$ then X is NBU(2) (resp. NWU(2)).*

Remark 7.11.1 Corollary 7.11.1(d) is given in Chen (1994), (e) is due to Li et al. (2000) whereas (f) may be obtained in Li and Kochar (2001).

7.11.1 Characterizations in Terms of Residual Life Functions

Below we characterize s-IFR class and its dual in terms of residual life of a used item through s-FR order. The results may be obtained in Hu et al. (2001).

Theorem 7.11.2 *For $s \in \mathbb{N}$, the following conditions are equivalent:*

(a) *X is s-IFR (resp. s-DFR).*
(b) *$X_t \geq_{s-FR}$ (resp. $\leq_{s-FR}$) $X_{t'}$ for all $t \leq t'$.*
(c) *$X \geq_{s-FR}$ (resp. $\leq_{s-FR}$) X_t for all $t \geq 0$.*
(d) *$X + t \leq_{s-FR}$ (resp. $\geq_{s-FR}$) $X + t'$ for all $t \leq t'$.*

Proof Suppose (a) holds. Then, on using Lemma 7.8.9, we have, for $0 \leq t \leq t'$,

$$\frac{\bar{T}_s(X_t, x)}{\bar{T}_s(X_{t'}, x)} = \frac{\bar{T}_s(X, x+t)}{\bar{T}_s(X, x+t')} \cdot \frac{\bar{T}_s(X, t')}{\bar{T}_s(X, t)}.$$

Let us write $t' - t = u$. Then the above expression can be rewritten as

$$\frac{\bar{T}_s(X_t, x)}{\bar{T}_s(X_{t'}, x)} = \frac{\bar{T}_s(X, x+t)}{\bar{T}_s(X, x+t+u)} \cdot \frac{\bar{T}_s(X, t')}{\bar{T}_s(X, t)},$$

7.11 Some Results in Terms of Excess Lifetime 241

which is increasing (resp. decreasing) in x if and only if

$$\frac{\bar{T}_s(X, x)}{\bar{T}_s(X, x+u)} \text{ is increasing (resp. decreasing) in } x, \qquad (7.11.77)$$

for all $u \geqslant 0$. Since (a) is assumed to be true, (7.11.77) holds by Theorem 7.9.1. This proves (b). Now, taking $t = 0$, we see that (b) implies (c). Again, suppose (c) holds. Then

$$\frac{\bar{T}_s(X_t, x)}{\bar{T}_s(X, x)} \text{ is decreasing (resp. increasing) in } x,$$

which, on using Lemma 7.8.9, can be rewritten as

$$\frac{\bar{T}_s(X, x+t)}{\bar{T}_s(X, x)} \text{ is decreasing (resp. increasing) in } x. \qquad (7.11.78)$$

Note that

$$\frac{\bar{T}_s(X+t, y)}{\bar{T}_s(X+t', y)} = \frac{\bar{T}_s(X, y-t)}{\bar{T}_s(X, y-t')} = \frac{\bar{T}_s(X, x+t^*)}{\bar{T}_s(X, x)}, \qquad (7.11.79)$$

where $x = y - t'$ and $t^* = t' - t$. Hence (d) follows from (7.11.78). Suppose (d) holds. Then (a) immediately follows from (7.11.79). □

Taking $s = 1, 2, 3$ we get the following corollary.

Corollary 7.11.2 *1. The following statements are equivalent:*

(a) X is IFR (resp. DFR).
(b) For all $t \leqslant t'$, $X_t \geqslant_{hr}$ (resp. $\leqslant_{hr}$) $X_{t'}$.
(c) For all $t \geqslant 0$, $X \geqslant_{hr}$ (resp. $\leqslant_{hr}$) X_t.
(d) For all $t \leqslant t'$, $X + t \leqslant_{hr}$ (resp. $\geqslant_{hr}$) $X + t'$.

2. The following statements are equivalent:

(a) X is $DMRL$ (resp. $IMRL$).
(b) For all $t \leqslant t'$, $X_t \geqslant_{mrl}$ (resp. $\leqslant_{mrl}$) $X_{t'}$.
(c) For all $t \geqslant 0$, $X \geqslant_{mrl}$ (resp. $\leqslant_{mrl}$) X_t.
(d) For all $t \leqslant t'$, $X + t \leqslant_{mrl}$ (resp. $\geqslant_{mrl}$) $X + t'$.

3. The following statements are equivalent:

(a) X is $DVRL$ (resp. $IVRL$).
(b) For all $t \leqslant t'$, $X_t \geqslant_{vrl}$ (resp. $\leqslant_{vrl}$) $X_{t'}$.
(c) For all $t \geqslant 0$, $X \geqslant_{vrl}$ (resp. $\leqslant_{vrl}$) X_t.
(d) For all $t \leqslant t'$, $X + t \leqslant_{vrl}$ (resp. $\geqslant_{vrl}$) $X + t'$.

Remark 7.11.2 The first two results of the above corollary may be obtained in Shaked and Shanthikumar (2007). □

In the following theorem, we characterize the condition that, between two used items, the remaining life of more used items is probabilistically less than that of a less used item which supports our usual understanding.

Theorem 7.11.3 *Let $s \in \mathbb{N} \cup \{0\}$. Then, for all $t \leqslant t'$,*

$$X_t \geqslant_{s-ST} (resp. \leqslant_{s-ST}) X_{t'}$$

if and only if X is s-IFR (resp. s-DFR).

Proof Suppose $X_t \geqslant_{s-ST} X_{t'}$, for all $t \leqslant t'$. Then, on using Lemma 7.8.9, this can equivalently be written as

$$\frac{\bar{T}_s(X; x+t)}{\bar{T}_s(X; t)} \text{ is decreasing in } t.$$

This gives that X is s-IFR. This proves one part of the theorem. The proof for the parenthetical statement is similar. □

Taking different values of s we get the following corollary:

Corollary 7.11.3 *(a) $X_t \geqslant_{wlr} (resp. \leqslant_{wlr}) X_{t'}$ for all $t \leqslant t'$ if and only if X is ILR (resp. DLR).*
(b) $X_t \geqslant_{st} (resp. \leqslant_{st}) X_{t'}$ for all $t \leqslant t'$ if and only if X is IFR (resp. DFR).
(c) $X_t \geqslant_{hmrl} (resp. \leqslant_{hmrl}) X_{t'}$ for all $t \leqslant t'$ if and only if X is DMRL (resp. IMRL). □

Similar to the above theorem, below we give a characterization of s-NBU class and its dual. Here we compare one used item with another new item in contrary to what is done in the above theorem where two used items are compared.

Theorem 7.11.4 *X is s-NBU (resp. s-NWU) if and only if $X \geqslant_{s-ST}$ (resp. $\leqslant_{s-ST}) X_t$, for all $t \geqslant 0$.*

Proof Suppose $X \geqslant_{s-ST} X_t$, for all $t \geqslant 0$. This means that, for all $t \geqslant 0$,

$$\frac{\bar{T}_s(X; x)}{\bar{T}_s(X_t; x)} \geqslant \frac{\bar{T}_s(X; 0)}{\bar{T}_s(X_t; 0)},$$

which, on using Lemma 7.8.9, can equivalently be written as

$$\frac{\bar{T}_s(X; x)}{\bar{T}_s(X; x+t)} \geqslant \frac{\bar{T}_s(X; 0)}{\bar{T}_s(X; t)},$$

for all $t \geqslant 0$. This means that X is s-NBU. The proof for the parenthetical statement is similar. □

Taking $s = 1$ in the above theorem, we immediately get that X is NBU (resp. NWU) if and only if $X \geqslant_{st}$ (resp. $\leqslant_{st}$) X_t, for all $t \geqslant 0$. Some more ageing classes are discussed in Sect. 8.5.

7.12 A Unified Study of Some Stochastic Orders

So far we have discussed generalized stochastic orders, called s-FR order, for $s \in \mathbb{N} \cup \{0\}$. In this section, we discuss the generalized stochastic order where s is not restricted to be an integer rather can take any positive real value. The detailed discussion can be obtained in Hu et al. (2004b). Note here that the random variables are not necessarily restricted to be non-negative.

Let X be a random variable having cdf F_X. For $s > 0$, write

$$\gamma^{(s)}(x) = \begin{cases} \frac{(-x)^{s-1}}{\Gamma(s)}, & x < 0 \\ 0, & x \geq 0 \end{cases}$$

and

$$\bar{\Phi}_s^X(x) = E[\gamma^{(s)}(x - X)] = \int_{-\infty}^{\infty} \gamma^{(s)}(x - x_0) dF_X(x_0),$$

for all $x \in (-\infty, \infty)$ with $\bar{\Phi}_0^X(x) = f_X(x)$ if the pdf f_X of X exists. Here s may be any non-negative real number. For $s \in \mathbb{N}$, $\bar{\Phi}_s^X(x)$ becomes

$$\bar{\Phi}_s^X(x) = \int_x^{\infty} \int_{x_{s-1}}^{\infty} \cdots \int_{x_1}^{\infty} dF_X(x_0) dx_1 \cdots dx_{s-1},$$

for all $x \in (-\infty, \infty)$.

For $s > 0$, we call $\bar{\Phi}_s^X(x)$ the generalized iterated integral of the survival function of X. For a continuous random variable X (resp. Y), let l_X (resp. l_Y) be the left endpoint of the support of X (resp. Y), and u_X (resp. u_Y) be the right endpoint of the support of X (resp. Y), which could be even infinity.

Definition 7.12.1 Let X and Y be two continuous random variables as above, and let $s > 0$. Then X is said to be smaller than Y in the (unrestricted) s-FR order, denoted as $X \leqslant_{s-fr} Y$, if

$$\frac{\bar{\Phi}_s^Y(x)}{\bar{\Phi}_s^X(x)} \text{ is increasing in } x \in (-\infty, \max(u_X, u_Y)).$$

In the above definition, we have used the convention $a/0 = \infty$, when $a > 0$, for the case $u_X < u_Y$. It can be easily seen that $X \leqslant_{s-fr} Y \Rightarrow u_X \leqslant u_Y$. This may be compared with discussion in Sect. 7.3. However, $X \leqslant_{s-fr} Y \not\Rightarrow l_X \leqslant l_Y$, in general, as can be seen in the following counterexample.

Counterexample 7.12.1 *Let* $X \sim U(1, 2)$ *and* $Y \sim U(0, 3)$. *Then*

$$\frac{\bar{\Phi}_2^X(t)}{\bar{\Phi}_2^Y(t)} = \begin{cases} \frac{3(3-2t)}{(3-t)^2}, & t \in [0, 1) \\ \frac{3(2-t)^2}{(3-t)^2}, & t \in [1, 2) \\ 0, & t \in [2, 3) \end{cases}$$

which is decreasing in $t \in [0, 3)$ *implying that* $X \leqslant_{mrl} Y$, *although* $l_X > l_Y$. □

The following theorem generalizes Theorem 7.8.5. The proof may be obtained in Hu et al. (2004b).

Theorem 7.12.1 *Let* $s > 0$. *Then* $X \leqslant_{s-fr} Y \Rightarrow X \leqslant_{t-fr} Y$, *for any* $t \geqslant s + 1$. □

From the above theorem, one may wonder whether the above theorem holds for $t \in (s, s + 1)$. This is answered in negative via the following counterexample.

Counterexample 7.12.2 *Let* $X \sim U(0, \theta)$ *and* $Y \sim (1, 1 + \theta)$, $\theta \in (0, 1)$. *Note that, for* $t \in (0, 1)$,

$$\lim_{x \to -\infty} \frac{\bar{\Phi}_t^Y(x)}{\bar{\Phi}_t^X(x)} = 1 \quad \text{and} \quad \frac{\bar{\Phi}_t^Y(0)}{\bar{\Phi}_t^X(0)} = \frac{(1+\theta)^t - 1}{\theta^t} < 1,$$

implying that $X \not\leqslant_{t-fr} Y$ *for* $t \in (0, 1)$, *although* $X \leqslant_{lr} Y$. □

Now we define s-IFR order for any real $s > 00$.

Definition 7.12.2 *Let* $s > 0$. *A continuous random variable* X *is said to belong to the* s-ifr *class if* $X - x \leqslant_{s-fr} X$, *for all* $x \geqslant 0$. □

The following theorem due to Hu et al. (2004b) generalizes quite a few theorems available in the literature.

Theorem 7.12.2 *Let us consider two random variables* X *and* Y. *Suppose* $i, j > 0$ *be such that* $i + j = s$, *and let* $Z \in j$-ifr *class, independent of* X *and* Y.

(i) *For* $j \geq 1$, *if* $X \leqslant_{(i+1)-fr} Y$ *then* $X + Z \leqslant_{s-fr} Y + Z$.
(ii) *If* $X \leqslant_{i-fr} Y$ *then* $X + Z \leqslant_{s-fr} Y + Z$.

Corollary 7.12.1 *Let* X *and* Y *be two independent random variables as above.*

(a) *If, for* $i > 0, j \geqslant 1$, $X \in (i + 1)$-ifr *and* $Y \in j$-ifr *then* $X + Y \in (i + j)$-ifr.
(b) *If, for* $i > 0, j > 0$, $X \in i$-ifr *and* $Y \in j$-ifr *then* $X + Y \in (i + j)$-ifr.

Chapter 8
Classes of Life Distributions

8.1 Relevance of Classification

Classification of life distributions based on important properties like failure rate and its averages, residual life and its summary measures, properties of the distribution or density function, coordinate sub-tangent of the survival curve, etc. is required for a non-parametric or semi-parametric approach in several problems in reliability and survival analysis. Thus, to find bounds on reliability and to work out optimal maintenance policies, a realistic assumption likely to hold good in many situations would be that the underlying failure time distribution has an increasing failure rate.

The ageing classes are generated mainly based on three concepts:

(i) Based on failure rate.
(ii) Based on conditional survival probability.
(iii) Based on conditional mean remaining life.

$IFR, IFRA, NBUC, NBUFR, NBUAFR$[1] classes and their duals are defined in terms of failure rate of a distribution, NBU and its dual are defined based on conditional survival probabilities whereas $DMRL, DVRL, NBUE, HNBUE$ classes and their duals are defined based on the conditional mean remaining lives. Among the classes mentioned above, some are based on monotonicity of reliability functions. To be specific, IFR and $IFRA$, and their duals are based on monotonicity of some functions of failure rate. These are called monotone ageing classes. However, there are distributions which have non-monotone ageing (see Fig. 8.1). A diagram similar to Fig. 8.1 may be obtained in Sengupta (2013).

Following the works of Barlow et al. (1963) and Bryson and Siddiqui (1969), various non-parametric classes of life distributions have been introduced in reliability literature. In recent years, attempts have been made to find out larger and larger

---

[1] A distribution with failure rate $r(t)$ is said to be new better than used in failure rate average (see Loh, 1984) if $tr(0) \leqslant \int_0^t r(x)\,dx$ for all $t > 0$. This is also sometimes abbreviated as $NBAFR$ or $NBUFRA$.

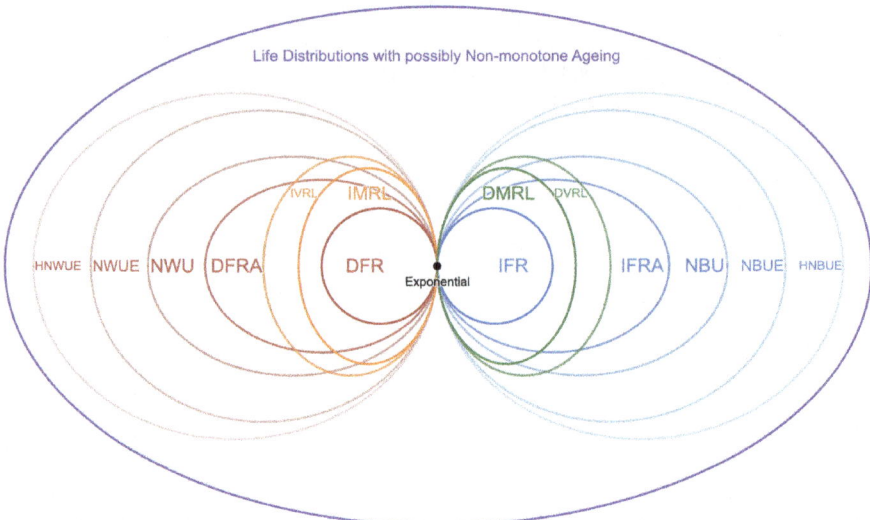

Fig. 8.1 Representation of different ageing classes

classes, and their inter-relations have been studied. Of course, only a few of these are truly useful in practice, others are mathematically interesting and motivate further studies.

An important consideration in developing a class of life distributions with some common properties is to examine if the class is closed under certain operations like formation of coherent systems, or of convolutions or in respect of mixtures. Even closure in respect of convergence to limiting forms is also desirable. In fact, one of the seminal contributions in the area of classification was the work of Block and Savits (1976) on closure of the $IFRA$ class with respect to convolution. Life distribution classes lacking closure properties at least in regard to some operations are not that useful.

It may be incidentally pointed out that a complete specification of some reliability characteristic of a life distribution may lead to its characterization, while an incomplete or partial knowledge may lead to the assignment of the life distribution to a particular class. In this sense, classification may be looked upon as incomplete characterization. For example, considering failure rate as a characteristic property of a life distribution, we note that a complete knowledge of the functional form of the FR function uniquely determines the distribution and thus provides a characterization of the distribution. On the other hand, a simple knowledge about increasing or decreasing behaviour of the FR function leads to a classification as IFR or DFR. Similar is the case with the MRL function or the mean inactivity time function.

Dealing with any life distribution class, one is interested to develop some lower (upper) bound for reliability, which could be desirably sharp. Bounds may take into account the mission time, at least as being greater (smaller) than the mean time-to-failure. Similarly, relations among various classes are also of interest to both researchers and practitioners.

While most of the classes are based on ageing properties, a class of distributions dominated by the exponential distribution in Laplace transform order has also been studied. Similarly, log-concave class of distribution functions has also some interesting features interpretable in terms of ageing properties.

In the present chapter, we review some of the important classes of life distributions—both continuous and discrete, explore their inter-relationships, and examine their preservation under certain reliability operations. Section 8.2 deals with classes based on failure rate and reversed hazard rate, while the next section considers classification based on residual life and inactivity time distributions. Section 8.4 is a brief exposition of classes based on coordinate sub-tangent. Classes based on some dominance relations involving the exponential distribution, viz., the $\mathscr{L}$-class, the $\mathscr{M}$-class, and the $\mathscr{LM}$-class have been discussed in Sect. 8.5. The next section deals with preservation of class properties with respect to common reliability operations like formation of coherent systems, convolutions, and mixtures. The extant literature on classification of life distributions is quite exhaustive and several interesting findings in this regard might have escaped our attention. Further, newer—and in most cases wider—classes are being proposed almost continuously and any account of life distribution classes is likely to miss out some contemporary developments, in terms of classes as also their properties.

8.2 Classes Based on Ageing Properties

8.2.1 Use of Failure Rate

Quite a few life distributions in common use have monotonic failure rates and we can conveniently put them in either the IFR class or the DFR class, with the exponential distribution showing a constant failure rate as a borderline between the two. Although the IFR class is closed in respect of convolution (standby system formation), the DFR class is not. It is true, however, that the IFR class is not closed in respect of coherent system formation with independent components, as mentioned in Sect. 2.4.1. It is noted that the series system with independent IFR components will have IFR distributions whereas this is not true for parallel system (cf. Esary and Proschan, 1963). And quite a few life distributions are found to have non-monotonic failure rates, e.g., when the failure rate curve is unimodal or bathtub shaped or inverted bathtub shaped.

This led to a search for wider classes based on averages of the failure rate over an interval, usually $(0, t)$ assuming that such an average behaves in the same manner for all t. We can think of the arithmetic, the geometric, and the harmonic means of the failure rate and depending on the behaviour of these means define the following classes of life distributions.

As defined in Sect. 2.4.1, a distribution F with corresponding failure rate $r(x)$ is said to belong to the $IFRA$ ($DFRA$) class if

$$A(t) = \frac{1}{t} \int_0^t r(x)dx = -\frac{\ln \bar{F}(t)}{t}$$

is increasing (decreasing) in $t > 0$. A distribution F is alternatively defined as $IFRA$ ($DFRA$) if $\bar{F}(\alpha x) \geq (\leq) \bar{F}^\alpha(x)$ for all $0 < \alpha < 1$ and for all $x \geq 0$. It should be noted that the IFR class is a sub-class of the $IFRA$ class and that the $IFRA$ class is the smallest class which is closed in respect of coherent system formation, convolution, and taking limits of sequences of distributions. In fact, the $IFRA$ class has been accepted as a sufficiently large class of life distributions.

A distribution F with corresponding failure rate $r(x)$ is said to belong to the increasing geometric mean failure rate ($IGMFR$) class if the geometric failure rate over $(0, x)$,

$$G(x) = e^{\frac{1}{x} \int_0^x \ln r(t) dt},$$

is increasing in x.

A distribution F with corresponding failure rate $r(x)$ is said to belong to the increasing harmonic mean failure rate ($IHMFR$) class if the harmonic mean of failure rate over $(0, x)$,

$$H(x) = \left(\frac{1}{x} \int_0^x \frac{dt}{r(t)} \right)^{-1},$$

is increasing in x. The $DGMFR$ and $DHMFR$ classes can be defined analogously. The following result due to Roy (1984) is interesting in this context.

> IFR class of distributions is a subset of $IGMFR$ and $IHMFR$ classes of life distributions and DFR class is a subset of $DGMFR$ and $DHFR$ classes.

It is easily seen that for $GMFR$ class the distribution function G gives

$$G'(x) = \frac{G(x)}{x} [\ln r(x) - \ln G(x)]$$

which is increasing (decreasing) as $r(x) > (<) G(x)$. Similarly,

$$H'(x) = \frac{H(x)}{x} \left[1 - \frac{H(x)}{r(x)} \right]$$

and hence $H(x)$ is increasing (decreasing) as $g(x) > (<) H(x)$. A similar result connects $DFRA$, $DGMFR$, and $DHMFR$ classes.

8.2 Classes Based on Ageing Properties

It can be easily shown that IFR property implies $IFRA$ property and, similarly, DFR property implies $DFRA$ property. However, for a distribution with non-monotone failure rate, different segments of the distribution may reveal deviations. For example, the Inverse Rayleigh distribution has three parts with $(IFR, IFRA)$, $(DFR, IFRA)$, and $(DFR, DFRA)$ properties (as discussed in Sect. 3.10.1).

8.2.2 Classification of Discrete Life Distributions

Classification of discrete life distributions has been attempted by several workers including Nanda and Sengupta (2005). Defining the reversed hazard rate for a discrete distribution with distribution function F over the set $\{0, 1, 2, \ldots\}$ as its support in terms of the ratio $[F(k) - F(k-1)]/F(k)$, many common distributions like Binomial, Poisson, Geometric, Hyper-geometric, Negative Binomial, Hyper-Poisson, Yule and Zeta distributions belong to the discrete Decreasing Reversed Hazard Rate (d-DRH) class. Nanda and Sengupta (2005) proved that the following statements are identical for any discrete random variable X.

(a) F is d-DRH;
(b) $F(n+k)/F(k)$ is decreasing in k for all n in the support in $\{0, 1, 2, \ldots\}$;
(c) $G_k(n) = P(k - X \leqslant n | X \leqslant k)$ is stochastically increasing in k.

It may be noted that G_k is the distribution function of the time since failure, at time k.

They also prove that a discrete distribution is d-DRH if and only if it is the discretized version of a continuous log-concave distribution. Here discretization is considered by taking integral part of the continuous random variable.

To note the closure properties of discrete distribution classes, it has been shown that if X and Y are independent discrete-IFR random variables, then $X + Y$ is also discrete IFR. For the somewhat lengthy proof, one may refer to Nanda and Sengupta (2005). The following results proved by the same authors are also quite important.

Theorem 8.2.1 (a) *If a sequence of d-DRH distributions converges to a limiting distribution, the limiting distribution is also d-DRH.*
(b) *If the components of a parallel system have independent d-DRH life distributions, then the system life distribution is also d-DRH.*
(c) *If the components of a k-out-of-n system have independent and identical d-DRH distributions, then the system life distribution is also d-DRH.*
(d) *Convolution of two d-DRH distributions produces a d-DRH distribution.* □

In this connection, it is noted that the life distribution of a series system with independent d-DRH component life distributions need not be discrete-DRH. An example given in Nanda and Sengupta (2005) considers two d-DRH distributions given by

$$F(k) = \left(\frac{1}{2}\right)^{4-k} \quad \text{and} \quad G(k) = \left(\frac{1}{4}\right)^{4-k}, \quad k = 0, 1, 2, 3, 4,$$

both of which are d-DRH. However, $1 - (1 - F)(1 - G)$ is not d-DRH. The authors also provide an example to point out that a mixture of two d-DRH distributions need not be d-DRH. If F is d-DRH with mean μ then they proved the following which is related to an upper bound for reliability

$$P(X > k) = \begin{cases} 1, & \text{if } k < [\mu] \\ \max_{i=k,k+1,\ldots} \left(1 - r_i^{i-k}\right), & \text{if } k \geqslant [\mu] \end{cases},$$

where $[\mu]$ is the integer part of μ and r_i is the unique solution to the equation

$$\frac{1 - r_i^{i+1}}{1 - r_i} = i + 1 - \mu.$$

It may be pointed out that unlike in the continuous case where the bound for $k \geqslant [\mu]$ is a function of time divided by mean life, this bound in the discrete set-up cannot be computed only from the knowledge of k/μ or $(k+1)/\mu$. Further, since the DFR class is a subset of the d-DRH class and the lower bound to reliability for d-DRH class is zero, the same for DFR class with the same mean life is also zero.

8.3 Classes Based on Residual Life and Inactivity Time

Depending on the 'work-hardening' or the 'wearing-out' phenomenon, reflected in decreasing and increasing rates of failure, respectively, we expect the remaining life of a unit of age t to be increasing or decreasing. This will also be linked to the property of a new (unused) unit to be considered as better than a used one in a probabilistic sense.

Let $X_t = (X - t | X \geqslant t)$ denote residual life at age t with the distribution function $F_{X_t}(x) = P(X_t \leqslant x)$ and the mean $\mu_X(t) = E(X_t)$, assumed to exist. Similarly, $_t X = (t - X | X \leqslant t)$ is used to denote the inactivity time (time since failure) at time t. Mean and variance of $_t X$ and some related results have been considered in Chaps. 5 and 7.

Wang (1996), and independently Belzunce et al. (1995), defined a new ageing class of life distributions based on the Laplace transform order, called the New Better than Used in Laplace Ordering ($NBUL$). A non-negative random variable X is said to be $NBUL$ if and only if, for all $t \geqslant 0$,

$$\int_0^\infty e^{-sx} \bar{F}(x+t)\,dx \leqslant \int_0^\infty e^{-sx} \bar{F}(x)\,dx$$

with $s > 0$.

Deshpande et al. (1986) proposed the $NBU(2)$ class (New Better than Used of second order) in which $X > X_t$ in the increasing concave ordering (denoted as

8.3 Classes Based on Residual Life and Inactivity Time

$X \geqslant_{icv} X_t$) for all $t \geqslant 0$. Cao and Wang (1991) proposed another class, viz., $NBUC$ (New Better than Used in Convex ordering) in terms of the relation $X \geqslant_{icx} X_t$ for all $t \geqslant 0$. Thus, F is $NBU(2)$ if, for all $t, y \geqslant 0$,

$$\int_0^y \bar{F}(x+t)\,dx \leqslant \bar{F}(t) \int_0^y \bar{F}(x)\,dx.$$

The dual concept of a New Worse than Used of second-order ($NWU(2)$) life distribution is defined by the reversed inequality.

Direct integration of both sides of the above inequality leads to F exhibiting the $NBUE$ property. Hence we have

$$NBU \Rightarrow NBU(2)\ (NBUC) \Rightarrow NBUE.$$

It is known that F is new better (worse) than used (NBU or NWU) if

$$\bar{F}(x+y) \leqslant (\geqslant) \bar{F}(x)\bar{F}(y),\ \text{for all}\ x, y \geqslant 0,$$

which gives that $\bar{F}(x) - \bar{F}(x+y) \geqslant (\leqslant) \bar{F}(x)(1 - \bar{F}(y))$. Thus

$$\frac{1}{\bar{F}(x)} \lim_{y \to 0} \frac{\bar{F}(x) - \bar{F}(x+y)}{y} \geqslant (\leqslant) \lim_{y \to 0} \frac{F(y)}{y}.$$

This means

$$\frac{f(x)}{\bar{F}(x)} \geqslant (\leqslant) f(0),\ x \geqslant 0.$$

If $f(0) > 0$, we get that F is New Better (Worse) than Used in Failure Rate ($NBUFR$ ($NWUFR$)) if $r(x) \geqslant (\leqslant) r(0)$ for all $x \geqslant 0$ (see also Sect. 7.9). Integrating both sides of this inequality, we have

$$-\frac{\ln \bar{F}(x)}{x} \geqslant (\leqslant) r(0),\ x > 0. \tag{8.3.1}$$

This leads to the definition of F being New Better (Worse) than Average Failure Rate ($NBAFR$ ($NWAFR$)) (also called New Better (Worse) than Used in Failure Rate Average and is written as $NBUFRA$ ($NWAFR$)) (as defined in Sect. 8.1), which can equivalently be written as

$$\bar{F}(x) \leqslant (\geqslant) e^{-f(0)x},\ x > 0.$$

These relations were proposed by Loh (1984). We have the following chain of implications

$$NBU \Rightarrow NBUFR \Rightarrow NBAFR$$

and the dual class
$$NWU \Rightarrow NWUFR \Rightarrow NWAFR.$$

As usual, exponential is the borderline distribution.

Noting that the ageing behaviour of a life distribution may change after some age t_0, Zahui and Xiaohu (1998) defined the NBU-t_0 and $IFRA$-t_0 classes, which were independently proposed and studied by Park and Boring (2023). A distribution F (or equivalently the corresponding random variable X) is said to belong to the NBU-t_0 (new better than used at t_0) class if

$$\bar{F}(x+t_0) \leq \bar{F}(x)\bar{F}(t_0),$$

for all $x \geq 0$ and some $t_0 \geq 0$. If this inequality is reversed we say that F is new worse than used at t_0 (NWU-t_0).

The NBU-t_0 class contains and is larger than the NBU class. While the only life distribution that belongs to both IFR and DFR, $IFRA$ and $DFRA$, NBU and NWU, $DMRL$ and $IMRL$, and $NBUE$ and $NWUE$ classes is the exponential, the boundary between NBU-t_0 and NWU-t_0 includes the following distributions (cf. Marsaglia and Tubilla (1975) and Park (2003)).

- $\bar{F}_1(x) = e^{-\lambda x}$, $x \geq 0$.
- $\bar{F}_2(x) = \bar{G}(x)$, $x \geq 0$, where $\bar{G}$ is a survival function with $\bar{G}(0) = 1$ and $\bar{G}(t_0) = 0$.
- $\bar{F}_3(x) = \begin{cases} \bar{G}(x), & \text{if } 0 \leq x < t_0 \\ \bar{G}(t_0)\bar{G}(x-t_0), & \text{if } t_0 \leq x < 2t_0 \\ \bar{G}^2(t_0)\bar{G}(x-2t_0), & \text{if } 2t_0 \leq x < 3t_0 \\ \vdots \\ \bar{G}^n(t_0)\bar{G}(x-nt_0), & \text{if } nt_0 \leq x < (n+1)t_0 \\ \vdots \end{cases}$ where $\bar{G}(x)$, $x \geq 0$, is a survival function.

It is to be noted that if $\bar{G}$ has a density over $[0, t_0]$, then $\bar{F}_3$ is periodic with period t_0.

Ahmad et al. (2005) introduced a class called new better than used in Total Time on Test Transform ($NBUT$). A random variable X having cdf F is said to belong to the $NBUT$ class if

$$\int_0^{F^{-1}(p)} \bar{F}(x+t)\,dx \leq \bar{F}(t) \int_0^{F^{-1}(p)} \bar{F}(x)\,dx,$$

for all $p \in (0, 1)$. They have shown that the $NBUT$ class is closed under formation of series system of iid components.

Incidentally, and more interestingly, a characterization of the more common classes of life distributions, in terms of scaled Total Time on Test Transform, $\phi(p)$, has been indicated as follows, where

$$\phi(p) = \frac{\int_0^{F^{-1}(p)} \bar{F}(u)\,du}{\int_0^\infty \bar{F}(u)\,du},$$

for $p \in [0, 1]$ (see Sect. 5.6).

(a) F is IFR (DFR) if and only if $\phi(p)$ is concave (convex) in $p \in [0, 1]$.
(b) F is $IFRA$ ($DFRA$) if and only if $\phi(p)/p$ is decreasing (increasing) in $p \in (0, 1)$.
(c) F is $DMRL$ ($IMRL$) if and only if $(1 - \phi(p))/(1 - p)$ is decreasing (increasing) in $p \in [0, 1)$.
(d) F is $NBUE$ ($NWUE$) if and only if $\phi(p) \geq (\leq) p$ for $p \in [0, 1]$.
(e) F is $HNBUE$ ($HNWUE$) if and only if $\phi(p) \geq (\leq) 1 - e^{-F^{-1}(p)/\mu}$ for $p \in [0, 1]$.
(f) F is BT (UBT, upside-down bathtub failure rate) if ϕ has only one reflection point u_0 such that $0 < u_0 < 1$ and it is convex (concave) on $[0, u_0]$ and concave (convex) on $[u_0, 1]$.

Classification based on properties of residual (remaining) life, particularly the mean and the variance of the residual life, has led to interesting bounds (lower and/or upper) for reliability. The more important classes are the $DMRL$ and its dual $IMRL$ classes with the exponential distribution having a constant mean residual life defining the border. Similar is the classification in terms of the variance of residual life, viz., $IVRL$ and $DVRL$. Stoyanov and Al-Sadi (2004) showed that the $DVRL$ class is not closed in respect of convolution or mixing and the $IVRL$ class is not closed under formation of coherent systems.

8.4 Use of Coordinate Sub-tangent

The coordinate sub-tangent (CST) to a curve $y = f(x)$ at the point (x_0, y_0) is given by $(x^* - x_0)$ where $(x^*, 0)$ is the point at which the tangent to the curve at point (x_0, y_0) meets the x-axis. Thus, we have

$$x^* = x_0 - \frac{f(x_0)}{f'(x_0)}.$$

Writing CST as a function of $T(\cdot)$ we have $T(x) = -f(x)/f'(x)$. The reciprocal of $T(x)$ is called 'reciprocal coordinate sub-tangent'. Thus, the CST to the survival function $\bar{F}(x)$ is just $1/r(x)$. It is just evident that $T(x)$ for the exponential distribution with failure rate λ is the constant $1/\lambda$. For the Weibull distribution, having survival function $e^{-\lambda x^\alpha}$, we have

$$T(x) = \frac{x}{\lambda \alpha x^\alpha - (\alpha - 1)}$$

and for the Pearsonian Type XI distribution (discussed in Sect. 9.6)

$$T(x) = \frac{\alpha}{q-1}\left(1 + \frac{x}{\alpha}\right).$$

A distribution is said to be Increasing CST ($ICST$) if the CST function $T(\cdot)$ is increasing. Similarly the class of decreasing CST ($DCST$) of distributions can be defined. The following is an interesting result.

Theorem 8.4.1 *If X is IFR (DFR) with decreasing (increasing) rate of failure rate, then X is $DCST$ ($ICST$).* □

The result follows from the fact that

$$r'(x) = \frac{f'(x)}{\bar{F}(x)} + \frac{f^2(x)}{\bar{F}^2(x)}$$

implying

$$\frac{1}{T(x)} = r(x) - \frac{r'(x)}{r(x)}.$$

Hence, if $r(x)$ is increasing and $r'(x)$ is decreasing, then $T(x)$ must be decreasing. Similar is the simple proof for the dual property.

It is quite easy to prove these results. Mukherjee and Roy (1989) also showed that a non-negative random variable X has a density which is convex ordered with respect to the exponential density if and only if X has a $DCST$ distribution. This result provides alternative definitions of the $ICST$ and $DCST$ classes. And these provide some linkages between increasing harmonic mean CST (decreasing harmonic mean CST), written as $IHCST$ ($DHCST$), given in the following theorem due to Roy (1988). Before that we need to define harmonic mean CST ($HCST$).

Definition 8.4.1 The harmonic mean CST of a curve $y = f(x)$ is defined as

$$HCST(x) = \left(\frac{1}{x}\int_0^x \frac{du}{T(u)}\right)^{-1} = \frac{x}{\int_0^x \frac{du}{T(u)}}.$$

Theorem 8.4.2 *X is $DHCST$ ($IHCST$) if and only if $(f(x)/f(0))^{1/x}$ is decreasing (increasing) in x.* □

It can be easily shown that $DCST$ is a sub-class of $DHCST$ and $ICST$ is a sub-class of $IHCST$. Further, the exponential distribution is the only continuous distribution which is both $DHCST$ and $IHCST$. The CST function links the parent distribution class to the class of the equilibrium distributions. This is indicated in the following theorem.

Theorem 8.4.3 *A random variable $X \geq 0$ with density $f(x)$ is $DHCST$ ($IHCST$) implies*

8.4 Use of Coordinate Sub-tangent

$$f(x+y)f(0) \leq (\geq) f(x)f(y)$$

for all $x, y \geq 0$.

Proof Note that

$$HCST(x) = x \left[\int_0^x \frac{du}{T(u)} \right]^{-1}$$

$$= x \left[-\ln\left(\frac{f(x)}{f(0)}\right) \right]^{-1}$$

$$= -\left[\ln\left(\frac{f(x)}{f(0)}\right)^{1/x} \right]^{-1}.$$

Thus, X to be $DHCST$, $(f(x)/f(0))^{1/x}$ is decreasing in x. Hence, for any $x, y \geq 0$,

$$\left[\frac{f(x+y)}{f(0)} \right]^{1/(x+y)} \leq \left(\frac{f(x)}{f(0)} \right)^{1/x}$$

yielding

$$\left[\frac{f(x+y)}{f(0)} \right]^{x/(x+y)} \leq \frac{f(x)}{f(0)}.$$

Similarly,

$$\left[\frac{f(x+y)}{f(0)} \right]^{y/(x+y)} \leq \frac{f(y)}{f(0)}.$$

A combination results in

$$\frac{f(x+y)}{f(0)} \leq \frac{f(x)}{f(0)} \cdot \frac{f(y)}{f(0)}$$

and hence is the proof. □

The class of $DHCST$ ($IHCST$) distributions can now be widened in terms of satisfying the above result. This may be called extended $DHCST$ ($IHCST$) or $EDHCST$ ($EIHCST$) class. It follows that

$$DCST \Rightarrow DHCST \Rightarrow EDHCST.$$

The dual relation is also true.

Theorem 8.4.4 *If G is the distribution function for the equilibrium distribution corresponding to the original distribution F, then*

(a) *G is $DCST$ implies and is implied by the fact that F is IFR;*
(b) *G is $ICST$ implies and is implied by the fact that F is DFR;*

(c) G is $DHCST$ implies and is implied by the fact that F is $IFRA$;
(d) G $IHCST$ is equivalent to F is $DFRA$;
(e) G is $EDHCST$ is equivalent to F is NBU;
(f) G is $EIHCST$ is equivalent to F is NWU.

Some characterization results based on CST are discussed in Sect. 9.8

8.5 Classes Based on Stochastic Dominance

In this section, we consider some non-parametric classes which are generated by considering different stochastic orders.

8.5.1 The $\mathscr{L}$-Class

The most widely used life distribution classes, viz., IFR, $IFRA$, $DMRL$, NBU, and $NBUE$ along with their duals have a common characteristic that is Laplace order domination by the exponential distributions with identical means. Thus, if F represents a life distribution with a finite mean and belongs to this class denoted as $\mathscr{L}$ (introduced by Klefsjö, 1983), then

$$L_F(s) \leqslant L_G(s),$$

where G stands for the distribution function of the exponential distribution and $L_F(s) = \int_0^\infty e^{-st} \bar{F}(t) dt$.

Thus, a distribution F with mean μ is in the $\mathscr{L}$ class if

$$\int_0^\infty e^{-tx} \bar{F}(x)\, dx \geqslant \frac{\mu}{1+t\mu}$$

for all $t \geqslant 0$. We also have the dual of the $\mathscr{L}$ class if the above inequality is reversed. This is denoted by $\bar{\mathscr{L}}$. A similar result holds in respect of discrete life distributions with the above properties compared to the geometric distribution.

Reliability operation	$\mathscr{L}$-class	$\bar{\mathscr{L}}$-class
Formation of coherent structures	Not preserved	Not preserved
Mixture	Not preserved	Preserved
Convolution	Preserved	Not preserved

Klefsjö (1983) (see also Miziula, 2012) presents results regarding closure of the class $\mathscr{L}$ and its dual under different reliability operations given in the following table.

Regarding finiteness of moments, it is known that distributions in the $HNBUE$ class and hence in all hierarchically smaller classes of life distributions have finite moments of all orders. However, for the $\mathscr{L}$ class it could be shown that moments of order r with $-1 < r \leqslant 2$ only are finite (Bhattacharjee and Sengupta, 1996 and Basu and Mitra, 1998). Bhattacharjee and Sengupta (1996) further showed that

(i) if $X \in \mathscr{L}$ then $C(X) \leqslant 1$;
(ii) if $X \in \bar{\mathscr{L}}$ then $C(X) \geqslant 1$.

8.5.2 The $\mathscr{M}$-Class and the $\mathscr{L}\mathscr{M}$-Class

The $\mathscr{L}$-class is the largest of well-known classes of life distributions. However, it is so large as to include distributions which can hardly be said to exhibit positive ageing property. Klar (2002) gives an example of a distribution belonging to this class with an infinite third moment, where mean remaining life at age t tends to infinity as t tends to infinity and the failure rate tends to zero. This led Klar and Müller (2003) to suggest an alternative $\mathscr{M}$-class which is obtained by replacing the Laplace transform order with the moment generating function order.

A distribution F with mean μ is said to be in the $\mathscr{M}$-class if $F \leqslant_{mgf} Exp(1/\mu)$, where $Exp(1/\mu)$ denotes the exponential distribution with mean μ and $X \leqslant_{mgf} Y$ if $E\left(e^{tX}\right) \leqslant E\left(e^{tY}\right)$ for all $t \geqslant 0$ with $E\left(e^{t_0 Y}\right) < \infty$ for some $t_0 > 0$. Thus, F is in the $\mathscr{M}$-class if and only if

$$E\left(e^{tX}\right) \leqslant \frac{1}{1-\mu t}, \text{ for all } 0 \leqslant t < 1/\mu.$$

It may be noted that $\mathscr{M}$-class is closed under convolution.

A distribution F belongs to the $\mathscr{L}\mathscr{M}$-class if it belongs to the $\mathscr{L}$-class as also to the $\mathscr{M}$-class. It is important to note that distributions in the $\mathscr{L}$-class (and hence in any smaller class like $IFR, NBU, NBUE, HNBUE$, etc.) cannot have point masses at the origin. The $\mathscr{M}$-class has this property and hence it is wider than any other class. The following theorem due to Klar and Müller (2003) provides the necessary support.

Theorem 8.5.1 *The $\mathscr{M}$-class contains all random variables X with $P(a \leqslant X \leqslant b) = 1$ and*

$$E(X) \geqslant \frac{a+b}{2}, \text{ for some } 0 \leqslant a < b.$$

Proof Consider first a two-point distribution with masses $p \leqslant 1/2$ and $1 - p$ at points $a \geqslant 0$ and $b > a$, respectively. Obviously, $E(X) = pa + (1 - p)b$ and this distribution belongs to the $\mathscr{M}$-class if and only if

$$f_p(x) = \frac{1}{1 - [pa + (1-p)b]x} - pe^{ax} - (1-p)e^{bx}$$

for $0 \leqslant x \leqslant 1/E(X)$. Expanding $f_p(x)$ in Taylor series, we have

$$f_p(x) = \sum_{k=0}^{\infty} a_k(p) x^k,$$

where

$$a_k(p) = [pa + (1-p)b]^k - p \cdot \frac{a^k}{k!} - (1-p) \cdot \frac{b^k}{k!}.$$

Hence we need to show that $a_k(p) \geqslant 0$ for all $p \leqslant 1/2$ and all $k = 0, 1, 2, \ldots$. For $p = 1/2$, it is just evident and for $p < 1/2$,

$$\frac{d}{dp} a_k(p) = a'_k(p) = k[pa + (1-p)b]^{k-1}(a-b) - \frac{a^k}{k!} + \frac{b^k}{k!}$$

$$\leqslant \frac{1}{k!}(b^k - a^k) - k(b-a)\left(\frac{a+b}{2}\right)^{k-1}$$

$$= (b-a) \sum_{i=0}^{k-1} a^i b^{k-1-i} \left(\frac{1}{k!} - \frac{k}{2^{k-1}} \binom{k-1}{i}\right)$$

$$\leqslant 0.$$

Thus, $a_k(p) \geqslant 0$ for all $p \leqslant 1/2$ and all $k = 0, 1, 2, \ldots$

Now, suppose $P(a \leqslant X \leqslant b) = 1$ and $E(X) \geqslant (a+b)/2$. Define Y such that

$$P(Y = a) = \frac{b - E(X)}{b - a} = 1 - P(Y = b).$$

It follows that $X \leqslant_{cx} Y$ and hence $X \leqslant_{mgf} Y$ also. According to the first part, $Y \leqslant_{mgf} Exp(1/E(X))$. Hence X belongs to the $\mathcal{M}$-class. $\square$

The condition on the mean relates to a notion of ageing, namely, that the unit is more likely to fail near the end of its potential lifetime than at its beginning. It can be verified that all symmetric distributions with bounded support fulfil conditions of the theorem and we state that any symmetric life distribution is contained in the $\mathcal{M}$-class.

Noting that $HNBUE$ property of a life distribution F with mean μ may be characterized by $F \leqslant_{cx} Exp(1/\mu)$, it follows that the $\mathcal{M}$-class and the $\mathcal{LM}$-class contain all $HNBUE$ distributions. A class in between the $HNBUE$ and the $\mathcal{M}$-classes can be defined in terms of the relation

$$F \leqslant_{mom} Exp(1/\mu),$$

where $X \leqslant_{mom} Y$ if $E(X^r) \leqslant E(Y^r)$, for all $r \in \mathbb{N}$.

8.6 Preservation of Class Properties

While some indications about closure of some classes with respect to some of the important reliability operations have been given in a scattered fashion here and there, a more or less consolidated picture is presented in Table 8.1. Some of the results can be easily proved, while some others may need some ingenuous effort. Interested readers may refer to the relevant articles or books (as indicated in References and Suggested Reading) for detailed discussions.

The $NBUC$ class of life distributions is closed in respect of parallel system formation (Li et al., 2000) but is not closed under series system formation (Hendi

Table 8.1 Preservation class properties under reliability operations

Class	Reliability operations		
	Coherent systems	Convolution	Mixture
ILR	Not preserved	Preserved	Not preserved
IFR	Not preserved	Preserved	Not preserved
$IFRA$	Preserved	Preserved	Not preserved
NBU	Preserved	Preserved	Not preserved
$NBUE$	Not preserved	Preserved	Not preserved
$NBUC$	Not preserved	Preserved	Not preserved
$NBUFR$	Preserved	Preserved	Not preserved
$NBAFR$	Preserved	Preserved	Not preserved
$DMRL$	Not preserved	Not preserved	Not preserved
$HNBUE$	Not preserved	Preserved	Not preserved
NBU-t_0	Preserved	Not preserved	Not preserved
$\mathscr{L}$	Not preserved	Preserved	Not preserved
DFR	Not preserved	Not preserved	Preserved
DLR	Not preserved	Not preserved	Preserved
$DFRA$	Not preserved	Not preserved	Preserved
NWU	Not preserved	Not preserved	Not preserved
$NWUE$	Not preserved	Not preserved	Not preserved
$NWUC$	Not preserved	Not preserved	Not preserved
$NWUFR$	Not preserved	Not preserved	Not preserved
$NWAFR$	Not preserved	Not preserved	Preserved
$IMRL$	Not preserved	Not preserved	Preserved
$HNWUE$	Not preserved	Not preserved	Preserved
NWU-t_0	Not preserved	Not preserved	Not preserved
BT	Not preserved	Not preserved	Not preserved
$\bar{\mathscr{L}}$	Not preserved	Not preserved	Preserved
$DRHR$	Not preserved	Preserved	Not preserved

et al., 1993). Thus the $NBUC$ class property is not preserved under formation of coherent systems.

The survival function of a series system with independently distributed component lives having $NBU(2)$ distributions with finite means has the $NBU(2)$ property (cf. Franco et al., 2001). The closure under convolution property of $NBU(2)$ class may be obtained in Hu and Xie (2002). It may be added that both $NBUE$ and $NBUC$ classes are closed under convolutions (cf. Marshall and Proschan, 1972; Hu and Xie, 2002). As pointed out by Li and Kochar (2001), it is not known whether the $NBU(2)$ class is closed under parallel system formation. However, $NBU(2)$ class is not closed under mixtures, since mixtures of some exponential life distributions often belong to the DFR class. This $NWU(2)$ class is not closed under convolution (Barlow and Proschan, 1981), and is not closed under the formation of parallel systems (Franco et al., 2001) although it is closed under the formation of series system of independent components.

Chapter 9
Characterizations of Life Distributions

9.1 Introduction

Choosing an appropriate probability model for the underlying random variable(s), based on a sample of observations on the variable(s), is a basic task in any reliability analysis exercise, and particularly so for any reliability inference exercise. In fact, this model-selecting inference problem has been of great interest in statistical inference.

Going by the traditional approach, we assume a model based on some pattern revealed by the data (e.g., in a frequency diagram) and then carry out a test for goodness of fit after estimating the parameters in the assumed model to find out if the assumed model provides a reasonably good fit or not. In the latter case, we will pick up another model and repeat the exercise. Eventually, we will settle with some model that provides the best possible fit. Alternatively, we can find out some measure of the Information Content provided by some model about the underlying true distribution by considering the maximized likelihood of the sample observations under the assumed model and penalizing the likelihood by a function of the number of free parameters in the model to be estimated from the model. Thus, we have the Akaike Information Criterion (AIC) or Bayes Information Criterion (BIC) among several other such measures which are used to identify the model with the highest information content. With the goodness-of-fit approach, the choice of class-intervals is solved somewhat subjectively and with the Information Criterion we have to find out the criterion value for several competing models and, assuming the true model to be one among these, to choose one that maximizes the criterion value. Thus, a unique choice of the true model is difficult.

Characterization results, particularly those which can be verified from the given sample, can provide a unique answer. It is true that many characterization results for some of the well-known probability distributions involve sampling distributions and hence cannot be verified from a given sample. (An indirect possibility often tried out in case of a large sample would be to break up the sample into a number of smaller sub-samples and to work out approximate sampling distribution for the statistic(s) required, or to verify the result for each of these sub-samples). Once a characterization result for a known distribution is verified on the basis of the observed sample, we can accept that distribution as a unique model for the underlying random variable, within limits of uncertainty associated with the verification exercise. Thus, characterization results play an important role in selecting probability models.

Characterization results may be based on

(1) properties of the probability distribution under consideration like moment-ratios;
(2) ageing properties like failure rate, hazard function, ageing intensity, and residual life distribution properties;
(3) order statistics and relations connecting these;
(4) entropy or information, including residual entropy, past entropy, etc.;
(5) dominance relations between the distribution under reference and some other reference distribution like the Exponential distribution, its own equilibrium distribution, its failure rate distribution, etc.;
(6) other atypical properties of the distribution.

Whenever, questions of stochastic independence or dominance or even some relation involving the distribution of some function of sample observations come in, we require sampling distributions which are not generally available. Verification of a characterization result need not always be based on sampling distributions. It may suffice to verify the result for a small number of time points or of subsets of observations.

Another useful contribution of characterizations is the fact that once a characterization result for a univariate life distribution can be extended, uniquely or otherwise, to the multivariate case, we can work out the multivariate extension of the initial univariate distribution. In some cases, the characterization result may admit of several extensions and can lead to several forms of multivariate distributions, each being characterized by one such form.

In this chapter, we will discuss characterizations of some probability models commonly used in reliability analysis, *viz*. Exponential, Weibull, Gamma, Rayleigh, and some Finite-range distributions. In fact, most random variables involved in reliability analysis are continuous, e.g., time to failure or time between failures or time to repair or down time or stress encountered and strength built in and so on and so forth. We will emphasize on simple and conveniently verifiable characterizations, though other known results will also be covered. In most cases, detailed proofs will be avoided. Results derived by the authors and their associates have been accorded some emphasis, as expected.

A discussion on characterization can proceed along one of the two alternative ways. We can consider a few oft-used probability distributions and present their characterizations based on some properties or characteristics. Otherwise, we can indicate how some of the reliability properties help us in characterizing life distributions. We have primarily followed the first track here. We must mention here that quite a few results of this chapter are borrowed from Roy (1988).

9.2 Characterizations of the Exponential Distribution

The Exponential distribution with its many simple and interesting properties has been studied extensively by a whole host of authors and we have a deluge of characterization results for this distribution. In fact, some of its simple characterization results have motivated a lot of characterizations for some other distributions. This distribution also acts as a reference distribution that has led to several results characterizing some of the well-known distributions. The loss-of-memory property and the associated constancy of failure rate are somewhat like household statements in reliability parlance.

Simply stated, the loss-of-memory property implies that the conditional probability of survival up to or beyond $x + t$ given survival up to x is the same as the unconditional probability of surviving up to or beyond time t. This property directly leads to a constant failure rate as a characterization of the Exponential distribution. In terms of a life-test with n copies of a product put on test, noting the numbers surviving up to several time points $t_1, t_2, t_3, \ldots$ and the number of failures in the windows $(t_k, t_k + \delta t)$ where δt is small but allows at least one failure in a window, we can have nonparametric estimates of the failure rate r at these time points. We can find out if the failure rate is sensibly constant or not.

The Exponential distribution has been characterized by Reinhardt (1968) in terms of the mean residual life $m_X(u) = C$ (a constant) for all u. This implies that $m_X(u) = m_X(0) = E(X)$ and hence $m_X(u) - m_X(0) = 0$ or a zero 'memory' characterizes this distribution (see Sect. 2.4.7 for definition and discussion on memory). Guerrieri (1965) has shown that a constant variance of residual life

$$V_X(u) = Var(X - u | X > u)$$
$$= \frac{2}{\bar{F}_X(u)} \int_u^\infty x \bar{F}_X(x) dx - 2u m_X(u) - m_X^2(u)$$

characterizes the Exponential distribution. It may be pointed out that a more general form for the loss-of-memory property of the Exponential distribution has been reported as

$$\bar{F}_X(x + y) = \bar{F}_X(x) . \bar{F}_X(y), \text{ for all } x, y \geqslant 0.$$

The requirement of this Cauchy-type equation to be valid for all x and y was relaxed by Marsaglia and Tubilla (1975) who needed the equation to be true for only two values of y such that their ratio is an irrational number. In fact it is quite likely that a complete knowledge about the functional forms of $m_X(u)$ and $V_X(u)$ may determine the underlying distribution uniquely as they are related to $\bar{F}_X(x)$ through integral forms. It is more interesting to examine ratios of moments for characterization of distributions as there is a possibility that while taking ratios some common factors may get eliminated. As the coefficient of variation is a widely used measure of relative dispersion, its use may be emphasized here. Writing $C_X(u)$ for the coefficient of variation, $C_X(u) = \sqrt{V_X(u)}/m_X(u)$, it has been observed that unlike $m_X(u)$ the functional form of $C_X(u)$ cannot uniquely determine the underlying distribution. For example, the constancy of $C_X(u)$ with respect to u is a feature of three possible distributions including the Exponential distribution as one. A further specification of the value of this constant, however, ensures characterization. We have the following.

Theorem 9.2.1 *For a continuous non-negative random variable X, $C_X(x) = 1$ for all x if and only if X follows Exponential distribution.*

Proof We prove the 'only if' part since the 'if part' is easy to prove. Given that $C_X(x) = 1$ for all x. This means that $V_X(x) = m_X^2(x)$ for all x. It is known that

$$V_X(x) = \frac{2}{\bar{F}(x)} \int_x^\infty t\bar{F}(t)\,dt - m_X^2(x) - 2x\, m_X(x).$$

Since $C_X(x) = 1$, the above expression reduces to

$$m_X^2(x) = \frac{1}{\bar{F}(x)} \int_x^\infty t\bar{F}(t)\,dt - x\, m_X(x) \tag{9.2.1}$$

which gives

$$\frac{1}{\bar{F}(x)} \int_x^\infty t\bar{F}(t)\,dt = m_X(x)(m_X(x) + x). \tag{9.2.2}$$

Differentiating both sides of (9.2.1) with respect to x we get

$$2m_X(x)\, m_X'(x) = -m_X(x) - x\, m_X'(x) - x - \frac{\bar{F}'(x)}{\bar{F}^2(x)} \int_x^\infty t\bar{F}(t)\,dt. \tag{9.2.3}$$

Again,

$$m_X(x) = \frac{1}{\bar{F}(x)} \int_x^\infty \bar{F}(t)\,dt$$

which, after differentiating with respect to x, reduces to

$$m_X'(x) = -1 - \frac{\bar{F}'(x)}{\bar{F}(x)} m_X(x). \tag{9.2.4}$$

On using (9.2.2) and (9.2.4), (9.2.3) gives

$$m_X(x) \frac{\bar{F}'(x)}{\bar{F}(x)} + 1 = 0.$$

On using (9.2.4), this expression reduces to $m_X'(x) = 0$ for all x giving $m_X(x)$ is constant for all $x > 0$, which proves the result. □

Instead of considering the mean or the variance or the coefficient of variation of residual life, one may consider similar measures for some functional transforms

9.2 Characterizations of the Exponential Distribution

of the original variable X to obtain interesting characterization results. One such characterization of the Exponential distribution is given below.

Theorem 9.2.2 *A random variable X follows Exponential distribution if and only if*
$$Var\left(e^{-X} - e^{-u}|X > u\right) = bE^2\left(e^{-X} - e^{-u}|X > u\right),$$
where $b < 1$ is a constant. □

Before we prove this theorem we need a couple of lemmas given one after another.

Lemma 9.2.1 *For a non-negative random variable X, define*
$$H_1(u) = E(h(X) - h(u)|X > u), \quad (9.2.5)$$
$$H_2(u) = V(h(X) - h(u)|X > u)$$
for a differentiable function h such that, for $r = 1, 2$, $h^r(x)\bar{F}_X(x) \to 0$, as $x \to \infty$. Then we have
$$H_1(u) = \frac{1}{\bar{F}_X(u)} \int_u^\infty h'(x)\bar{F}_X(x)\,dx, \quad (9.2.6)$$
$$H_2(u) = \frac{2}{\bar{F}_X(u)} \int_u^\infty h'(x)h(x)\bar{F}_X(x)\,dx - H_1^2(u) - 2h(u)H_1(u). \quad (9.2.7)$$

Proof Note that
$$H_1(u) = -\frac{1}{\bar{F}_X(u)} \int_u^\infty (h(x) - h(u))\,d\bar{F}_X(x)$$
$$= -h(u) - \frac{1}{\bar{F}_X(u)} \int_u^\infty h(x)\,d\bar{F}_X(x)$$
$$= \frac{1}{\bar{F}_X(u)} \int_u^\infty h'(x)\bar{F}_X(x)\,dx.$$

The third equality follows by integration by parts and the fact that $h(x)\bar{F}_X(x) \to 0$ as $x \to \infty$. Again,
$$H_2(u) = E(h^2(X)|X > u) - E^2(h(X)|X > u)$$
$$= -\frac{1}{\bar{F}_X(u)} \int_u^\infty h^2(x)\,d\bar{F}_X(x) - (H_1(u) + h(u))^2$$
$$= \frac{2}{\bar{F}_X(u)} \int_u^\infty h'(x)h(x)\bar{F}_X(x)\,dx - H_1^2(u) - 2h(u)H_1(u).$$

The second equality follows from (9.2.5) whereas the last equality follows using integration by parts and the fact that $h^2\bar{F}_X(x) \to 0$ as $x \to \infty$. This completes the proof.

Lemma 9.2.2 If $H_2(u) = bH_1^2(u)$ where b is a constant then

$$\frac{(1-b)\bar{F}_X'(u)}{\bar{F}_X(u)} = \frac{2bH_1'(u)}{H_1(u)}$$

where H_1 is as defined in Lemma 9.2.1.

Proof Since $H_2(u) = bH_1^2(u)$, on using (9.2.7) we have

$$(b+1)H_1^2(u) = \frac{2}{\bar{F}_X(u)} \int_u^\infty h'(x)h(x)\bar{F}_X(x)\,dx - 2h(u)H_1(u). \qquad (9.2.8)$$

Differentiating the above expression with respect to u we get

$$(b+1)H_1(u)H_1'(u) = -h'(u)h(u) - \frac{\bar{F}_X'(u)}{\bar{F}_X^2(u)} \int_u^\infty h'(x)h(x)\bar{F}_X(x)\,dx$$
$$-h'(u)H_1(u) - H_1'(u)h(u). \qquad (9.2.9)$$

Differentiating (9.2.6) with respect to u we get

$$H_1'(u) = -h'(u) - \frac{\bar{F}_X'(u)H_1(u)}{\bar{F}_X(u)}. \qquad (9.2.10)$$

Plugging this expression in (9.2.9) we get

$$-(b+1)H_1(u)\left[h'(u) + \frac{\bar{F}_X'(u)H_1(u)}{\bar{F}_X(u)}\right] = -\frac{\bar{F}_X'(u)}{\bar{F}_X^2(u)} \int_u^\infty h'(x)h(x)\bar{F}(x)\,dx - h'(u)H_1(u)$$
$$+h(u)H_1(u)\frac{\bar{F}_X'(u)}{\bar{F}_X(u)} \qquad (9.2.11)$$

which, with the help of (9.2.8), reduces to

$$-(b+1)H_1(u)\left[h'(u) + \frac{\bar{F}_X'(u)H_1(u)}{\bar{F}_X(u)}\right] = -\frac{(b+1)\bar{F}_X'(u)}{2\bar{F}_X(u)}H_1^2(u) - h'(u)H_1(u).$$

After simplification this reduces to

$$-(b+1)\frac{\bar{F}_X'(u)}{\bar{F}_X(u)} = 2b\frac{h'(u)}{H_1(u)} \qquad (9.2.12)$$
$$= \frac{2b}{H_1(u)}\left[-H_1'(u) - \frac{\bar{F}_X'(u)}{\bar{F}_X(u)}H_1(u)\right],$$

where the last equality follows from (9.2.10). This, when simplified, gives the required result.

9.2 Characterizations of the Exponential Distribution

Lemma 9.2.3 *As per the conditions on h and the definitions of H_1 and H_2 given in Lemma 9.2.1, $H_2(u) = b\, H_1^2(u)$ if and only if*

$$\bar{F}_X(u) = k(H_2(u))^{b/(1-b)}, \tag{9.2.13}$$

where k and b$(\neq 1)$ are constants.

Proof Since $b \neq 1$, we get from Lemma 9.2.2,

$$\frac{\bar{F}'_X(u)}{\bar{F}_X(u)} = \frac{2b}{1-b} \cdot \frac{H'_1(u)}{H_1(u)}$$

which is equivalent to

$$\frac{d}{du} \ln \bar{F}_X(u) = \frac{2b}{1-b} \frac{d}{du} \ln H_1(u).$$

This, when integrated, reduces to

$$\begin{aligned}\bar{F}_X(u) &= k_1(H_1(u))^{2b/(1-b)} \\ &= k_2(H_2(u))^{b/(1-b)},\end{aligned} \tag{9.2.14}$$

where k_1 and k_2 are constants. The last equality follows from the given relation between H_1 and H_2. This proves 'only if' part. The 'if part' follows as given below.

Given that

$$\bar{F}_X(u) = k(H_2(u))^{b/(1-b)}$$

which gives

$$\frac{\bar{F}'_X(u)}{\bar{F}(u)} = \frac{b}{1-b} \cdot \frac{H'_2(u)}{H_2(u)}. \tag{9.2.15}$$

From (9.2.7), we have

$$\begin{aligned}H_2(u) &= \frac{2}{\bar{F}_X(u)} \int_u^\infty h'(x)h(x)\bar{F}_X(x)\,dx - H_1^2(u) - 2h(u)H_1(u) \\ &= L(u) - H_1^2(u) - 2h(u)H_1(u),\end{aligned} \tag{9.2.16}$$

where

$$L(u)\bar{F}_X(u) = 2\int_u^\infty h'(x)h(x)\bar{F}_X(x)\,dx$$

which after differentiation reduces to

$$L'(u) = -\frac{2h'(u)h(u)\bar{F}_X(u) + L(u)\bar{F}'_X(u)}{\bar{F}_X(u)}. \tag{9.2.17}$$

Differentiating (9.2.16) with respect to u we get

$$H_2'(u) = L'(u) - 2H_1'(u)H_1(u) - 2h'(u)H_1(u) - 2h(u)H_1'(u)$$
$$= -\frac{2h'(u)h(u)\bar{F}_X(u) + L(u)\bar{F}_X'(u)}{\bar{F}_X(u)} + 2H_1(u)\left(h'(u) + \frac{\bar{F}_X'(u)H_1(u)}{\bar{F}_X(u)}\right)$$
$$- 2h'(u)H_1(u) + 2h(u)\left(h'(u) + \frac{\bar{F}_X'(u)H_1(u)}{\bar{F}_X(u)}\right).$$

The second equality follows by substituting the value of $L'(u)$ from (9.2.17) and $H_1'(u)$ from (9.2.10). Multiplying both sides of the above expression by $\bar{F}_X(u)$ we get

$$H_2'(u)\bar{F}_X(u) = \bar{F}_X'(u)\left(2H_1^2(u) + 2h(u)H_1(u) - L(u)\right)$$
$$= \bar{F}_X'(u)\left(H_1^2(u) - H_2(u)\right).$$

The second equality follows from (9.2.16). Thus, we have

$$\frac{\bar{F}_X'(u)}{\bar{F}_X(u)} = \frac{H_2'(u)}{H_1^2(u) - H_2(u)}. \tag{9.2.18}$$

Comparing (9.2.15) and (9.2.18) we get the required result.

Lemma 9.2.4 *As per the conditions on h and the definitions of H_1 and H_2 given in Lemma 9.2.1, if $H_2(u) = b\, H_1^2(u)$ then, for $b \neq 1$,*

$$h(u) = k_1 \left(\bar{F}_X(u)\right)^{\frac{1-b}{2b}} + k_2,$$

where k_1 and k_2 are constants.

Proof Since $H_2(u) = b\, H_1^2(u)$ we immediately get from Lemma 9.2.3 that

$$H_1(u) = C(\bar{F}_X(u))^{\frac{1-b}{2b}}. \tag{9.2.19}$$

From (9.2.10) we get

$$h'(u) = -H_1'(u) - \frac{\bar{F}_X'(u)H_1(u)}{\bar{F}_X(u)}$$

which gives

$$h(u) = -H_1(u) - \int \frac{\bar{F}_X'(u)}{\bar{F}_X(u)} H_1(u)\, du$$
$$= -C(\bar{F}_X(u))^{\frac{1-b}{2b}} - C \int (\bar{F}_X(u))^{\frac{1-b}{2b}-1} d\bar{F}_X(u)$$

9.2 Characterizations of the Exponential Distribution

$$= -C(\bar{F}_X(u))^{\frac{1-b}{2b}} - \frac{2bC}{1-b}(\bar{F}_X(u))^{\frac{1-b}{2b}} + k_2$$

$$= k_1 \left(\bar{F}_X(u)\right)^{\frac{1-b}{2b}} + k_2,$$

where k_1 is a contact (function of b and C) and k_2 is a constant of integration, and the second equality follows from (9.2.19). □

Now we are in a position to prove Theorem 9.2.2.

Proof of Theorem 9.2.2 Write $k(X, u) = \left(e^{-X} - e^{-u} | X > u\right) < 0$.
Only If Part: We note that if X is distributed Exponentially with failure rate λ, then

$$E[k(X, u)] = -\frac{e^{-u}}{\lambda + 1},$$

$$V[k(X, u)] = \frac{\lambda}{(\lambda + 2)(\lambda + 1)^2} e^{-2u}.$$

Combining we get

$$V[k(X, u)] = \frac{\lambda}{\lambda + 2} E^2[k(X, u)] = bE^2[k(X, u)],$$

where $b = \frac{\lambda}{\lambda+2} < 1$.

If Part: Let $b < 1$ and $V[k(X, u)] = bE^2[k(X, u)]$. Then taking $h(u) = e^{-u}$, we have $H_2(u) = bH_1^2(u)$. Then, by Lemma 9.2.4, we have

$$e^{-u} = k_1[\bar{F}(u)]^{\frac{1-b}{2b}} + k_2,$$

where k_1, k_2 are constants. Allowing $u \to \infty$ in the above expression and noting the fact that $b < 1$, we get $k_2 = 0$ and hence

$$e^{-u} = k_1[\bar{F}(u)]^{\frac{1-b}{2b}}.$$

Again, putting $u = 0$ in the above expression, we get $k_1 = 1$ so that

$$e^{-u} = [\bar{F}(u)]^{\frac{1-b}{2b}}$$

which gives $\bar{F}(u) = e^{-\lambda u}$ where $\lambda = 2b/(1 - b)$ giving $b = \lambda/(\lambda + 2)$. Thus, X follows the Exponential distribution. □

Let $X_1, X_2, \ldots, X_n$ be n random sample observations on X and let $Y_1 \leqslant Y_2 \leqslant \ldots \leqslant Y_n$ be the corresponding order statistics. Srivastava (1967) observed that the conditional expectation $E(Y_{m+1} - Y_m | Y_m = y)$ is independent of y if and only if X follows Exponential distribution. Govindarajulu (1966) has shown that independence of $(Y_2 - Y_1, Y_3 - Y_1, \ldots, Y_n - Y_1)$ and Y_1 is both necessary and sufficient for X

to have Exponential distribution. He has also stated that independence of Y_1 and $\sum_{j=2}^{n}(Y_j - Y_1)$ characterizes the Exponential law.

Writing

$$Z_k = \frac{1}{n-k} \sum_{j=k+1}^{n} (Y_j - Y_k),$$

for $1 \leqslant k \leqslant n-1$ and

$$W_k = \frac{1}{k-1} \sum_{i=1}^{k-1}(Y_k - Y_i),$$

for $2 \leqslant k \leqslant n$, we can restate Govindarajulu's result through independence of Y_1 and Z_1. Dallas (1973) has characterized Exponential distribution through $E(Z_1|Y_1 = y)$ = constant, independent of y. Wang and Srivastava (1980) have characterized a few distributions through

$$E(Z_k|Y_k = y) = \alpha y + \beta, \qquad (9.2.20)$$

where α and β are constants, as follows.

- The Exponential distribution

$$\bar{F}_X(x) = e^{-(x-\mu)/\theta}, \quad -\infty < \mu < x; \ \theta > 0$$

is characterized by $\alpha = 0$, and $\theta = \beta > 0$.
- The Pearsonian distribution given by

$$\bar{F}_X(x) = \left(\frac{\nu - x}{\nu - \mu}\right)^\theta, \quad -\infty < \mu < x < \nu < \infty, \theta > 0$$

is characterized for $\alpha \in (-1, 0)$, $\theta = -(\alpha + 1)/\alpha > 0$, $\mu = (1 - \beta)/\alpha < \nu = -\beta/\alpha$.
- Pareto distribution having survival function

$$\bar{F}_X(x) = \left(\frac{\mu + \nu}{x + \nu}\right)^\theta, \quad -\infty < \mu < x; \theta > 0, \ \nu > -\infty$$

is characterized for $\alpha > 0$, $\theta = (\alpha + 1)/\alpha > 0$, $\mu = (1 - \beta)/\alpha$, $\nu = -\beta/\alpha$. □

They have shown that, for $\alpha \leqslant -1$ there does not exist any distribution having the property given in (9.2.20).

In place of the random variable X if we take $Y = -X$, then the above distributions (as distribution of Y) may be characterized by $E(W_k|Y_k = y) = \alpha y + \beta$ exactly in the same way (cf. Wang and Srivastava, 1980). These distributions have been characterized by Beg and Kirmani (1974) through $E(Z_1|Y_1 = y) = \alpha y + \beta$. □

9.2 Characterizations of the Exponential Distribution

Observing the similarity between these results and those based on mean residual life, we have examined the suitability of coefficient of variation in place of expectation for characterization of life distributions. The following result is an outcome of that study.

Theorem 9.2.3 *If, for an absolutely continuous non-negative random variable X, $Y_1, Y_2, \ldots, Y_n$ are the ordered observations with Z_k as defined earlier, the conditional coefficient of variation $C(Z_k|Y_k = y) = \frac{1}{\sqrt{n-k}}$ if and only if X follows Exponential distribution.*

Proof Write $S = \{1, 2, \ldots, n\}$. Following the unordering technique of Wang and Srivastava (1980), let $s \in S$ and σ be a subset of size k of the set S with $s \in \sigma$. Let σ_0 be the set $\{1, 2, \ldots, k\}$. Write

$$A_{\sigma,s} = \{(X_1, X_2, \ldots, X_n) : X_j < y, \text{ for } j \in \sigma \text{ (excluding element } s); X_s = y, X_j > y \text{ for } j \notin \sigma\}.$$

Then

$$P(A_{\sigma,s}|Y_k = y) = \frac{1}{n\binom{n-1}{k-1}}.$$

From the given condition we get

$$E(Z_k^2|Y_k = y) = \frac{n-k+1}{n-k} E^2(Z_k|Y_k = y),$$

which gives

$$E\left[\left(\sum_{i=k+1}^{n}(Y_i - Y_k)\right)^2 \bigg| Y_k = y\right] = \frac{n-k+1}{n-k} E^2\left[\left(\sum_{i=k+1}^{n}(Y_i - Y_k)\right) \bigg| Y_k = y\right].$$

Thus, we have

$$\sum_{\sigma,s} E\left[\left(\sum_{j\notin\sigma}(X_j - y)\right)^2 \bigg| A_{\sigma,s}\right] P(A_{\sigma,s}|Y_k = y) = \frac{n-k+1}{n-k}\left[\sum_{\sigma,s} E\left(\sum_{j\notin\sigma}(X_j - y)\bigg| A_{\sigma,s}\right) P(A_{\sigma,s}|Y_k = y)\right]^2.$$

The above expression can be rewritten as

$$E\left[\left(\sum_{j\notin\sigma_0}(X_j - y)\right)^2 \bigg| A_{\sigma_0,s}\right] = \frac{n-k+1}{n-k}\left[E\left(\sum_{j\notin\sigma_0}(X_j - y)\bigg| A_{\sigma_0,s}\right)\right]^2$$

which is equivalent to

$$E\left[\left(\sum_{j=k+1}^{n}(X_j-y)\right)^2 \bigg| A_{\sigma_0,s}\right] = \frac{n-k+1}{n-k}\left[E\left(\sum_{j=k+1}^{n}(X_j-y)\bigg| A_{\sigma_0,s}\right)\right]^2.$$

This can be rewritten as

$$E\left[\sum_{j=k+1}^{n}(X_j-y)^2 + \sum_{j\neq j'}(X_j-y)(X_{j'}-y)\bigg| A_{\sigma_0,s}\right] = \frac{n-k+1}{n-k}\left[E\left(\sum_{j=k+1}^{n}(X_j-y)\bigg| A_{\sigma_0,s}\right)\right]^2.$$

Since X_j are independent copies of X, the above expression reduces to

$$(n-k)E\left[(X_n-y)^2|X_n>y\right] + (n-k)(n-k-1)E^2(X_n-y|X_n>y)$$
$$= \frac{n-k+1}{n-k}(n-k)^2 E^2(X_n-y|X_n>y)$$

which can be simplified to

$$E\left[(X_n-y)^2|X_n>y\right] = 2E^2(X_n-y|X_n>y)$$

giving
$$V(X_n-y|X_n>y) = E^2(X_n-y|X_n>y)$$

so that $V_{X_n}(y) = m_{X_n}^2(y)$. Now the result follows from Theorem 9.2.1. □

A similar result can be proved for W_k also as given below.

Theorem 9.2.4 *For a continuous random variable X, with ordered observations $Y_1 < Y_2 < Y_3 < \ldots < Y_n$, and G_k and H_k defined as*

$$G_k = \prod_{i=k+1}^{n}(Y_i - Y_k),\ 1 \leq k \leq n-1$$

and

$$H_k = \prod_{i=1}^{k-1}(Y_k - Y_i),\ 2 \leq k \leq n$$

(a) $E(G_k|Y_k = y) = \alpha^{n-k}$ *if and only if X follows Exponential distribution, and*
(b) $E(H_k|Y_k = y) = \alpha^{k-1}$ *if and only if X follows Exponential distribution,*

where $\alpha > 0$ is any constant.

Proof We prove (a) only since the proof of (b) is similar. The notations used below are same as those in Theorem 9.2.3. Note that

$$E(G_k|Y_k = y) = \sum_{\sigma,s} E\left[\prod_{j\notin\sigma}(X_j - y)\bigg| A_{\sigma.s}\right] P(A_{\sigma,s}|Y_k = y)$$

$$= E\left[\prod_{j\notin\sigma_0}(X_j - y)\bigg| A_{\sigma_0.s}\right]$$

$$= E\left[\prod_{j=k+1}^{n}(X_j - y)\bigg| A_{\sigma_0.s}\right]$$

$$= \prod_{j=k+1}^{n} E(X_j - y|X_j > y)$$

$$= [E(X_n - y|X_n > y)]^{n-k}.$$

So, $E(G_k|Y_k = y) = \alpha^{n-k}$ if and only if $E(X_n - y|X_n > y) = \alpha$. This gives $m_{X_n}(y)$ is constant so that the underlying distribution is Exponential.

9.3 Characterization Through Record Values

Chandler (1952) has introduced and studied many basic properties of records. Since then these have been extensively studied in the literature. Let X_i, $i = 0, 1, \ldots, n$, be iid non-negative random variables having absolutely continuous survival function $\bar{F}(\cdot)$ and density function $f(\cdot)$ having finite mean. Then X_n is called an upper record value (which is commonly called record) if $X_n > X_i$, for every $i < n$. It can be noted that X_0 is a trivial record. The random variables $\{L(n), n \geq 0\}$, defined by

$$L(n) = \min\{j : X_j > X_{L(n-1)}\}, \quad n \geq 1, \tag{9.3.21}$$

with $L(0) = 0$, give the sequence of epochs at which these record values occur. The sequence of record values corresponding to $\{X_n, n \geq 0\}$ is defined by $R_n^X = X_{L(n)}$, $n \geq 0$. Here R_n^X is known as n-th upper record. An analogous definition deals with lower record values (cf. Kundu and Nanda, 2010a). See also Sect. 5.7.2.

The earliest characterization through record values is due to Tata (1969) who has proved that $X_{L(0)}$ and $X_{L(1)} - X_{L(0)}$ are independently distributed when and only when X is Exponentially distributed. Subsequently, the Exponential distribution has been characterized through independence of $X_{L(r)}$ and $X_{L(r+1)} - X_{L(r)}$ by Ahsanullah (1981) and Pfeifer (1982). The more restrictive condition of distributional independence has been replaced by the independence of conditional moments from the conditioning variate by several authors. It has been shown by Nagaraja (1977) and Ahsanullah (1978) that $E\left(X_{L(r+1)} - X_{L(r)}|X_{L(r)} = y\right)$ is independent of y if and only if $\bar{F}_X$ has the Exponential form. Similarly, based on second-order central moment, Ahsanullah (1981) has proved that the independence of

$$Var\left(X_{L(r+1)} - X_{L(r)} | X_{L(r)} = y\right)$$

with respect to y is both necessary and sufficient condition for the variables X_i, $i = 0, 1, \ldots, n$, to follow Exponential distribution.

The striking similarity between these latter results and those based on order statistics and residual life has led to two unification works—one is due to Deheuvels (1984) and the other is due to Gupta (1984). While Deheuvels has drawn one's attention to the Markovian nature of record values, Gupta has noted that $\{X_{L(r+1)} - X_{L(r)} | X_{L(r)} = y\}$ behaves in the same way as $\{X - y | X > y\}$. Gupta has further obtained a general result characterizing the Exponential distribution through independence of

$$E\left[\left(X_{L(r+1)} - X_{L(r)}\right)^k | X_{L(r)} = y\right]$$

with respect to y for a given $k \geq 1$.

However, all these above-mentioned studies are restricted to those on two consecutive record values to exploit, knowingly or unknowingly, the Markovian nature or the residual life equivalence. The only exception to these results is the work of Nayak (1981) who has extended the result of Tata (1969) by characterizing the Exponential distribution through independence of

$$\left[X_{L(t)} - X_{L(s)} | X_{L(r)} = u\right]$$

with respect to u in the following manner.

Theorem 9.3.1 *For a sequence $\{X_n\}$ of non-negative random variables distributed independently with a common density function $f(\cdot)$ and belonging to either the NBU or the NWU class of distributions, $E\left[\left(X_{L(t)} - X_{L(s)}\right)^k | X_{L(r)} = u\right]$ is independent of u if and only if X_n follows Exponential distribution for some integers r, s and t such that $0 \leq r \leq s < t$, where $\{L(n)\}$ is the sequence of epochs defined in (9.3.21).* □

In this context it may be mentioned that while record values have been used by most of the authors for characterizing the Exponential distribution, it is possible to present a general characterization result as follows.

Theorem 9.3.2 *For a sequence of iid non-negative random variables $\{X_n\}$ with a common survival function $\bar{F}(\cdot)$,*

$$E\{X_{L(r+1)} - X_{L(r)} | X_{L(r)} = u\} = h(u) \tag{9.3.22}$$

if and only if

$$\bar{F}_X(x) = \frac{h(0)}{h(x)} e^{-\int_0^x \frac{du}{h(u)}},$$

where r is an integer with $0 \leq r < n$ and $\{L(n)\}$ is the sequence of epochs as defined in (9.3.21).

Proof Following the line of proof of Theorem 3.1 of Gupta (1984), (9.3.22) can equivalently be written as

$$\frac{1}{\bar{F}_X(u)} \int_u^\infty \bar{F}_X(x)\,dx = h(u).$$

Multiplying both sides of the above expression by $\bar{F}_X(u)$ and then differentiating with respect to u we get

$$\frac{\bar{F}'_X(u)}{\bar{F}_X(u)} = -\frac{1 + h'(u)}{h(u)}.$$

Integrating both sides in $(0, x)$ and on using the boundary condition that $\bar{F}_X(0) = 1$, the solution of the above differential equation gives the required result. □

Different feasible choice of $h(u)$ will characterize different distribution. For example, $h(u) = c$, a constant, characterizes Exponential distribution.

9.4 Characterizations of the Weibull Distribution

We represent the Weibull distribution, which can be regarded as a generalization of the Exponential distribution with a wide range of applications in situations that reveal increasing failure rate as also decreasing failure rate behaviour, by the symbol $W(\alpha, \lambda)$. The pdf of the two-parameter Weibull distribution is given by

$$f_X(x; \alpha, \lambda) = \alpha\lambda x^{\alpha-1} e^{-\lambda x^\alpha}, \quad x > 0,$$

with α and λ as the shape and the scale parameters, respectively. The class of Weibull distributions can be divided into three sub-classes according to the values of the shape parameter α. A value $\alpha > 1$ implies an IFR distribution and a value $\alpha < 1$ corresponds to a DFR distribution, while $\alpha = 1$ reduces to the Exponential distribution.

An early characterization of the distribution is due to Dubey (1967) stating that the minimum order statistic follows the Weibull distribution if and only if the lifetime distribution is Weibull. Shimizu and Davies (1981) characterized the Weibull distribution through functional equations and covered the stable distribution in general.

The two-parameter Weibull distribution as a generalization of the one-parameter Exponential distribution can be characterized by a generalized loss-of-memory property as follows.

Theorem 9.4.1 *A non-negative random variable X which satisfies the property*

$$P[X^\alpha \geq x^\alpha + z^\alpha | X \geq z] = P[X \geq x]$$

has the Weibull distribution with shape parameter α and vice versa. □

It is clear from the definition of hazard function that a complete knowledge of $R(x) = \int_0^x r(t)dt$ determines the lifetime distribution uniquely. However, if $R(x)$ is known to satisfy certain functional relation only, we may like to examine the possibility of determining the distribution from the same. Thus, famous loss-of-memory property that characterizes the Exponential distribution can be expressed as a functional relation in $R(x)$ as

$$R(x+y) = R(x) + R(y), \quad x, y > 0.$$

A product relationship yields the following result.

Theorem 9.4.2 *For a non-negative random variable X,*

$$R(xy)R(1) = R(x)R(y) \text{ for all } x > 0, y > 0 \qquad (9.4.23)$$

if and only if X follows Weibull distribution, where $R(1) > 0$. □

Following a result on loss-of-memory property by Marsaglia and Tubilla (1975) we modify the above theorem as follows.

Theorem 9.4.3 *For a non-negative random variable X, (9.4.23) holds true for all $x > 0$ and for two values y_1 and y_2 of $y > 0$ such that $(\ln y_2 / \ln y_1)$ is irrational, where $R(1) > 0$, if and only if X follows Weibull distribution.* □

We point out here another variant of these results.

Theorem 9.4.4 *A function $F(\cdot)$, with support on $[0, \infty)$, is a Weibull distribution function if and only if $R(\cdot)$ satisfies the conditions:*

(a) $R(xy)R(1) = R(x)R(y)$, *with $R(1) > 0$, for at least one value of $y > 0$, $y \neq 1$, and for all $x > 0$,*
(b) $xr(x)/R(x)$ *is non-decreasing in $x > 0$.* □

Proof If the distribution is Weibull, the proof is straight forward. Conversely, suppose (a) and (b) are given. Define $L(x) = R(x)/R(1)$ and $\phi(u) = L(e^u)$ with $u = \ln x$ and $v = \ln y$. Then (a) can be rewritten as

(a') $\phi(u+v) = \phi(u).\phi(v)$ for all u and at least one $v (\neq 0)$.

Also write $\psi(u) = [\phi(v)]^{-u/v}\phi(u)$. Then clearly $\psi(u+v) = \psi(u)$ giving that $\psi(u)$ is periodic in u with a period of v so that $\ln \psi(u)$ is also periodic in u. Note that

$$\frac{d \ln \psi(u)}{du} = -\frac{1}{v} \ln \phi(v) + \frac{\phi'(u)}{\phi(u)}. \qquad (9.4.24)$$

Again, $\phi(u) = L(e^u) = R(e^u)/R(1)$ gives

$$\frac{\phi'(u)}{\phi(u)} = \frac{R'(e^u)e^u}{R(e^u)}. \qquad (9.4.25)$$

9.4 Characterizations of the Weibull Distribution

Combining (9.4.24) and (9.4.25) we get

$$\frac{d \ln \psi(u)}{du} = -\frac{1}{v} \ln \phi(v) + \frac{R'(e^u)e^u}{R(e^u)}$$

which is increasing in u, by (b). Since $\ln \psi(u)$ is periodic and increasing we have $\ln \psi(u) = constant = \ln C$, say, which gives

$$-\frac{u}{v} \ln \phi(v) + \ln \phi(u) = \ln C$$

so that

$$\phi(u) = Ce^{u \ln \phi(v)/v} = c \left(e^u\right)^{\ln \phi(v)/v}.$$

This gives

$$R(e^u) = CR(1) \left(e^u\right)^{\ln \phi(v)/v}$$

giving $R(x) = CR(1)x^\alpha$ where $\alpha = \ln \phi(v)/v$. Thus, (a) becomes

$$CR^2(1)x^\alpha y^\alpha = C^2 R^2(1) x^\alpha y^\alpha.$$

So, $C = 0$ or $C = 1$. Since $R(1) > 0$, C cannot be zero and hence $C = 1$. Thus, we have $R(x) = R(1)x^\alpha$ giving $r(x) = R(1)\alpha x^{\alpha-1}$. Hence, $X \sim W(\alpha, R(1))$. This completes the proof.

It may be pointed out that, in terms of failure rate $r(x)$ satisfying the functional equation $r(xy)r(1) = r(x)r(y)$, one may characterize the Weibull distribution.

For notational convenience we shall denote by Z, the failure rate transform (FRT) from Y with probability density function and distribution function of Z as $h(\cdot)$, $H(\cdot)$ and those of Y as $g(\cdot)$, $G(\cdot)$, respectively, where Y is the FRT of X. Then, by the definition,

$$Y = \frac{f(X)}{\bar{F}(X)} \quad \text{and} \quad Z = \frac{g(Y)}{\bar{G}(Y)},$$

where $\bar{G} \equiv 1 - G$.

We now have the following characterization results.

Theorem 9.4.5 *(a) If a random variable X follows DFR Weibull distribution, then Y follows Inverse Weibull distribution.*
(b) Let X belong to the class of DFR distributions with $y \to \infty$ as $x \to 0$. Then Y follows Inverse Weibull distribution only if X is DFR Weibull.

Proof Let X have a DFR Weibull distribution $W(\alpha, \lambda)$ so that $\alpha < 1$. Then

$$G(y) = P(Y \leq y) = P\left(X \geq \left(\frac{y}{\lambda\alpha}\right)^{1/(\alpha-1)}\right) = e^{-\lambda(\lambda\alpha)^\beta y^{-\beta}} = e^{-\delta y^{-\beta}},$$

which is the distribution function of an Inverse Weibull distribution, where $\beta = \alpha(1-\alpha)^{-1} > 0$ and $\delta = \lambda(\lambda\alpha)^\beta$. This proves (a). To prove (b), let Y follow an Inverse Weibull distribution so that

$$P(r(X) \leq y) = e^{-\delta y^{-\beta}}$$

with $\beta, \delta > 0$. This implies

$$\bar{F}\left(r^{-1}(y)\right) = P(X \geq r^{-1}(y)) = e^{-\delta y^{-\beta}}$$

since X belongs to the DFR class. This, writing $y = r(t)$, reduces to

$$\ln \bar{F}(t) = -\delta(r(t))^{-\beta}$$

which after differentiation gives

$$(r(t))^{-\beta-1} \frac{dr(t)}{dt} = -\frac{r(t)}{\beta\delta}.$$

Writing $(r(t))^{-\beta-1} = z$, we get

$$\frac{dz}{dt} = \frac{\beta+1}{\beta\delta}$$

giving

$$z = \frac{\beta+1}{\beta\delta} t + A,$$

where A is the constant of integration. Substituting the value of z we get

$$(r(t))^{-\beta-1} = \frac{\beta+1}{\beta\delta} t + A.$$

On using boundary condition, we get $A = 0$ so that

$$(r(t))^{-\beta-1} = \frac{\beta+1}{\beta\delta} t$$

so that

$$r(t) = \left(\frac{\beta\delta}{\beta+1}\right)^{1/(\beta+1)} . t^{-1/(\beta+1)}.$$

This proves (b). □

A similar result is the following.

9.4 Characterizations of the Weibull Distribution

Theorem 9.4.6 (a) *If a random variable X follows IFR Weibull distribution then Y also follows IFR Weibull distribution.*
(b) *If X belongs to the class of IFR distributions with $y \to 0$ for $x \to 0$ then Y follows IFR Weibull distribution only if X follows IFR Weibull distribution.*

Proof Since X follows an IFR Weibull distribution $W(\alpha, \lambda)$ with $\alpha > 1$,

$$P(Y \geq y) = P\left(X \geq \left(\frac{y}{\lambda\alpha}\right)^{1/(\alpha-1)}\right) = e^{-\lambda' y^{\alpha'}},$$

where $\lambda' = \lambda(\lambda\alpha)^{-\alpha'}$ and $\alpha' = \alpha(\alpha-1)^{-1}$. Since $\alpha' > 1$ and $\lambda' > 0$, we conclude that Y has an IFR Weibull distribution. This proves (a).

To prove (b), let Y have an IFR Weibull distribution with survival function given by

$$\bar{G}(y) = P(r(X) \geq y) = e^{-\lambda y^{\alpha}}, \quad \alpha > 1, \ \lambda > 0.$$

Since X is *IFR*, writing $r(t) = y$, the above relation reduces to

$$\bar{F}(t) = P(X \geq t) = e^{-\lambda(r(t))^{\alpha}}.$$

Taking log on both sides and then differentiating with respect to t we get

$$(r(t))^{\alpha-2}\frac{dr(t)}{dt} = \frac{1}{\alpha\lambda}.$$

Proceeding similarly as above and using the boundary conditions, we get

$$r(t) = Ct^{1/(\alpha-1)},$$

where C is a constant independent of t. This proves the result. □

An associated characterization result is the following.

Theorem 9.4.7 *For a non-negative random variable X,*

(i) *X follows Rayleigh distribution if and only if Y is proportional to X, and*
(ii) *in the IFR class, X follows IFR Weibull distribution if and only if Z is proportional to X.*

Proof If X follows Rayleigh distribution with pdf given by

$$f(x) = 2\lambda x e^{-\lambda x^2},$$

it is obvious that

$$r(X) = Y = 2\lambda X,$$

and hence Y is proportional to X. Conversely, let Y be proportional to X as $Y = kX$ where k is a non-negative constant. Then

$$\bar{F}(x) = e^{-kx^2/2}$$

and this implies that X follows Rayleigh distribution. This proves (i). To prove (ii), let X follow IFR Weibull distribution $W(\alpha, \lambda)$ with $\alpha > 1$. Then, it is known that $Y = \alpha \lambda X^{\alpha-1}$. Since, by Theorem 9.4.6, it is known that Y has a Weibull distribution with parameters α', λ' we have the failure rate transform of Y as

$$Z = \alpha' \lambda' Y^{\alpha'-1}$$

which eventually simplifies to

$$Z = \alpha' \lambda' (\alpha \lambda)^{\alpha'-1} X$$

or $Z = kX$ where k is a constant involving α and λ.

To prove the converse, let X be IFR such that $Z = kX$. From the IFR property, it follows that $G(y) = F(q(y))$ where $q(\cdot) = r^{-1}(\cdot)$ is the inverse function of $r(\cdot)$ so that $g(y) = f(q(y))q'(y)$ and hence the failure rate function of Y is given by $r_Y(y) = r(q(y))q'(y) = yq'(y)$. Since

$$q'(y) = \left[\left(\frac{f'(x)}{f(x)} \right) r(x) + r^2(x) \right]^{-1},$$

we finally get

$$z = \left(\frac{f'(x)}{f(x)} + r(x) \right)^{-1}.$$

Thus, the condition $Z = kX$ implies that

$$kX = \left(\frac{f'(X)}{f(X)} + r(X) \right)^{-1},$$

which gives

$$r(X) = \frac{1}{kX} - \frac{f'(X)}{f(X)}.$$

Hence, for every realization x of X, we get

$$r(x) = \frac{1}{kx} - \frac{f'(x)}{f(x)}$$

yielding, on integration, $-\ln \bar{F}(x) = (1/k) \ln x - \ln f(x) - \ln A$ which gives

$$\bar{F}(x) = A f(x) x^{-1/k}$$

so that

$$\frac{f(x)}{\overline{F}(x)} = \frac{1}{A} x^{1/k},$$

where A is a constant of integration. Thus, finally we have $r(x) = r(1)x^{1/k}$ which translates into

$$f(x) = \alpha \lambda x^{\alpha-1} e^{-\lambda x^\alpha},$$

where $\lambda = r(1)/\alpha$ and $\alpha = 1 + 1/k$. And α being greater than 1 as $k > 0$, X follows IFR Weibull distribution. □

Roy and Mukherjee (1986) characterized the Weibull distribution in terms of Fisher Information within the family of probability distributions with a scale parameter λ-dependent density function $\lambda f(\lambda x)$ satisfying the following conditions. Let $\mathcal{F}$ be the class of density functions $\mathcal{F}_\alpha$ with support in $(0, \infty)$ such that

$$\mathcal{F} = \{\mathcal{F}_\alpha : 0 < \alpha < \infty\},$$

where $\mathcal{F}_\alpha$ is defined by the collection of the density functions $f(\cdot)$ satisfying the following conditions:

(i) f is continuously differentiable on the support,
(ii) $xf(x) \to 0$ as $x \to 0+$ and $x^{1+\alpha} f(x) \to 0$ as $x \to \infty$,
(iii) $E(X^\alpha) = 1$ and $E(X^{2\alpha}) = 2$

so that coefficient of variation of X^α equals 1. Among all members of this class, Fisher information is minimized by $W(\alpha, 1)$. Then we have the following.

Theorem 9.4.8 *Among all members of F_α, the Fisher information is minimized by $W(\alpha, 1)$.*

Result to establish minimum Fisher information for the Weibull distribution $W(\alpha, \lambda)$ can be similarly constructed.

9.5 Characterizations of the Generalized Gamma Distribution

The generalized gamma distribution with the density

$$f(x : \lambda, \alpha, \beta) = \frac{\lambda}{\alpha^\beta \Gamma(\beta/\lambda)} x^{\beta-1} e^{-(x/\alpha)^\lambda}, \quad x \geqslant 0, \tag{9.5.26}$$

as given in (3.8.3), also known as the Stacy distribution, is symbolically represented by $GG(\alpha, \beta, \lambda)$. Different shapes of this distribution are shown in Fig. 9.1.

It is noted that if $X_i \sim GG(\alpha_i, \beta_i, \lambda)$, for $i = 1, 2$ independently then, for $T = X_1/X_2$,

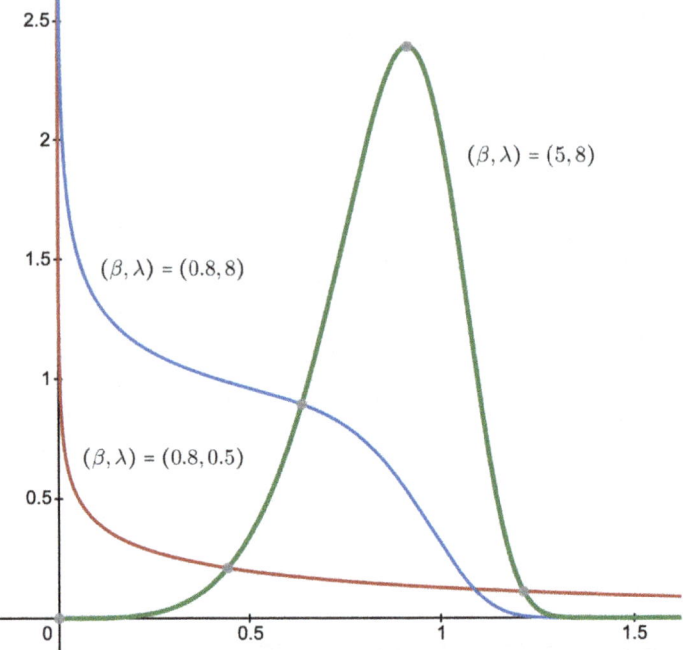

Fig. 9.1 Different shapes of the Generalized Gamma Distributions for different values of the parameters β and λ when the scale parameter α is taken to be unity

$$W = \frac{T^\lambda}{T^\lambda + \left(\frac{\alpha_1}{\alpha_2}\right)^\lambda} \sim Beta\left(\frac{\beta_1}{\lambda}, \frac{\beta_2}{\lambda}\right).$$

Patil and Seshadri (1964) characterized several distributions through conditional pdf constructed out of two independent random variables by considering their sum. Roy (1984) presented a variant of that result by considering ratio of two independent random variables, and the conditional densities constructed there from.

Lemma 9.5.1 *If X and Y are two independent non-negative random variables with positive probability density functions $f(\cdot)$ and $g(\cdot)$ such that the conditional density of X given X/Y, denoted by $c(x, x/y)$, satisfies the relation*

$$\frac{c\left(\frac{x}{y}, \frac{x}{y}\right).c\left(1, \frac{1}{y}\right)}{c\left(\frac{1}{y}, \frac{1}{y}\right).c\left(x, \frac{x}{y}\right)} = \frac{h\left(\frac{x}{y}\right)h(1)}{h(x).h\left(\frac{1}{y}\right)}$$

for some arbitrary positive function $h(\cdot)$, then

$$f(x) = f(1)h(x)x^a \quad \text{and} \quad g(y) = g(1)k(y)y^{-a-1},$$

9.5 Characterizations of the Generalized Gamma Distribution

where a is an arbitrary constant, $f(1)$ and $g(1)$ are suitable normalizers and

$$k(y) = \frac{h\left(\frac{1}{y}\right)}{h(1)} \cdot \frac{c\left(1, \frac{1}{y}\right)}{c\left(\frac{1}{y}, \frac{1}{y}\right)}.$$

Proof Being independent, the joint distribution of X and Y is given by $f_{X,Y}(x, y) = f(x)g(y)$. Let us write $W = X/Y$. Then the joint distribution of X and W is given by

$$f_{X,W}(x, w) = f(x)g\left(\frac{x}{w}\right)\frac{x}{w^2}.$$

Since $c(x, w)$ is the conditional density of X given W, we have

$$c(x, w).\eta(w) = f(x)g\left(\frac{x}{w}\right) \cdot \frac{x}{w^2}, \qquad (9.5.27)$$

where $\eta(w)$ is the density of W. We now replace x by w in (9.5.27) so that

$$c(w, w).\eta(w) = f(w)g(1).\frac{1}{w}. \qquad (9.5.28)$$

If we replace w by $1/y$ in (9.5.28) we get

$$c\left(\frac{1}{y}, \frac{1}{y}\right).\eta\left(\frac{1}{y}\right) = f\left(\frac{1}{y}\right)g(1)y. \qquad (9.5.29)$$

Now, in (9.5.27), we replace x by 1 and w by $1/y$ so that

$$c\left(1, \frac{1}{y}\right).\eta\left(\frac{1}{y}\right) = f(1)g(y)y^2. \qquad (9.5.30)$$

Now, if the product of (9.5.28) and (9.5.30) is divided by that of (9.5.27) and (9.5.29) and nothing that fact that $x/y = w$ we get

$$\frac{c(w, w).c\left(1, \frac{1}{y}\right)}{c\left(\frac{1}{y}, \frac{1}{y}\right).c(x, w)} = \frac{f(w)f(1)}{f\left(\frac{1}{y}\right)f(x)}.$$

From the given condition (and noting that $w = x/y$), the above equation reduces to

$$\frac{f(w)f(1)}{f\left(\frac{1}{y}\right)f(x)} = \frac{h(w)h(1)}{h(x)h\left(\frac{1}{y}\right)}$$

which can be rewritten as

$$\frac{f(w)h(1)}{h(w)f(1)} = \frac{f\left(\frac{1}{y}\right)h(1)}{f(1)h\left(\frac{1}{y}\right)} \cdot \frac{f(x)h(1)}{f(1)h(x)} \qquad (9.5.31)$$

Now writing $\lambda(x) = [f(x)h(1)]/[h(x)f(1)]$ we immediately get from (9.5.31)

$$\lambda(w) = \lambda(x)\lambda\left(\frac{1}{y}\right).$$

This means

$$\lambda\left(\frac{x}{y}\right) = \lambda(x)\lambda\left(\frac{1}{y}\right)$$

for all $x, y > 0$. This gives $\lambda(x) = x^a$ for some arbitrary constant a so that

$$f(x) = \frac{f(1)}{h(1)}.h(x)x^a.$$

Again, from (9.5.29) and (9.5.30), we get

$$\frac{c\left(\frac{1}{y}, \frac{1}{y}\right)}{c\left(1, \frac{1}{y}\right)} = \frac{f\left(\frac{1}{y}\right)g(1)}{f(1)g(y)y}$$

which gives

$$g(y) = g(1) \cdot \frac{f\left(\frac{1}{y}\right)}{f(1)y} \cdot \frac{c\left(1, \frac{1}{y}\right)}{c\left(\frac{1}{y}, \frac{1}{y}\right)}$$

$$= g(1)k(y)y^{-a-1}.$$

The last equality follows because we have

$$\frac{f(x)h(1)}{f(1)h(x)} = \lambda(x) = x^a.$$

This completes the proof. □

Based on the above result, we prove a few characterization results for $GG(\alpha, \beta, \lambda)$.

Theorem 9.5.1 *Let X and Y be two independent non-negative random variables with $c(\cdot, z)$ as the conditional pdf of X given $Z = X/Y$. If $c(x, z)$ is factorized in the form*

9.5 Characterizations of the Generalized Gamma Distribution

$$c(x, z) = c_1(x) c_2(z) c_3\left(\frac{x}{z}\right),$$

then

(i) X follows a generalized Gamma distribution if and only if $c_1(x)$ is proportional to e^{-bx^a}, for $a > 0$ and $b > 0$;

(ii) Y follows a generalized Gamma distribution if and only if $c_3(x)$ is proportional to e^{-dx^c}, for $c > 0$ and $d > 0$.

Proof We prove (i) only since the proof of (ii) is similar. From Lemma 9.5.1 and the given condition, we have

$$\frac{c\left(\frac{x}{y}, \frac{x}{y}\right) . c\left(1, \frac{1}{y}\right)}{c\left(\frac{1}{y}, \frac{1}{y}\right) . c\left(x, \frac{x}{y}\right)} = \frac{c_1\left(\frac{x}{y}\right) c_1(1)}{c_1(x) c_1\left(\frac{1}{y}\right)}$$

$$= \frac{c_1\left(\frac{x}{y}\right)/c_1(1)}{(c_1(x)/c_1(1))\left(c_1\left(\frac{1}{y}\right)/c_1(1)\right)}$$

$$= \frac{h\left(\frac{x}{y}\right) h(1)}{h(x) h\left(\frac{1}{y}\right)},$$

where $h(x) = c_1(x)/c_1(1)$.

Given that $c_1(x) \propto e^{-bx^a}$. So, $h(x) = e^{b(1-x^a)}$. Hence by Lemma 9.5.1, we have

$$f(x) = f(1) x^\alpha e^{b(1-x^a)},$$

where α is an arbitrary constant. Hence X follows generalized gamma distribution. The proof of the converse is straight forward. □

It may be noted that knowledge about the functional forms of $c_2(\cdot)$ and $c_3(\cdot)$ is not necessary to characterize the distribution of X, and that of $c_1(\cdot)$ and $c_2(\cdot)$ is not necessary to characterize the distribution of Y. In fact a suitable combination of results (i) and (ii) of the above theorem will lead to the following result. The proof of the 'if part' is straight forward whereas that of 'only if part' can be constructed using Lemma 9.5.1. For details see Roy (1988).

Theorem 9.5.2 *Let X_1 and X_2 be two independent non-negative random variables. The conditional distribution of X_1 given $W = X_1/X_2$ is*

$$GG\left(\left(\frac{1}{\alpha_1^\lambda} + \frac{1}{w^\lambda \alpha_2^\lambda}\right)^{-1/\lambda}, \beta_1 + \beta_2, \lambda\right)$$

if and only if X_1 and X_2 have Generalized Gamma distributions of the form $GG(\alpha_1, \beta_1, \lambda)$ and $GG(\alpha_2, \beta_2, \lambda)$, respectively.

Below is one important theorem.

Theorem 9.5.3

(a) If we denote by $I(\cdot, z)$ the conditional information function for the random variable X given $Z = X/Y$ then $I(x, z) = -\ln c(x, z)$.

(b) If X and Y are two non-negative random variables with conditional information function of X given $Z = X/Y$ as $I(\cdot, z)$ then

$$I\left(\frac{x}{y}, \frac{x}{y}\right) + I\left(1, \frac{1}{y}\right) - I\left(\frac{1}{y}, \frac{1}{y}\right) - I\left(x, \frac{x}{y}\right) = H\left(\frac{x}{y}\right) - H(x) - H\left(\frac{1}{y}\right),$$

for all x and y, where $H(xy)H(1) = H(x)H(y)$, for all x and two values y_1 and y_2 of y such that $(\ln y_1 / \ln y_2)$ is irrational if and only if both X and Y follow generalized gamma distribution with the same shape parameter. □

The following is a generalization of the characterization result for Generalized Gamma distribution. Before we give this result, we state the following lemma (cf. Kagan et al., 1973).

Lemma 9.5.2 *Let X be a random variable having density $f(x)$ with support $S \subseteq \mathbb{R}$. Suppose $h_1, h_2, \ldots$ are integrable functions defined on S such that*

$$\int_S h_i(x) f(x) dx = g_i, \quad i = 1, 2, \ldots \qquad (9.5.32)$$

Then the maximum entropy is attained by distributions with the density of the form

$$f(x) = e^{a_0 + a_1 h_1(x) + a_2 h_2(x) + \ldots} \qquad (9.5.33)$$

(and only by them), if there exist $a_0, a_1, a_2, \ldots$ such that the density (9.5.33) satisfies (9.5.32).

Theorem 9.5.4 *Within the class of non-negative random variables with*

(a) *a given value G of the geometric mean, and*
(b) *a given value $1/\lambda$ for a raw moment of arbitrary but fixed order $\alpha(> 0)$,*

maximum entropy is attained if and only if X follows the Generalized Gamma distribution with shape parameter α.

Proof Condition (a) gives

$$\int_0^\infty \ln x f(x) dx = \ln G \qquad (9.5.34)$$

and (b) gives

$$\int_0^\infty x^\alpha f(x) dx = \frac{1}{\lambda} \qquad (9.5.35)$$

for some $\alpha > 0$. So, following Lemma 9.5.2, we have

$$\ln f(x) = \ln k + b \ln x - m x^\alpha$$

so that

$$f(x) = k x^b e^{-m x^\alpha}$$

which is the density of a Generalized Gamma distribution given in (9.5.26). The parameters m and b can be obtained from the two given conditions (9.5.34) and (9.5.35) and k as normalizing constant.

9.6 Characterizations of Pearsonian Type XI Distribution

A random variable X is said to follow Pearsonian Type XI distribution if its pdf is of the form:

$$f(x; a, q) = \frac{q}{a}\left(1 + \frac{x}{a}\right)^{-q-1}, \quad x > 0, \ a > 0, \ q > 1.$$

We denote this as $X \sim P(a, q)$.

As a particular case, we note the following result in Arnold (1983).

Proposition 9.6.1 *For a non-negative random variable X,*

$$m_X(t) = E(X - t | X > t) = E(X) + (B - 1)t,$$

where $B > 1$, for all $t > 0$ if and only if $X \sim P(a, q)$. $\square$

Note that, in the above Proposition, we have $a = E(X)/(B - 1)$ and $q = B/(B - 1)$ with $E(X) = a/(q - 1)$. The distribution has the survival function given by

$$\bar{F}_X(x) = \left(1 + \frac{x}{a}\right)^{-q}$$

and the failure rate

$$r_X(x) = \left(\frac{q}{a}\right)\left(1 + \frac{x}{a}\right)^{-1},$$

which is Pareto Type II distribution. However, a more interesting result can be presented in terms of the coefficient of variation $C_X(u)$ of residual life at elapsed time u. A similar characterization for the Exponential distribution has been indicated earlier.

Theorem 9.6.1 *For an absolutely continuous non-negative random variable X, coefficient of variation $C_X(u)$ of residual life is a constant greater than unity if and only if $X \sim P(a, q)$. If X belongs to the $IMRL$ class, $C_X(u)$ is a constant if and only if $X \sim P(a, q)$.* $\square$

A characterization of Pearsonian Type XI distribution through $m_X(u)r_X(u)$ can be obtained from the following theorem.

Theorem 9.6.2 *For a non-negative random variable X, $m_X(u)r_X(u)$ is a constant greater than unity for all u if and only if $X \sim P(a, q)$.* □

The following results provide some more characterizations of $P(a, q)$.

Theorem 9.6.3 (a) *If, for absolutely continuous non-negative random variable X, $Y_1, Y_2, \ldots, Y_n$ are ordered observations then the conditional coefficient of variation*

$$C(Z_k|Y_k = y) = \frac{\delta}{\sqrt{n-k}},$$

where $\delta > 1$, if and only if X follows $P(a, q)$, for suitably chosen a and q.
(b) *If, for absolutely continuous non-negative random variable X, $Y_1, Y_2, \ldots, Y_n$ are ordered observations then the conditional coefficient of variation*

$$C(W_k|Y_k = y) = \frac{\delta}{\sqrt{k-1}},$$

where $\delta > 1$, if and only if X follows $P(a, q)$, for suitably chosen a and q.
(c) *For non-negative random variable X with ordered observations as $Y_1, Y_2, \ldots, Y_n$,*

  (i) $E(G_k|Y_k = y) = (\alpha y + \beta)^{n-k}$, *where $\alpha > 0$, if and only if X follows $P(a, q)$*
  (ii) $E(H_k|Y_k = y) = (\alpha y + \beta)^{k-1}$, *where $\alpha > 0$, if and only if X follows $P(a, q)$,*

  for suitably chosen a and q.
(d) *For a non-negative absolutely continuous random variable X,*

  (i) $\bar{F}(a, x) = \bar{F}(1, x/a)$, $a > 0$ (i.e., a is a scale parameter), and
  (ii) $\bar{F}(a, x + y) = \bar{F}(a, x)\bar{F}(a + x, y)$ *for all x and for all y in the neighbourhood of zero if and only if X follows $P(a, q)$.*

(e) *If, for a non-negative random variable X, failure rate $r(x) > 0$, for $x > 0$, admits differential coefficient and if $\lim_{x \to 0+} r(x)$ exists, then the failure rate at the arithmetic mean of two time points s and t is equal to the harmonic mean of the failure rates at s and t for all s and $t (> 0)$ if and only if X follows $P(a, q)$.*
(f) *Let X be a random variable with $X_{(1)}$ as the minimum order statistic based on a random sample (of size n) observations drawn on X. Then X follows $P(a, q)$ if and only if $X_{(1)}$ follows $P(a, nq)$.* □

In (a), (b), and (c) of the above theorem, symbols Z_k, W_k, G_k, and H_k are as defined in Sect. 9.2. This may be compared with Theorem 9.2.3.

9.7 Characterizations Through Ageing Intensity Function

Let X be a random variable having density function f_X, survival function $\bar{F}_X$, and failure rate function r_X. Then its ageing intensity function L_X is defined as

$$L_X(t) = \frac{-tf_X(t)}{\bar{F}_X(t)\ln \bar{F}_X(t)} = \frac{tr_X(t)}{\int_0^t r_X(u)du}.$$

The different ageing classes, based on monotonicity of the ageing intensity function, have been studied in Nanda et al. (2007) and Bhattacharjee et al. (2013a) whereas Sunoj and Rasin (2017) have studied quantile-based ageing intensity function. For some more properties of ageing intensity function one may refer to Goodarzi et al. (2017), Kundu and Ghosh (2017), Misra and Bhattacharjee (2018), Szymkowiak (2018, 2019), to name a few. It is noted in Giri et al. (2023) that, for $t > 0$, and any $c > 0$,

$$\bar{F}(t) = \exp\left\{\ln \bar{F}(c) \exp\left(\int_c^t \frac{L_X(u)}{u} du\right)\right\}. \tag{9.7.36}$$

It is not difficult to see that random variables having proportional hazard rates have the same ageing intensity function.

Considering the sum of two failure rates as the failure rate of a distribution, called additive Weibull distribution by Xie and Lai (1995), its survival function is given by

$$\bar{F}_X(t) = \exp(-\alpha t^\theta - \beta t^\gamma), \ t \geq 0, \ \alpha, \beta, \theta, \gamma \geq 0.$$

We write this as $X \sim AW(\theta, \gamma, \alpha, \beta)$. If Y is a random variable having survival function $\bar{F}_Y(t) = [\bar{F}_X(t)]^A$, for some $A > 0$, we call this as Exponentiated Additive Weibull distribution and write this as $Y \sim EAW(\theta, \gamma, \alpha, \beta, A)$. It is easy to see that $EAW(\theta, \gamma, \alpha, \beta, A) \equiv AW(\theta, \gamma, A\alpha, A\beta)$. Thus, $EAW(\theta, \gamma, \alpha, \beta, A)$ and $AW(\theta, \gamma, \alpha, \beta)$ have the same ageing intensity function. The following theorem characterizes the EAW distribution in terms of its AI function. The 'only if' part of the above theorem is easy to prove. 'If part' of the theorem follows by using (9.7.36).

Theorem 9.7.1 *A non-negative random variable X follows $EAW(\theta, \gamma, \alpha, \beta, A)$ if and only if its AI function is given by*

$$L_X(t) = \frac{\alpha\theta t^\theta + \beta\gamma t^\gamma}{\alpha t^\theta + \beta t^\gamma}, \ t > 0,$$

for any $A > 0$. □

Exponentiated Weibull distribution, introduced by Mudholkar and Srivastava (1993), has the survival function

$$\bar{F}_X(t) = 1 - \left(1 - e^{-\alpha t^\theta}\right)^\beta, \ t > 0,$$

with $\alpha, \theta, \beta > 0$. Giri et al. (2023) have shown that the above distribution or a distribution having survival function

$$\left[1 - \left(1 - e^{-\alpha t^\theta}\right)^\beta\right]^A$$

with $A > 0$ is characterized by its ageing intensity function

$$L_X(t) = \frac{-\alpha\beta\theta t^\theta e^{-\alpha t^\theta}(1 - e^{-\alpha t^\theta})^{\beta-1}}{\{1 - (1 - e^{-\alpha t^\theta})^\beta\}\ln\{1 - (1 - e^{-\alpha t^\theta})^\beta\}}.$$

The Modified Weibull Extension distribution, proposed by Xie et al. (2002), having survival function

$$\bar{F}_X(t) = \exp\left\{\lambda\alpha\left(1 - e^{\left(\frac{t}{\alpha}\right)^\beta}\right)\right\}, t \geq 0$$

is characterized by the ageing intensity function

$$L_X(t) = \frac{\beta(t/\alpha)^\beta e^{(t/\alpha)^\beta}}{e^{(t/\alpha)^\beta} - 1}, t > 0.$$

Nadarajah and Kotz (2005) proposed a Weibull distribution which has its survival function given by

$$\bar{F}_X(t) = \exp\left\{-at^b\left(e^{ct^d} - 1\right)\right\} = \left[\exp\left\{-t^b\left(e^{ct^d} - 1\right)\right\}\right]^a, t \geq 0$$

with $a > 0, b \geq 0, c \geq 0, d > 0$. This can easily be characterized by using (9.7.36) in terms of its ageing intensity function

$$L_X(t) = \frac{cdt^d e^{ct^d} - b + be^{ct^d}}{e^{ct^d} - 1}.$$

Inverse Weibull distribution (also known as Reversed Weibull distribution or Reciprocal Weibull distribution) due to Mudholkar and Kollia (1994) has its survival function

$$\bar{F}(t) = 1 - e^{-\alpha t^{-\theta}}, \ t > 0, \alpha > 0, \theta > 0.$$

The corresponding exponentiated distribution is known as Kumaraswamy Inverse Weibull distribution having survival function

$$\bar{F}^*(t) = \left(1 - e^{-\alpha t^{-\theta}}\right)^b, \ \alpha > 0, \theta > 0,$$

for any real $b > 0$. This distribution is characterized in terms of its ageing intensity function (cf. Giri et al., 2023)

9.7 Characterizations Through Ageing Intensity Function

$$L_X(t) = \frac{-\alpha\theta e^{-\alpha t^{-\theta}}}{t^\theta (1 - e^{-\alpha t^{-\theta}}) \ln(1 - e^{-\alpha t^{-\theta}})}.$$

The Beta Inverse Weibull distribution due to Hanook et al. (2013) has its survival function given by

$$\bar{F}_X(t) = 1 - I_{e^{-t^{-\theta}}}(a, b),$$

where

$$I_x(a, b) = \frac{1}{B(a, b)} \int_0^x t^{a-1}(1-t)^{b-1} dt.$$

We may write this as $X \sim HW(\theta, a, b)$. The Exponentiated HW distribution, denoted as $EHW(\theta, a, b, A)$, has its survival function given by

$$\bar{F}_X^*(t) = \left(1 - I_{e^{-t^{-\theta}}}(a, b)\right)^A,$$

for any $A > 0$. The EHW distribution may be characterized through its AI function given by

$$L_X(t) = \frac{-\theta e^{-at^{-\theta}} \left(1 - e^{-t^{-\theta}}\right)^{b-1}}{t^\theta B(a, b) \left\{1 - I_{e^{-t^{-\theta}}}(a, b)\right\} \ln\{1 - I_{e^{-t^{-\theta}}}(a, b)\}}.$$

A non-negative continuous random variable is said to follow Exponentiated Truncated Log-Weibull distribution if its survival function is given by

$$\bar{G}_X(t) = C^{-A} \exp\left\{-A e^{\frac{t-a}{b}}\right\},$$

for any $A > 0$. This distribution may be characterized in terms of its ageing intensity function

$$L_X(t) = \frac{t e^{\frac{t-a}{b}}}{b \left(e^{\frac{t-a}{b}} + \ln C\right)}.$$

A non-negative continuous random variable X is said to follow Exponentiated Generalized Weibull distribution, denoted by $X \sim EGW(\alpha, \theta, \lambda, A)$, if its survival function is given by

$$\bar{G}_X(t) = (1 - \alpha\lambda t^\theta)^{A/\lambda},$$

for any $A > 0$. The AI function given by

$$L_X(t) = \frac{-\alpha\theta\lambda t^\theta}{(1 - \alpha\lambda t^\theta) \ln(1 - \alpha\lambda t^\theta)}$$

characterizes the $EGW(\alpha, \theta, \lambda, \cdot)$ distribution.

Ghitany et al. (2005) have developed the Extended Weibull distribution (using the semi-parametric method developed by Marshall and Olkin (1997)), denoted by $EW(\alpha, \lambda, \beta)$, having survival function given by

$$\bar{F}_X(t) = \frac{\alpha e^{-(\lambda t)^\beta}}{1 - \bar{\alpha} e^{-(\lambda t)^\beta}}, \quad t > 0,$$

with $\bar{\alpha} = 1 - \alpha$, and $\alpha, \lambda, \beta > 0$. A random variable X is said to follow Exponentiated Extended Weibull distribution, denoted by $X \sim EEW(\alpha, \lambda, \beta, A)$, if its survival function is given by

$$\bar{G}_X(t) = \left(\bar{F}_X(t)\right)^A,$$

for some $A > 0$. The EEW family of distributions is characterized by its AI function given by

$$L_X(t) = \frac{\beta \lambda^\beta t^\beta e^{(\lambda t)^\beta}}{\left\{e^{(\lambda t)^\beta} - \bar{\alpha}\right\} \ln \left\{\frac{e^{(\lambda t)^\beta} - \bar{\alpha}}{\alpha}\right\}}.$$

On using power transformation, Ghitany et al. (2013) proposed the Power Lindley distribution obtained from a Lindley random variable proposed by Lindley (1958). A non-negative random variable X is said to follow Power Lindley distribution, denoted by $X \sim PL(\alpha, \beta)$, if its survival function is given by

$$\bar{F}_X(t) = \left(1 + \frac{\beta t^\alpha}{\beta + 1}\right) e^{-\beta t^\alpha}, \quad t \geq 0,$$

with $\alpha, \beta > 0, t \geq 0$. Giri et al. (2023) have shown that, if any one of the following is true, the failure rate function of the distribution is decreasing.

(i) $0 < \alpha \leq \frac{1}{2}, \beta > 0$.
(ii) $1/2 < \alpha < 1, \beta \geq \frac{1 - 2\sqrt{\alpha(1-\alpha)}}{\alpha}$.

They have also shown that if any one of the conditions

(a) $\alpha = 1, 0 < \beta < 1$
(b) $\alpha > 1, \beta > 0$

holds good, then the distribution is unimodal. The roller-coaster behaviour of the failure rate of the distribution is also studied by them. A random variable X is said to follow Exponentiated Power Lindley distribution, denoted by $X \sim EPL(\alpha, \beta, A)$, if its survival function is given by

$$\bar{G}_X(t) = \left(1 + \frac{\beta t^\alpha}{\beta + 1}\right)^A e^{-A\beta t^\alpha}, \quad t \geq 0,$$

9.7 Characterizations Through Ageing Intensity Function

which is characterized by its AI function given by

$$L_X(t) = \frac{\alpha\beta^2(1+t^\alpha)t^\alpha}{(1+\beta+\beta t^\alpha)\left\{\beta t^\alpha - \ln(\frac{1+\beta+\beta t^\alpha}{1+\beta})\right\}}.$$

Generalized Power Weibull distribution, proposed by Nikulin and Haghighi (2006), is a non-negative continuous random variable having survival function given by

$$\bar{F}_X(t) = \exp\left\{1 - (1+\alpha t^\theta)^{\frac{1}{\lambda}}\right\}.$$

Its exponentiated version, called Exponentiated Generalized Power Weibull distribution, has the survival function given by

$$\bar{G}_X(t) = \exp\left\{A\left(1 - (1+\alpha t^\theta)^{\frac{1}{\lambda}}\right)\right\}, \ t \geq 0.$$

This is characterized in terms of its AI function

$$L_X(t) = \frac{\alpha\theta t^\theta(1+\alpha t^\theta)^{\frac{1}{\lambda}-1}}{\lambda[(1+\alpha t^\theta)^{1/\lambda} - 1]}.$$

Odd Weibull distribution, defined by Cooray (2006), is a non-negative random variable having survival function

$$\bar{F}_X(t) = \{1 + (e^{\alpha t^\theta} - 1)^\lambda\}^{-1}.$$

A random variable X is said to follow Exponentiated Odd Weibull distribution if its survival function is given by

$$\bar{G}_X(t) = \{1 + (e^{\alpha t^\theta} - 1)^\lambda\}^{-A}, \ t \geq 0, A > 0.$$

This can be characterized by its AI function given by

$$L_X(t) = \frac{\alpha\theta\lambda t^\theta e^{\alpha t^\theta}(e^{\alpha t^\theta} - 1)^{\lambda-1}}{\{1 + (e^{\alpha t^\theta} - 1)^\lambda\}\ln\{1 + (e^{\alpha t^\theta} - 1)^\lambda\}}.$$

A continuous non-negative random variable follows Flexible Weibull Extension distribution due to Bebbington et al. (2007) if its survival function is given by

$$\bar{F}_X(t) = \exp\left\{-e^{\alpha t - \beta/t}\right\}, \ \alpha, \beta > 0, t \geq 0,$$

which can be easily shown to be characterized by its AI function

$$L_X(t) = \alpha t + \frac{\beta}{t}.$$

Exponentiated Kumaraswamy Weibull distribution, a five-parameter distribution due to Lemonte et al. (2013), has its survival function given by

$$\bar{F}_X(t) = 1 - \left[1 - \left\{1 - \left(1 - e^{-\delta t^\theta}\right)^\alpha\right\}^\beta\right]^\gamma, \quad \delta, \theta, \alpha, \beta, \gamma > 0, t > 0.$$

A random variable Y is said to follow Exponentiated Lemonte distribution if its survival function is given by $\bar{F}_Y(t) = [\bar{F}_X(t)]^A$ for some $A > 0$. This can be characterized by its AI function given by

$$L_X(t) = \frac{e^{-\delta t^\theta}\left(1 - e^{-\delta t^\theta}\right)^{\alpha-1}\left\{1 - \left(1 - e^{-\delta t^\theta}\right)^\alpha\right\}^{\beta-1}\left[1 - \left\{1 - \left(1 - e^{-\delta t^\theta}\right)^\alpha\right\}^\beta\right]^{\gamma-1} t^\theta \delta \alpha \beta \gamma \theta}{\left[1 - \left[1 - \left\{1 - \left(1 - e^{-\delta t^\theta}\right)^\alpha\right\}^\beta\right]^\gamma\right] \ln\left[1 - \left[1 - \left\{1 - \left(1 - e^{-\delta t^\theta}\right)^\alpha\right\}^\beta\right]^\gamma\right]}.$$

Kumaraswamy Modified Weibull distribution, introduced by Cordeiro et al. (2014), has its survival function given by

$$\bar{F}_X(t) = \left[1 - \left(1 - e^{-\alpha t^\theta e^{\lambda t}}\right)^a\right]^b,$$

with $\alpha, \theta, \lambda, a, b > 0, t \geq 0$, which may be characterized by its AI function

$$L_X(t) = \frac{a\alpha e^{\lambda t} e^{-\alpha t^\theta e^{\lambda t}}\left(1 - e^{-\alpha t^\theta e^{\lambda t}}\right)^{a-1} t^{\theta-1}(\theta + \lambda t)}{\left(1 - \left(1 - e^{-\alpha t^\theta e^{\lambda t}}\right)^a\right) \ln\left(1 - \left(1 - e^{-\alpha t^\theta e^{\lambda t}}\right)^a\right)}.$$

A non-negative random variable X having survival function given by

$$\bar{F}_X(t) = \exp\left(1 - a^{t^\alpha}\right), a > 1, \alpha > 0, t > 0$$

is known as Loglog distribution, a Weibull distribution due to Pham (2002). This has a bathtub-shaped failure rate. This family of distribution is characterized in terms of its AI function

$$L_X(t) = \frac{\alpha t^\alpha a^{t^\alpha} \ln a}{a^{t^\alpha} - 1}.$$

9.8 Characterizations Through Coordinate Sub-tangent

For a non-negative random variable X with coordinate sub-tangent (CST) $T(x)$, suppose $T(x)r(x) = \alpha$, a constant. Then we have the following theorem (see Sect. 8.4 for discussion on CST).

9.8 Characterizations Through Coordinate Sub-tangent

Theorem 9.8.1 *For any non-negative random variable X,*

(i) $T(x)r(x) = 1$ *if and only if X has an Exponential distribution;*
(ii) $0 < T(x)r(x) < 1$ *if and only if X has a Pearsonian Type XI distribution;*
(iii) $T(x)r(x) < 0$ *or $T(x)r(x) > 1$ if and only if X has the Finite-range distribution with survival function given by $\bar{F}(x) = \left(1 - \frac{x}{R}\right)^\theta$, $0 < x < R$ and $\theta > 1$.*

Proof *Only if Part*: Suppose $T(x)r(x) = \alpha$, a constant. Given that $\alpha = 1$. So, we get

$$\frac{f(x)}{\bar{F}(x)} = -\frac{f'(x)}{f(x)}$$

yielding $-\ln \bar{F}(x) = -\ln f(x) + \ln \lambda$, for any arbitrary constant λ. Hence $f(x)/\bar{F}(x) = \lambda$, which gives

$$f(x) = \lambda e^{-\lambda x},$$

for any arbitrary constant $\lambda > 0$. If $\alpha \neq 1$, we get

$$f(x) = k\bar{F}^{1/\alpha}(x),$$

where $k > 0$ is any arbitrary constant. Integrating both sides we get

$$\int_0^t \frac{f(x)}{\bar{F}^{1/\alpha}(x)} dx = kt + c.$$

If $\alpha \in (0, 1)$, we have $1 - 1/\alpha < 0$. Let us write, for some $q > 0$, $1 - 1/\alpha = -1/q$. Then the above integral reduces to

$$-q\left[1 - \left(\bar{F}(t)\right)^{-1/q}\right] = kt + c.$$

The initial condition gives $c = 0$ so that

$$\bar{F}(t) = \left(1 + \frac{t}{a}\right)^{-q},$$

where $a = q/k$, $k, q > 0$, $a > 0$. So, X follows Pearsonian Type XI with $\alpha \in (0, 1)$. If $\alpha < 0$ or $\alpha > 1$, writing $q = -\theta$, $\theta > 0$, we get

$$\bar{F}(x) = \left(1 + \frac{x}{a}\right)^\theta = \left(1 - \frac{x}{R}\right)^\theta$$

where $R = -a = -q/k = \theta/k > 0$. Thus, X follows the Finite-range distribution over $(0, R)$.

If Part: Note that, for the Exponential distribution, $T(x)r(x) = 1$, for the Pearsonian Type XI distribution this works out to be $q/(q+1) < 1$ and, for the Finite-range distribution having survival function, as given in (iii), we get $T(x)r(x) = \theta/(\theta - 1)$ which is either greater than 1 if $\theta > 1$ and is less than 0 in case $\theta < 1$.

Chapter 10
Distributions of Special Interest

10.1 Introduction

Distributions of time-to-failure (which may be called life distributions, rightly so in the case of a non-repairable, continuous-duty device or unit) discussed in the previous chapter assume (indirectly) only two states of functioning of a device, viz. failed or functioning. No differentiation was attempted among different possible states of being functional, say, with different associated levels of performance and failure times could be exactly observed without any error. Real-life situations exist where we have to recognize multiple states of functioning as also where the failure time of a device can be subject to error of observation/measurement. And, in the latter case which can complicate analysis even in the first case, attempts have been made to make use of the 'fuzziness' concept, with appropriate choice of membership functions.

Quite a few situations calling for probabilistic models to represent failures in engineering structures subjected to operational and environmental stresses have led to the development of fatigue failure models. The models try to capture the failure mechanism in terms of existing theoretical and empirical knowledge of crack propagation and related phenomena and then incorporate adjustments for some of the parameters likely to vary randomly.

In conventional stress-strength analysis, observations/measurements on stress (X) and strength (Y) are considered as exact values and reliability is defined as $R = P(Y > X)$. However, measurement of stress (at failure) and possibly of strength as well is subject to some inherent imprecision as distinct from random variation in the measured stress (or strength) from one prototype to another or from one operation cycle to another.

Mukherjee and Maiti (1996b) considered the estimation of stress-strength reliability using damaged data. This chapter provides a brief outline of probability models which are of some special interest either in terms of the relevance in some typical situations or for taking into consideration some aspects of reality which usually go unnoticed in the classical models.

10.2 Distribution of Quality-Adjusted Life

Traditional reliability analysis takes note of two states for a product in use, viz 'functional' and 'failed'. Multi-state reliability analysis, of course, recognizes more than these two states, by considering different partly functional states. However, not much has been done to associate appropriate measures of efficiency (quality of performance) with the various states, to note the random duration of stay in each of these states and to define a quality-adjusted life (QAL) in terms of the weighted total of these durations, using corresponding measures of efficiency as weights. A somewhat analogous concept used in the context of 'Burden of Disease' is the Disability Adjusted Years of Life (DALY) proceeding to Disability Adjusted Life Expectancy (DALE). QAL can be regarded as the cumulative value of a critical parameter (which can be treated as an efficiency factor) that determines whether a product in use functions fully satisfactorily or not till the time its value $Y(t)$ remains above a threshold. (A value below the threshold will lead to product failure.)

In the simplest set-up, we can recognize only three states with a state of functioning with reduced efficiency in between the fully functioning and the failed states. The product can go directly from the fully functional to the failed state, with some probability p, say. And let the time-to-failure be T_1. With the complementary probability, it passes from the fully functional to the partly functional state after a random time, say T_2, and stays for time T_3 in that state before passing to the failed state. Mukherjee and Pal (2002) assumed the full efficiency factor as 1, regarded the durations of stay T_1, T_2 and T_3 to be independently and exponentially distributed with parameters λ_i, $i = 1, 2, 3$ and worked out the distribution of QAL defined as

$$T = \begin{cases} T_1, & \text{with probability } p \\ T_2 + bT_3, & \text{with probability } 1 - p \end{cases}$$

Later on, Mukherjee and Pal (2007) assumed T_2 and T_3 to follow a bivariate exponential distribution of the Farlie-Gumbel-Morgenstern type and worked out the distribution of QAL. In both the cases, failure rate and mean remaining life functions were studied.

10.3 Life Distribution from Imprecise Data

The exact or true life (time-to-failure for a non-repairable item) may not be available, in the absence of a continuously operating recording mechanism or because the data are intentionally under-reported by customers (users) or over-reported by dealers and distributors (to potential customers) in some cases. In the latter situation, the observed life or failure time X is composed of two components, viz. an unknown true value T and a reporting error U. In estimating the distribution of T from the observations on X and the assumed distribution for U, we use the multiplicative model $X = T.U$ more often than the additive model $X = T + U$.

In the multiplicative model, over-reporting implies $U > 1$ and under-reporting corresponds to the range $0 < U < 1$.

There exist situations where the time-to-failure cannot be exactly determined, except in terms of an interval between two consecutive times for inspection or unless the item is put on operation (maybe, after the last cycle of operation, in case of an intermittent duty item). Thus observed or recorded failure times may be treated as 'fuzzy'.

Alam et al. (1980) considered the case of a power law for the true failure time with density

$$f(t) = \frac{p}{\alpha}\left(\frac{t}{\alpha}\right)^{p-1}, \ t < \alpha, \ p > 0,$$

while reporting error factor follows the density

$$g(u) = \lambda\left(\frac{1}{u}\right)^{\lambda+1}, \ u \geqslant 1.$$

The authors worked out the distribution of X, though properties of the same were not reported by them.

10.4 Reliability from Fuzzy Stress and Strength

One way to characterize imprecise data is to introduce fuzzy numbers. A fuzzy number X with a membership or characterizing function $\mu(\cdot)$ is a measurable real function obeying $0 \leqslant \mu(x) \leqslant 1$ for all $x \in \mathbb{R}$ and $\{x \in \mathbb{R} : \mu(x) = 1\} \neq \emptyset$ such that $[x \in \mathbb{R} : \mu(x) > 0]$ is an interval. The idealization of an exact stress/strength observation X is the special case $\mu(\cdot) = I_{\{x\}}(\cdot)$.

Viertl (1988) considered non-parametric estimation of mission time reliability function based on fuzzy lifetime data using both the classical and the Bayesian approaches. Mukherjee and Maiti (1996b) considered the estimation of stress-strength reliability using damaged as well as fuzzy data. They also studied the problem of estimating reliability $R = P(X < Y)$, where X and Y are fuzzy random stress and strength, respectively. For this purpose, they considered the membership function proposed by Civanlar and Trussell (1986) which is, in a sense, data-based unlike most other formulations of membership function. This method finds the smallest fuzzy set which assigns high average membership values to those objects with the defining features distributed according to a given probability density function. In fact, this membership function $\mu(\cdot)$ is based on the following principles:

(1) $0 \leq \mu(x) \leq 1$,
(2) $E(\mu(X)|X$ is distributed according to the underlying pdf $p(\cdot)) \geq C$ (a constant),
(3) $\int \mu^2(x)\,dx$ should be minimized. [This condition is required to obtain a selective membership function so that the size of the set should be as small as possible.]

Here $C < 1$ is the confidence level to be chosen close to 1. Quantitatively, since the only information available is the pdf for the members of the set, the average membership values assigned to the values distributed according to the given pdf should be large.

Now the membership function is given by

$$\mu(x) = \begin{cases} \lambda p(x), & \text{if } \lambda p(x) < 1 \\ 1, & \text{if } \lambda p(x) \geq 1, \end{cases} \quad (10.4.1)$$

where $p(x)$ is the density and the constant λ is to be solved from

$$\lambda \int_{\{x:\lambda p(x)<1\}} p^2(x)\,dx + \int_{\{x:\lambda p(x)>1\}} p(x)\,dx - C = 0. \quad (10.4.2)$$

For a given pdf, $\lambda = \lambda(C)$ can be worked out from (10.4.2). In the particular case of the exponential pdf with failure rate α, λ comes out as $\frac{1}{2}(\alpha(1-C))^{-1}$ and the membership function is determined by the parameter k which is a function of C, as derived from (10.4.1) and given by

$$k = \max\left\{0, -\frac{1}{\lambda}\ln(2(1-C))\right\} = -\frac{1}{\lambda}\ln(2(1-C)).$$

Hence

$$\mu(x) = \begin{cases} \frac{1}{2(1-C)} e^{-\alpha x}, & \text{if } x > k \\ 1, & \text{if } x < k. \end{cases} \quad (10.4.3)$$

10.5 Reliability Under Exponential Stress and Exponential Strength

Assume that X has the membership function $\mu_1(\cdot)$ with underlying pdf exponential with failure rate α_1 and Y has the membership function $\mu_2(y)$ with underlying pdf exponential with failure rate α_2.

In conventional approach

$$R = \frac{\alpha_1}{\alpha_1 + \alpha_2}.$$

10.6 Fatigue Failure Models   301

When $\mu_1(x)$ and $\mu_2(y)$ are of type (10.4.3), we get

$$R = \begin{cases} \frac{\alpha_1}{\alpha_1+\alpha_2}\left[1 - W_1 e^{-\alpha_2 k_1}\right] + (D_2 W_2^2 - W_2)(1 - W_1) + 2D_1 \frac{\alpha_1}{2\alpha_1+\alpha_2}\left[W_1^2 e^{-\alpha_2 k_1} - W_2 e^{-2\alpha_1 k_2}\right] \\ + D_1\left(D_2 W_2^2 - W_2\right)\left(W_1^2 - e^{-2\alpha_1 k_2}\right) + \frac{\alpha_1}{\alpha_1+\alpha_2}\cdot D_1 D_2 W_2^2 e^{-2\alpha_1 k_2} & \text{if } k_1 < k_2 \\[6pt] \frac{\alpha_1}{\alpha_1+\alpha_2}\left(1 - W_2 e^{-\alpha_1 k_2}\right) + (D_2 W_2^2 - W_2)\left(1 - e^{-\alpha_1 k_2}\right) + D_2 \cdot \frac{\alpha_1}{\alpha_1+\alpha_2}\left(W_2^2 e^{-\alpha_1 k_2} - W_1 e^{-2\alpha_2 k_1}\right) \\ + \frac{\alpha_1}{\alpha_1+\alpha_2}\cdot D_1 D_2 W_1^2 e^{-2\alpha_2 k_1} & \text{if } k_1 > k_2, \end{cases}$$

where $D_1 = (4(1 - C_1))^{-1}$, $D_2 = (4(1 - C_2))^{-1}$, $W_1 = e^{-\alpha_1 k_1}$, and $W_2 = e^{-\alpha_2 k_2}$.

10.6 Fatigue Failure Models

Structures and components made of brittle as also non-brittle materials subjected to constant or occasional stresses of constant or varying intensity arising from different factors develop cracks which propagate in a random manner, causing eventual failures. Such stresses could be initially present because of manufacturing defects or caused during operation or even rest periods mainly due to inherent properties of the material or the environment. The consequence is that these structures or components fail at random points of time. Attempts to predict failure times and to estimate reliability of such products are based on models representing propagation of fatigue cracks relating length of the crack developed and the time elapsed to the number of load/stress cycles gone through, during three phases, viz. initial, stable growth and fast growth. Most of these models are deterministic, though some stochastic differential models taking care of randomness of some factors affecting crack length have also been proposed. From any such crack propagation model, the probability distribution of time-to-failure (number of load/stress cycles) can be worked out by making some reasonable assumptions. A comprehensive review of deterministic models has been provided in Beden et al. (2009), while some other authors have adopted the stochastic process approach.

Most of the failure-time distributions derived under plausible physical considerations are based on fatigue crack growth under variable stress or cycle load. However, distributions of fatigue life under a constant stress situation are also important. In these models, time-dependent growth of cracks continues until a critical size is reached after which fast fracture sets in. And failure takes place as soon as the crack length exceeds a critical size. Factors which affect the failure propagation process depend mostly on properties of the material and on measures associated with the stress cycle under which the structure or component operates. Static fatigue refers to the phenomenon describing sub-critical flaw (fatigue crack) growth to such critical dimensions (Cooray and Ananda, 2008).

Let us consider a model to represent the phenomenon of fatigue failure and to work out a model for time-to-failure of a structure subject to randomly varying (in intensity) stress in successive stress cycles or to a sustained constant pressure or stress (essentially caused by the designed operating environment like temperature). There can exist two strategies for predicting reliability. A simple strategy would be based directly on time-to-failure or, in this case, number of stress cycles to failure. The other takes into account growth of a crack due to stress as caused by different physical factors exceeding a critical level that depends on material properties. For a complex structure with a designed life which is pretty 'long' like the pier of a bridge, we cannot observe failure times directly. A long search into the past may provide sparse data at the best and we have to fall back upon a fatigue failure model based on empirical data on stress applied directly or otherwise. However, failure times can be directly observed in the case of structures or components made of brittle materials and usually subjected to a constant distribution of stress, implying mean stress along with material properties and surface geometry. Even in the latter case, fatigue failure models have been and are being used to predict failure time for a given specimen or to estimate reliability of a lot or brand of specimens. For the sake of simplicity, we assume that the structure or component under consideration did not have any flaw or crack initially.

10.6.1 Deterministic Models

Let us first consider the case of a set of brittle homogeneous specimens subjected to a constant sustained pressure at stress level σ due primarily to the environment, in terms of temperature or vibration or a similar operating condition. In reality, small variations in temperature and variations in vibration magnitudes and intervals are likely to be there. The requirement in the model is that the distribution of stress due to such factors will not change and the distribution can be well taken care of by its mean value. We will notice that the specimens (units) fail at random times, due primarily to variations in material properties and environmental impacts likely to hasten fatigue failure. We assume that no cracks existed at the beginning and will appear only because of the applied stress.

The empirical law of crack velocity connecting strength degradation to its factors may be stated as

$$\frac{dL}{dt} = AK_t^B,$$

where L is the crack length at time t, K_t is the crack tip stress intensity factor for Phase II, and A and B are constants.

If $K_t < K_0$ (material fracture toughness), the specimen is tough enough to withstand the effect of stress concentration around the crack.

Stress intensity K_t corresponding to a crack length L under a (mean) stress σ is given by

$$K_t = \sigma Y L^{1/2},$$

with a constant Y depending on crack geometry.

10.6 Fatigue Failure Models

We assume that the stress distribution remains constant, while the crack grows larger and that small cracks relative to the specimen dimensions do not obliterate the surface geometry of cracks, leaving Y a constant when crack length grows. We thus get crack length at time t as

$$L = ct^\alpha,$$

where $\alpha = 2/(2 - B)$, $B < 2$, and c depends on σ and crack geometry. From the crack geometry eventually developing, the dominant crack length (usually treated as an extreme value) is identified and fatigue failure results from the eventual extension of the dominant crack beyond a critical value. The specimen fails as soon as L exceeds L_c. Usually, a half-normal distribution is assumed for L. Under this assumption, probability of failure of the specimen is given by

$$P(L \leqslant L_c) = F_L(L_c) = 2\Phi(L_c) - 1, \ L_c \geqslant 0.$$

Subsequently, the distribution of time-to-failure as the value of t corresponding to $(L/c)^{1/\alpha}$ can be worked out. As Cooray and Ananda (2008) have shown, the resulting distribution function for failure time comes out as

$$F(x) = 2\Phi\left[\left(\frac{x}{\theta}\right)^\alpha\right] - 1, \ \alpha, \theta > 0, \ x \geqslant 0.$$

This distribution has been sometimes referred to as the generalized half-normal distribution.

It is interesting to note that this distribution has been found to provide better fit to several data sets considered by Cooray and Ananda (2008) over some commonly used distributions like gamma, Weibull, log-normal, and Birnbaum-Saunders. They illustrated this fact by taking a set of 101 data points representing stress-rupture life of Kevlar49 epoxy strands which were subjected to constant sustained pressure at the 90% stress level until all had failed. A similar finding came up with a set of 49 data points using 70% stress level.

Case of Metallic Structures under Varying Stress: Let us now attempt to model the phenomenon of fatigue failure of a metallic structure or component during operation under randomly varying stress in successive stress cycles. Unlike random variations in times-to-failure of similar such structures, this probability model is to be somewhat specific for this structure. What can be observed are not failure times of similar structures, but some measure of 'degradation' or 'fatigue' or its outcome like the length of some crack developing internally with cumulative stress encountered up to different points of time. Such a measure can be related to time-to-failure, may be in terms of a differential equation, and a probability model for time-to-failure can then be derived.

Let us now consider a model to represent the length of a crack after N cycles of the load, which we denote by $a(N)$. Given this, the lifetime of a component is easily defined. Since, near failure, the crack grows very quickly (in Phase III of the crack propagation process), one can specify lifetime as the time at which the length of a crack first exceeds a suitable threshold length A_{th}. Thus, the time-to-failure can be represented (in terms of the number of cycles) as

$$N_j = \inf\{N | a(N) > A_{th}\} \tag{10.6.4}$$

and, under the assumption that $a(N)$ is non-decreasing, the reliability function is

$$P(N_j > N) = P(a(N) \leqslant A_{th}). \tag{10.6.5}$$

In this section, we consider some deterministic models of fatigue crack growth based on material properties, details of the geometry, and pattern of stress under which the units operate. Quite a few models for crack length velocity have been proposed, some specific to certain phases of the crack propagation process. Two such models, viz. the Paris-Erdogan equation and the Forman equation have been quite often used in metallurgical research. The Paris-Erdogan equation is derived from empirical considerations and has no real theoretical basis. The equation models the relation between crack velocity and an abstract quantity called the 'stress intensity range', which characterizes the magnitude of stress at the tip of the crack. This range is denoted by Δk and is usually defined as $\Delta k = Q\Delta\sigma/\sqrt{\alpha}$, where Q is a parameter corresponding to crack geometry, $\Delta\sigma$ is the range of stress in a cycle, and α is a constant.

The form of the Paris-Erdogan equation is

$$\frac{da}{dN} = C(\Delta k)^n \quad \text{and} \quad a(0) = A_0. \tag{10.6.6}$$

We note that C and n are regarded as experimentally determined material constants, but they also depend upon factors such as frequency, temperature, and stress ratio. The Paris-Erdogan equation gives good results for long cracks when the material constants are known, but we note that a large effort is required to determine them, since they are functions of many variables. As remarked earlier, the Paris-Erdogan equation is valid only in the stable (crack) growth region and does not work well either in the threshold region or in the accelerated growth region. Further, this equation does not take into account the stress ratio $R = \sigma_{\min}/\sigma_{\max}$ (ratio of minimum to maximum stress in a cycle) which has an important effect on crack growth, according to Bannantine et al. (1990).

Walker (1970) improved upon the Paris-Erdogan model by incorporating the stress ratio by defining stress intensity factor in terms of a new parameter R' as $\Delta k' = K_{\max}(1-R')^w$ resulting in the law given by

$$\frac{da}{dN} = C_w \left(\frac{\Delta k}{1-R^{1-\gamma}}\right)^n, \tag{10.6.7}$$

where γ is a material-specific constant. This equation reduces to the Paris-Erdogan law as $R = 0$.

The Forman equation also accounts for the stress ratio and has the form

$$\frac{da}{dN} = \frac{C(\Delta k)^n}{(1 - R')K_c - \Delta k},$$

where K_c is a critical level for the stress intensity, corresponding to unstable fracture.

10.6.2 Stochastic Model

It is easily understood that crack propagation, being a random time-dependent process, should be modelled better in terms of stochastic model, rather than any of the deterministic ones. One can directly formulate a stochastic differential equation and proceed solving that. Alternatively, we can use a stochastic process to model the crack length by an N-dependent stochastic process. Another possibility is to starting with some chosen deterministic model like the Paris-Erdogan model and randomize the same in several possible ways. Thus,

(1) we can assume random model parameters θ. Since α is a function of θ, a distribution on crack length is implied. Since N_j is a function of α and therefore of θ, a lifetime distribution is also implied; or
(2) we can take some non-decreasing stochastic process (such as a birth process) indexed by N and specify process parameters such that the expected value of the process is $\alpha(N)$, or
(3) if $\alpha(N)$ has been defined in terms of a differential equation, one can form an equivalent stochastic differential equation whose solution is a stochastic process model for $\alpha(N)$.

We consider here the random parameter model. A simple random model can be generated by taking the Paris-Erdogan model and assigning probability distributions to the parameters, n and A_0. This approach to study fatigue failure has been illustrated in the case of ceramic materials by Paluszny and Nicholls (1978). For ceramics, the assumption of a constant stress σ, instead of a cyclic one, seems more reasonable and one indexes growth by time t and the equivalent to the Paris-Erdogan equation comes out as

$$\frac{d\alpha}{dt} = v_0 k^n, \tag{10.6.8}$$

where $k = Q\alpha\sqrt{\sigma}$ is the stress intensity, and v_0 and n are material-specific parameters.

In this approach, we use a model to represent the strength of a specimen (defined as the maximum load or stress that the specimen can bear) rather than the crack length. The strength $S(t)$ inherent in the specimen at time t is related to crack length by the equation

$$S(t) = \sqrt{\frac{B}{\pi a(t)}} \qquad (10.6.9)$$

for a material constant B; this equation is derived in a seminal work of Griffith (1921). Since strength is inversely related to crack length at time t, the specimen fails as soon as $S(t) < \sigma$, so that time-to-failure can be defined as

$$T = \inf\{t : S(t) < \sigma\}. \qquad (10.6.10)$$

Using Eqs. (10.6.8), (10.6.9), and (10.6.10), and given that $n > 10$, for ceramic materials, one can show that the lifetime is approximately

$$T \cong K[S(0)]^{n/2}, \qquad (10.6.11)$$

where K is a function of material constants and stress intensity. If $S(0)$ is assumed to follow Weibull distribution then the distribution of T can be shown to be Weibull distributed.

The fact that the initial strength $S(0)$ is determined by the size of the largest initial crack, which can be reasonably assumed to follow one of the three well-known extreme value distributions applicable to the maximum of a number of normalized *iid* random variables, provides justification for assuming a Weibull distribution for $S(0)$. There have been other approaches to work on a 'randomized' version of the Paris-Erdogan equation, e.g., by introducing a stochastic process $X(N, \theta)$ as a multiplier of $C(\Delta K)^n$ to take care of any deviation from the equation over time and to solve the stochastic differential equation. This is expectedly complicated and has been rarely tried out.

Chapter 11
Using an Appropriate Probability Model

The use of an appropriate model is not only important in all forms of statistical analysis but also it is central to any scientific enquiry. In fact, taking a broad view of the word 'data' as 'whatever are given' by way of figures, maps, signatures, images, and documents and the like to study a phenomenon and of the word 'model' to imply a construct that can be used to experiment on the phenomenon and to generate pertinent data to be used as a part of the premises in any exercise for inferencing about the phenomenon, scientists have often raised doubts about which to come first: the model or the data? The question is so basic that it has led to two distinct schools of thought, one arguing that a model be first considered and then used to generate data that can be analysed to throw light on the phenomenon, while in the other, data are first collected and subsequently analysed through an appropriately chosen model to make data-based conclusions valid beyond the observed data. In either case, use of a model remains important, though the acceptance of a model as 'appropriate' requires a different outlook in each of the two thoughts.

Any analysis of observed data to make inferences or predictions must use an appropriate probability distribution to represent the data-generating mechanism (from which the true or the underlying probability model can be developed) and to reveal the underlying features of the data which are relevant for the analysis and for the goals or objectives of the data-based analysis. We are most often interested in making some probability-based inductive inferences to proceed from the observed data to more generally applicable results. And in this process, the use of an appropriate probability model cannot be escaped. The question of 'appropriateness' is easier appreciated than stated formally in terms of certain verifiable requirements. In any case, appropriateness takes into account both 'performance' of the model judged in terms of closeness of predictions made under the model with 'actual' realizations later or by way of validity of inferences based on the model as inputs into an existing theory or even in terms of the nature and extent of explanation of some phenomenon associated with the data, as well as 'simplicity' in terms of number of parameters to be estimated from the sample data. It may be remembered that two identifiable goals for model selection are (a) understanding the data-generating process and hence of the underlying phenomenon, thus ensuring valid inferences about the phenomenon and (b) making predictions based on the given data. The two goals may not be achieved simultaneously.

In some simple situations, the underlying data-generating mechanism may be made out—at least to some extent—by the data analyst who can then introduce some tenable assumptions or postulates, so that the phenomenon under study can be represented in terms of a symbolic model like differential equation (partial or total) or even a system of such equations with some boundary conditions. Some parameter(s) in the equation or system of equations (in case of a complex phenomenon) may behave randomly from one occasion to another. This random variation can be taken care of by assuming some simple and reasonable probability distribution for each such unknown parameter. This finally results in a probability model that can be adopted for purposes of inferencing or prediction. An example of such a situation has been taken up earlier to derive the Weibull distribution in the context of fatigue failure through propagation of a crack inside a metallic structure. Several stochastic process models have been developed to consider the impact of randomly occurring shocks of (randomly varying) intensities on structures or other products leading to failures. This activity can be broadly referred to as the 'model development' approach in the present context. It must be added that availability of or access to scientific evidence about the underlying physics or mechanism that generates the observed values of some variable like time-to-failure or time-between-consecutive failures or time-to-repair as the 'response' of some experiment—maybe an implicit one—to the data analyst is often quite limited and inadequate for the purpose of formulating a mathematical representation of the underlying phenomenon.

In most other situations in practice, we can think of two alternative approaches to identifying the most appropriate model referred to as the model selection approach, wherein we start with a set of candidate models (with a known cardinality) and develop a criterion for selecting one element from this set as the most appropriate. Of course, the candidate models should be chosen carefully by making use of any existing knowledge about the underlying mechanism for data generation and/ or by matching the graphical data plot with plotted representations of some well-known distributions or by focusing on certain typical features of the observed data and matching those with corresponding features of some candidate models. In this connection, different information criteria which are essentially derived from the maximized likelihood of the observed data on the basis of a model, penalized for the number of independent parameters in the model to be estimated from the data are used. In the other approach, we may use some characterization result(s) for each of the probability models that appear (maybe as indicated by a suitable graphical display of the observed data) to be more or less appropriate, work out a sample (estimated) version of some characterization result (usually avoiding those which relate to sampling distributions derived from primary models) and check (if possible, by way of a statistical test) the closeness (or distance) between the sample version and the actual result characterizing a particular model. Thus, we may examine the failure rate behaviour, or memory, or mean remaining life, or some measure of entropy, or some relation between different averages of failure rate, etc., to choose an appropriate model.

And in both the approaches, we have to test for goodness-of-fit to check on the performance of a chosen model in the task of explaining, inferencing, and prediction. Given candidate models of equal explanatory or predictive power, the simplest model is most likely to be the 'best' choice. (This reminds us of the Occam's razor). However, the 'best' among several candidate models will be appropriate for further analysis to the extent indicated by the way the candidate models were chosen.

11.1 Model Selection

Statistical literature as also literature on signal processing is replete with many interesting articles on the model selection problem and one can conveniently go through several comprehensive reviews. Mention may be made of Ghosh and Samanta (2001), Sclove (1994), and Schöniger et al. (2014). Quite a few research articles have appeared in the area of Bayesian model selection and averaging as well as the related topic of Bayesian model evidence evaluation.

In the model selection approach, we start assuming that there exists a true model or probability distribution, denoted by P, and that we have a set of candidate models Q from which we can select one that provides the 'best' approximation to the unknown model P. Towards this objective, we make use of the Kullback-Leibler (K-L) measure of distance between P and Q and try to estimate Q that minimizes this distance. It is to be mentioned here that some people prefer to call this as a 'divergence measure' than a distance measure since this is not a distance in true sense of the term because it is not symmetric, i.e., distance between P and Q is not necessarily same as that between Q and P. We start with a class of models M_m which is a set of probability density functions to represent the observed data, with an associated parameter space. The cardinality or the number of models in the set may be fixed or may depend on the sample size. For each model set, there is a given number of parameters giving the dimension of the model. For each candidate model, we fit the observed data to that model by estimating the parameters involved. For obvious reasons, we use the method of maximum likelihood to estimate the parameters. Having found the maximized likelihood of the observed data for each of the candidate models, the next question is to select the 'best' candidate by using some selection criterion. The original basis for selection was to minimize the (cumulative) loss. A loss function or a scoring function commonly used is the negative of the logarithm of the candidate distribution and the choice of this loss function yields the maximum likelihood estimate of the distribution.

The first part of the K-L distance cannot be estimated, but remains the same for all the candidate models and hence does not influence the choice of Q. The second component can be estimated in terms of the entropy estimate that is related to the maximum likelihood estimation of the parameters in a candidate model.

Information criteria generally refer to model selection methods that are derived from likelihood functions and are applicable to parametric model-based procedures. Some of the well-known and oft-used information criteria are noted below:

- Akaike Information Criterion (AIC) is based on the idea to approximate the out-sample prediction loss by the sum of the in-sample prediction loss and a correction term. Using the logarithmic loss, the AIC procedure is to select the model M_m that minimizes
$$AIC = -2\ln L + 2k,$$
where L is the maximized likelihood and k is the number of independent parameters in the model required to be estimated. For small sample sizes, a correction factor has been suggested to yield
$$AIC_c = AIC + \frac{2k(k+1)}{n-k-1},$$
where n is the sample size. Here we assume that the model is univariate, is linear in its parameters and has normally distributed residuals. Evidently, the correction factor tends to 0 as n is quite large compared to k. It has been argued that AIC is a first-order estimate of the information loss, while AIC_c is a second-order estimate and is likely to be more accurate. However, its formula for some models may be quite complicated. A predecessor of AIC, suggested by Akaike himself, was the 'final prediction error criterion'. This criterion tests each candidate model on a different data set. The criterion is defined as
$$FPE = det\left\{\frac{1}{N}\sum_{t=1}^{n} e(t,\widehat{\theta}_N)e(t,\widehat{\theta}_N)'\right\}\left(\frac{1+d/N}{1-d/N}\right),$$
where N is the number of values in the estimation data set, $e(t,\widehat{\theta}_N)$ is the vector of prediction errors (for the N values in the test data set), $\widehat{\theta}_N$ is the vector of estimated parameters, and d is the number of estimated parameters. Smaller the FPE, better is the model.

An extension of AIC is the Takeuchi criterion that incorporates model misspecification, but is not widely used because of its computational complexity.

For any information criterion, consistency and efficiency are the yardsticks for performance comparison.

- In the Bayesian paradigm, we associate with each candidate model a prior probability of the candidate being the 'true' model. These prior probabilities may be given from historical data or from scientific reasoning. We obtain the posterior probabilities for the models, given the observed data set. The maximum a posteriori approach selects the model with the highest posterior probability. Introduced by Spiegelhalter et al. (2002), DIC can be regarded as the Bayesian analogue of AIC in the sense that it uses estimation based on the posterior mean, instead of the $MLEs$. This measure of model complexity and model fit is given by

11.1 Model Selection

$$DIC = p_D + \overline{D(\theta)} = D(\bar{\theta}) + 2p_D,$$

where $D(\theta) = -2\ln p(y|\theta) + C$, y stands for data, the effective number of parameters is $p_D = \overline{D(\theta)} - D(\bar{\theta})$, $\bar{\theta}$ being the prior expectation of θ. DIC can be looked upon as being equivalent to the natural model-robust version of the AIC.

Bayes factors are often computed to compare within a pair of candidate models. Bayes factors remove the impact of prior probabilities on the models. The Bayes Information Criterion (BIC) is taken as

$$BIC = -2\ln L + k\ln n,$$

where k and n are as mentioned earlier.

- The MLE and the model dimension are replaced, respectively, by the posterior mean and the effective number of parameters in defining Deviance Information Criterion (DIC). This criterion enjoys some computational advantage for comparing complex models whose likelihood functions may not admit of analytic forms. In Bayesian settings, Markov Chain Monte Carlo (MCMC) methods can be taken advantage of to simulate posterior distributions of each candidate model, which can be further used to efficiently compute DIC.
- A new model selection criterion, called the Bridge Criterion (BC), has been recently proposed to achieve both consistency in a parametric framework and asymptotic efficiency in both parametric and non-parametric set-ups. The bridge criterion combines the advantages of both AIC and BIC. Proposed in the context of auto-regression model, the criterion uses a penalty function for the maximized likelihood which is the harmonic number of order d_m (the number of parameters to be estimated under model m), i.e.,

$$\sum_{i=1}^{d_m} \frac{1}{i}.$$

This term increases with the model dimension but the rate of increase depends on the model specification. BC adaptively chooses the best model for the data, irrespective of the scenario—parametric or non-parametric—we are in. The criterion is explicitly given by

$$BC = -2\sum \ln p_{\hat{\theta}_m}(y_i) + n^{2/3}\sum_{k=1}^{d_m} \frac{1}{k}$$
$$= -2\ln L_m + n^{2/3}\sum_{k=1}^{d_m} \frac{1}{k},$$

BC imposes a heavy penalty like BIC for a range of small models, but to alleviate the penalty for a large model if more evidence in supporting an infinite-dimensional true model.

Despite the fact that several different information criteria have been proposed over the years, basically starting from the maximum entropy principle, the most widely used criteria are the AIC and the BIC. It is also true that different criteria may result in different models being recommended as the 'best' approximations to the 'true' model. Model selection, howsoever done, remains an exploratory exercise and cannot be confirmatory. Nevertheless, good model selection tools can provide valuable and reliable information regarding explanation and prediction, and the latter are the primary objectives in most statistical analysis, in the context of reliability and survival analysis.

Several authors who have proposed some non-traditional probability models over the last few decades have sometimes used the model selection criteria to establish the 'superiority' of their models over existing traditional models, based on actual experimental data. They also come up with findings of a better fit by the proposed models to the observed data. However, small differences in the value of a criterion like AIC or BIC should not be taken as confirmatory evidence.

Model selection criteria have been generally examined for three important properties, viz. consistency in selection, asymptotic efficiency, and minimax-rate optimality. Selection consistency targets the goal of identifying the 'best' model on its own for scientific understanding. A model selection procedure is called 'consistent' if the 'best' model is selected with probability going to one if the sample size tends to be infinitely large. Asymptotic efficiency and minimax-rate optimality are concerned with the goal of prediction. AIC has been found to be asymptotically efficient in the non-parametric framework and is also minimax optimal. On the other hand, BIC is consistent and asymptotically efficient for the parametric framework. AIC is known to be inconsistent in the parametric framework where there exist at least two correct candidate models.

11.2 Testing Goodness of Fit

As mentioned earlier, model selection is sometimes carried out on a part of the observed data, usually referred to as the 'training sample', the remaining data being used to test the goodness-of-fit of the model selected to the data. The most common method of testing goodness-of-fit involves the chi-square test applied to the observed frequencies and the frequencies expected on the basis of the model in certain classes/groups into which the 'test' data can be divided. The test result may depend on the grouping adopted—the number of classes/groups and the location of class boundaries. It is also possible that the value of the chi-square statistic comes out the same for two competing models, making the choice between the two somewhat difficult. We can also go by the Kolmogorov-Smirnov statistic based on non-parametric estimates of cumulative distribution function for the selected model.

Another problem that is common to any distance-based procedure to test for goodness-of-fit by a particular probability model arises from the fact that the model-expected frequencies or cumulative proportions or other comparable measures depend on the estimates of the model parameters as derived from the observed data. And there do exist more than one method for such estimation. Any statistic summarizing the divergences between the observed and the expected quantities cannot be guaranteed to remain the same with the different methods. Moreover, even when we use a particular well-accepted method for estimating the parameters in the initially chosen model, we should probe whether the associated standard errors are relatively small or not. In fact, a robustness investigation may be taken up to find out if the same conclusion about fit of the model to the observed data set holds if we compute results based on the upper and the lower confidence limits for the estimate of each of the model parameters estimated.

Going by the use of an Information Criterion, a similar problem does arise in the sense that in most cases we can get the maximum likelihood of the observed data under any assumed model only approximately, that too following one out of several competing methods of approximation. Finally, given the fact that a whole host of distributions with a lot of similarity and some distinctiveness of each distribution, it may be quite problematic to try out each of these models to find out the best. Simplicity in estimating the model parameters and in using the model to derive model-based results of interest to a reliability analyst should be accorded due priority in choosing an appropriate—if not the 'best fitting' model.

11.3   Use of Characterization Results

The procedure to use characterization results for the purpose of model selection should desirably proceed on the basis of results relating to properties like the failure rate function or the entropy or the mean remaining life, etc., which admit of non-parametric estimates with desirable properties. In fact, quite a few such estimates are available for the failure rate function or the entropy, based only on times-to-failure of a sample of items put on test. To estimate the mean remaining life beyond certain chosen points is a slightly different exercise which also can be conveniently carried out. One can argue that the behaviour of any such property as estimated from a sample of data (in fact the whole sample or a part of it can be used) cannot be taken for granted and testing the observed behaviour for its validity in respect of the underlying phenomenon as a whole invites problems.

Characterization results can be put into several categories for the purpose of their uses to specify an underlying probability model, based on sample data. Thus, we have characterizations based on some properties of the life distribution that can be verified from sample data and can be conveniently tested for validity or otherwise. A simple relation between moments of two specified orders, e.g., a constant coefficient of variation equal to one within the class of continuous distributions is pretty useful. There are others which assume the underlying distribution to belong to a certain class or to have some specified property (distinct from the characterization result under consideration). For example, some entropy-based characterizations assume some given arithmetic or geometric mean or moment of some specified order. And, entropy-based results except in a very few cases are in terms of being the maximum within a class of models—something quite difficult to establish for a particular model. There exist other results which assume the true distribution to possess a property like IFR or NBU or $DMRL$ and the like. Then, there are characterizations based on independence of two or more random variables associated with the unknown true distribution. Similarly, characterizations based on order statistics or record values or even on failure rate transform or residual life are essentially based on sampling distributions and cannot obviously be applied conveniently.

Establishing exponentiality for the underlying model for a continuous random variable in terms of coefficient of variation equal to unity is not a difficult task. But verifying the result

$$E[X^k r(X)] = \frac{(k+1)\mu_k^2}{\mu_{k+1}}$$

for any natural number k or even for $k = 1$ is quite a problematic one, involving estimation of r and generating a sample of r values.

In this connection, we may note that statistical tests exist for verification of many class properties which assume exponentiality of the true model as the null hypothesis and, that way, indicate whether some class property can be taken for granted or should not be accepted. However, such tests are lacking with most of the well-known characterization results. This along with the other aforementioned problems circumscribes the application of characterization results for identifying the underlying or true probability model.

11.4 Some Examples

To exemplify the issues in selecting a model to analyse a data set and in testing for goodness-of-fit by a selected model, we refer below to a few relevant exercises.

Barbiero (2013) considered a data set containing the number of failures of a software observed over 62 weeks and assumed a discrete Type III Weibull distribution with the probability function given below as the underlying model.

$$P(x; c, \beta) = P(X = x) = e^{-c \sum_{j=1}^{x} j^\beta} \left(1 - e^{-c(x+1)^\beta}\right), \quad x = 0, 1, 2, \ldots$$

with $\sum_{j=1}^{x} j^\beta = 0$ for $x = 0$. He estimated the model parameters by the method of maximum likelihood, the method of moments and a method based on the proportion

11.4 Some Examples

of cases with $x = 0$ and $x = 1$ to get estimates of c as 0.361, 0.356, and 0.389, respectively. Estimates of β worked out as 0.068, 0.083, and -0.518. Thus, the method of maximum likelihood and the method of moments yielded close estimates of both the parameters, while the method using proportions provided an altogether different estimate for β, the estimate of c remaining more or less close to one another. It is found that the chi-square test for goodness-of-fit accepts the Weibull Type III model with the first two methods of estimation, giving values of the test statistic as 3.8843 and 3.8514 with p-values of 0.2742 and 0.2780, respectively. However, with the method of proportions these figures came out as 10.2323 and 0.0060, respectively. Thus, working at a 5% nominal level of significance, conclusions about the assumed model providing a good fit to the observed data depend on the method used for estimating the model parameters. Use of different groupings of the observed data could have given rise to further differences in results.

With more and more generalizations of some oft-used probability models to extend the versatility of the latter, it is interesting to find out if some sub-model (a particular case of a generalized model, involving fewer parameters) performs better than the full model, judged in terms of the information criteria generally used. In this context, we find that Pararai et al. (2014) proposed the generalized (in fact, gamma-generated) inverse Weibull distribution and considered a data set consisting of failure times (number of million revolutions before failure) of 23 ball bearings before failure to indicate its possible applications.

The pdf for the proposed model, denoted as $GIW(\beta, \lambda, \delta)$ is given by

$$f(x; \beta, \lambda, \delta) = \frac{\beta x^{-1}}{\Gamma \delta} \left(\lambda x^{-\beta}\right)^{\delta} e^{-\lambda x^{-\beta}}, \quad x > 0$$

and its failure rate function shows an upside-down bathtub shape. The authors took into account the sub-models of GIW like Inverse Weibull $(\beta, \lambda, 1)$, Frechet $(\beta, 1, 1)$, Inverse Rayleigh $(2, \lambda, 1)$, Generalized Inverse Exponential $(\beta, 1, \delta)$, Inverse Exponential $(\beta, 1, 1)$, and Generalized Inverse Rayleigh $(2, \lambda, \delta)$ to find the best (among those considered) model for the given data set.

Values of AIC, AIC_c, and BIC were computed for each of these sub-models as also for the full model. It was found that there was no significant difference in the criterion values between the Generalized Inverse Exponential and the Generalized Inverse Weibull distributions. Similar were the findings in respect of the Inverse Rayleigh and Generalized Inverse Rayleigh models. However the Generalized Inverse Exponential differed substantially from the Inverse Exponential model. Thus, it was argued that the Generalized Inverse Exponential could be selected as the appropriate model, of course among the set of models considered in this example.

References

Abadir, K. M. (2005). The mean-median-mode inequality. *Econometric Theory, 21*(2), 477–482.

Abe, S. I. (1993). An extension of the method of moments and its application to estimating Weibull parameters. Personal Communication to the first author.

Abdelaziz, M. A., Nofal, Z. M., & Afify, A. Z. (2024). The inverse-power Burr-Hatke-G family: Properties and inference with real-life applications. *Research Square.* https://doi.org/10.21203/rs.3.rs-4122305/v1.

Abdul-Moniem, I. B. (2012). Recurrence relations for moments of lower generalized order statistics form exponentiated Lomax distribution and its characterization. *Journal of Mathematical and Computational Science, 2*(4), 999–1011.

Abiodun, A. A., & Ishaq, A. I. (2022). On Maxwell-Lomax distribution: Properties and applications. *Aran Journal of Basic and Applied Sciences, 29*(1), 221–232.

Abouammoh, A., & Ahmed, A. N. (1988). The new better than used in failure rate class of life distribution. *Advances in Applied Probability, 20*(1), 237–240.

Abouammoh, A., & EL-Neweihi, E. (1986). Closure of the NBUE and DMRL classes under formation of parallel systems. *Statistics and Probability Letters, 4*, 223–225.

Afify, A. Z., & Butt, N. S. (2014). Transmuted complementary Weibull geometric distribution. *Pakistan Journal of Statistics and Operations Research, 10*(4), 435–454.

Afify, A. Z., Cordeiro, G. M., Butt, N. S., Ortega, E. M. M., & Suzuki, A. K. (2017). A new lifetime model with variable shapes for the hazard rate. *Brazilian Journal of Probability and Statistics, 31*(3), 516–541.

Ahmad, I. A., Ahmed, A., Elbatal, I., & Kayid, M. (2006). An aging notion derived from the increasing convex ordering: The NBUCA class. *Journal of Statistical Planning and Inference, 136*(3), 555–569.

Ahmad, A., Jallal, M., & Ahmad, A. (2022). A novel approach for constructing distributions with an example of the Rayleigh distribution. *Reliability: Theory and Applications, 17*(1), 52–64.

Ahmad, I. A., & Kayid, M. (2005). Characterizations of the RHR and MIT orderings and the DRHR and IMIT classes of life distributions. *Probability in the Engineering and Informational Sciences, 19*(4), 447–461.

Ahmad, I. A., Kayid, M., & Li, X. (2005). The NBUT class of life distributions. *IEEE Transactions on Reliability, 54*(3), 396–401.

Ahmad, Z., Mahmoudi, E., Alizadeh, M., Roozegar, R., & Afify, A. Z. (2021). The exponential T-X family of distributions: Properties and an application to insurance data. *Journal of Mathematics*, Article ID, *3058170*, 1–18.

Ahmed, A. N. (1988). Preservation properties of the mean remaining life function ordering. *Statistical Papers, 29*, 143–150.

Ahmed, A. N. (1991). Characterizations of beta, binomial and Poisson distributions. *IEEE Transactions on Reliability, 40*(3), 290–295.

Ahmed, A. N., Soliman, A. A., & Khider, S. E. (1996). On some partial ordering of interest in reliability. *Microelectronics and Reliability, 36*, 1337–1346.

Ahsanullah, M. (1978). Record values and the exponential distribution. *Annals of the Institute of Statistical Mathematics, 30*(1), 429–433.

Ahsanullah, M. (1981). On characterization of the exponential distribution by spacings. *Statistiche Hefte, 22*, 316–320.

Ahsanullah, M., & Alzaatreh, A. (2018). Some characterizations of the log-logistic distribution. *Stochastics and Quality Control, 33*(1), 23–29.

Aijaz, A., Jallal, M., Ain, S. Q. U., & Tripathi, R. (2020). The Hamza distribution with statistical properties and applications. *Asian Journal of Probability and Statistics, 8*(1), 28–42.

Alam, S. N., Yaqub, M., & Rizvi, S. G. A. (1980). On life estimation from over-reported data. *IAPQR Transactions, 5*(1), 1–8.

Al-Babtain, A. A., Ahmed, A. H. N., & Afify, A. Z. (2020). A new discrete analogue of the continuous Lindley distribution with reliability applications. *Entropy, 22*(6), 603, 18 pages.

Alexander, C., Cordeiro, G. M., Ortega, E. M. M., & Sarabia, J. M. (2012). Generalized beta-generated distributions. *Computational Statistics and Data Analysis, 56*(6), 1880–1897.

Al-Hussaini, E. K., & Abd-El-Hakim, N. S. (1989). Failure rate of the inverse Gaussian-Weibull mixture model. *Annals of the Institute of Statistical Mathematics, 41*, 617–622.

AL-Hussaini, E. K., Mohamad, A. M. A. M., & Sultan, K. S. (1997). Parametric and nonparametric estimation of $P(Y<X)$ for finite mixtures of lognormal components. *Communications in Statistics-Theory and Methods, 26*(5), 1269–1289.

Al-Hussaini, E. K., & Sultan, K. S. (2001). Reliability and hazard based on finite mixture models. In N. Balakrishnan & C. R. Rao (Eds.), *Handbook of Statistics* (Vol. 20, pp. 139–183). Elsevier.

Ali, M. M., Pal, M., & Woo, J. (2016). Skewed reflected distributions generated by the Laplace kernel. *Australian Journal of Statistics, 38*(1), 45–58.

Aljarrah, M. A., Lee, C., & Famoye, F. (2014). On generating T-X family of distributions using quantile functions. *Journal of Statistical Distributions and Applications, 1*(2), 1–17.

Almalki, S. J. (2014). Statistical Analysis of lifetime data using new modified Weibull distributions, *PhD Dissertation*, University of Manchester, England.

Almalki, S. J., & Nadarajah, S. (2014). Modifications of the Weibull distribution: A review. *Reliability Engineering and System Safety, 124*, 32–55.

Al-Masoud, T. A. (2013). A discrete general class of continuous distributions, *PhD Dissertation*, King Abdul Aziz University, Saudi Arabia.

Alnssyan, B. (2023). The modified-Lomax distribution: Properties, estimation methods, and application. *Symmetry, 15*(7), Article No. 1367. https://doi.org/10.3390/sym15071367.

Alzaatreh, A., Famoye F., Lee C. (2103a). Weibull-Pareto distribution and its applications. *Communications in Statistics-Theory and Methods, 42*(9), 1673–1691.

Alzaatreh, A., Lee, C., & Famoye, F. (2013). Weibull-Pareto distribution and its application. *Communications in Statistics-Theory and Methods, 42*, 1673–1691.

Alzaatreh, A., Lee, C., & Famoye, F. (2013). A new method for generating families of continuous distributions. *Metron, 71*(1), 63–79.

Alzaatreh, A., Famoye, F., & Lee, C. (2014). The gamma-normal distribution: Properties and applications. *Communications in Statistics-Theory and Methods, 69*, 67–80.

Alzaatreh, A., Lee, C., & Famoye, F. (2014). On generating T-X family of distributions using quantile functions. *Journal of Statistical Distributions and Applications, 1*(2), 1–17.

References

Anakha, K. K., & Chacko, V. M. (2020). On ageing properties of lifetime distributions. *Veritas Journal of Sciences, 1*(1), 107–139.

Andersen, P. K., Borgan, Ø., Gill, R. D., & Keiding, N. (1993). *Statistical Models Based on Counting Processes*. New York: Springer.

Andrade, C. (2017). Reliability analysis of corrosion onset: Initiation limit state. *Journal of Structural Integrity and Maintenance, 2*(4), 200–208.

Arnold, B. C. (1983). *Pareto Distributions*. USA: International Co-operative Publishing House.

Arnold, B. C., Balakrishnan, N., & Nagaraja, H. N. (1998). *Records*. Wiley.

Arnold, B. C., & Brockett, P. L. (1983). When does the βth percentile residual life function determine the distribution? *Operations Research, 31*(2), 391–396.

Arnold, B. C., & Strauss, D. (1988). Bivariate distributions with exponential conditionals. *Journal of the American Statistical Association, 83*(402), 522–527.

Arnold, B. C., & Zahedi, H. (1988). On multivariate mean residul life functions. *Journal of Multivariate Analysis, 25*, 1–9.

Arriaza, A., Sordo, M. A., & Suarez-Llorens, A. (2017). Comparing residual lives and inactivity times by transform stochastic orders. *IEEE Transactions on Reliability, 66*(2), 366–372.

Aryal, G. R., & Tsokos, C. P. (2011). Transmuted Weibull distribution: A generalization of the Weibull probability distribution. *European Journal of Pure and Applied Mathematics, 4*(2), 89–102.

Aryal, G. R., & Yousof, H. M. (2017). The exponentiated generalized-G Poisson family of distributions. *Stochastics and Quality Control, 32*(1), 7–23.

Asadi, M., & Berred, A. (2012). Properties and estimation of the mean past lifetime. *Statistics, 46*(3), 405–417.

Asadi, A., & Zohrevand, Y. (2007). On the dynamic cumulative residual entropy. *Journal of Statistical Planning and Inference, 137*(6), 1931–1941.

Astorga, J. M., Gómez, H. W., & Bolfarine, H. (2017). Slashed generalized exponential distribution. *Communications in Statistics-Theory and Methods, 46*(5), 2091–2102.

Astorga, J. M., Reyes, J., Santoro, K. I., Venegas, O., & Gómez, H. W. (2020). A reliability model based on the incomplete generalized integro-exponential function. *Mathematics, (MDPI), 8*(9), 1537, 12 pages.

Aubrun, G., & Nechita, I. (2009). Stochastic domination for iterated convolutions and catalytic majorization. *Annales de l'Institut Henri Poincaré-Probabilités et Statistiques, 45*, 611–625.

Awad, A. M. (1987). A statistical information measure. *Dirasat, XIV*(12), 7–20.

Azzalini, A. (1985). A class of distributions which include the normal. *Scandinavian Journal of Statistics, 12*, 171–178.

Badia, F. J., & Berrade, M. D. (2022). On the residual lifetime and inactivity time in mixtures. *Mathematics (MDPI), 10*(15), 2795, 20 pages.

Bagchi, S. B., & Das, A. (1985). On censored estimation of reliability of an over-reported distribution. *IAPQR Transactions, 10*, 103–113.

Bain, L. J. (1974). Analysis for the linear failure rate life-testing distribution. *Technometrics, 16*(4), 551–559.

Balakrishnan, N. (1992). *Handbook of the Logistic Distribution. A series of textbooks and monographs*. New York, USA: Marcel Dekker.

Balakrishnan, N., & Iliopoulos, G. (2009). Stochastic monotonicity of the MLE of exponential mean under different censoring schemes. *Annals of the Institute of Statistical Mathematics, 61*, 753–772.

Balakrishnan, N., & Kundu, D. (2018). Birnbaum-Saunders distribution: A review of models, analysis, and applications. *Applied Stochastic Models in Business and Industry, 35*(1), 1–46.

Bannantine, J. A., Comer, J. J., & Handrock, J. L. (1990). *Fundamentals of Metal Fatigue Analysis*. USA: Pearson Education.

Barbiero, A. (2010). A discretizing method for reliability computation in complex stress-strength models. *International Journal of Mathematical and Computational Sciences, 4*(11), 1395–1401.

Barbiero, A. (2013). Parameter estimation for the Type III discrete Weibull distribution: a comparative study. *Journal of Probability and Statistics*, Article ID 946562, 10 pages.

Bardwell, G. E., & Crow, E. L. (1964). A two-parameter family of hyper-Poisson distributions. *Journal of the American Statistical Association, 59*(305), 133–141.

Barlow, R. E. (1979). Geometry of the total time on test transform. *Naval Research Logistics Quarterly, 26*(3), 393–402.

Barlow, R. E., & Campo, R. (1975). Total time on test processes and applications to failure data analysis. In R. E. Barlow, R. Fussel, & N. D. Singpurwalla (Eds.), *Reliability and Fault Tree Analysis*. Society for Industrial and Applied Mathematics (pp. 451–481). Philadelphia.

Barlow, R. E., Marshall, A. E., & Proschan, F. (1963). Properties of probability distributions with monotone hazard rate. *Annals of Mathematical Statistics, 34*, 375–389.

Barlow, R. E., & Proschan, F. (1966). Inequalities for linear combinations of order statistics from restricted families. *Annals of Mathematical Statistics, 37*(6), 1574–1592.

Barlow, R. E., & Proschan, F. (1975). *Statistical Theory of Reliability and Life Testing: Probability Models*. Rinehart and Winston Inc: Holt.

Barlow, R. E., & Proschan, F. (1981). *Statistical Theory of Reliability and Life Testing: Probability Models*. Silver Spring, MD: To Begin With.

Basu, S. K., & Mitra, M. (1998). The class of life distributions that are Laplace order dominated by the exponential law and its ramifications. In A. P. Basu, S. K. Basu, & S. P. Mukhopadhyay (Eds.), *Frontiers in Reliability* (pp. 37–45). World Scientific.

Bebbington, M., Lai, C. D., & Zitikis, R. (2007). A flexible Weibull extension. *Reliability Engineering and System Safety, 92*, 719–726.

Beden, S. M., Abdullah, S., & Ariffin, A. K. (2009). Renew of fatigue crack propagation models for metallic components. *European Journal of Scientific Research, 28*(3), 364–397.

Beg, M. I., & Kirmani, S. N. U. A. (1974). On a characterization of exponential and related distributions. *Australian Journal of Statistics, 16*, 163–166.

Belzunce, F., Candel, J., & Ruiz, J. M. (1995). Ordering of truncated distributions through concentration curves. *Sankhyā, Series A, 57*(3), 375–383.

Belzunce, F., Candel, J., & Ruiz, J. M. (1998). Ordering and asymptotic properties of residual income distributions. *Sankhyā, Series B, 60*(2), 331–348.

Belzunce, F., Hu, T., & Khaledi, B. E. (2003). Dispersion-type variability orders. *Probability in the Engineering and Information Sciences, 17*(3), 305–334.

Belzunce, F., Lillo, R. E., Ruiz, J. M., & Shaked, M. (2001). Stochastic comparisons of nonhomogeneous processes. *Probability in the Engineering and Informational Sciences, 15*, 199–224.

Belzunce, F., Nanda, A. K., Ortega, E. M., & Ruiz, J. M. (2008). Generalized orderings of excess lifetimes of renewal processes. *TEST, 17*, 297–310.

Belzunce, F., Navarro, J., Ruiz, J. M., & del Aguila, Y. (2004). Some results on residual entropy function. *Metrika, 59*, 147–161.

Benduch-Fraszczak, M. (2010). Some properties of the proportional odds model. *Applicationes Mathematicae, 37*, 247–256.

Bennett, S. (1983). Analysis of survival data by the proportional odds model. *Statistics in Medicine, 2*, 273–277.

Bergeman, B., & Klefsjö, B. (1998). Recent applications of the TTT plotting technique. In A. P. Basu, S. K. Basu, & S. P. Mukhopadhyay (Eds.), *Frontiers in Reliability* (pp. 47–61). World Scientific.

Betsch, S., & Ebner, B. (2021). Fixed point characterizations of continuous univariate probability distributions and their applications. *Annals of the Institute of Statistical Mathematics, 73*(1), 31–59.

Bhat, V. A., & Pundir, S. (2022). Intervened exponential distribution: Properties and applications. *Pakistan Journal of Statistics and Operations Research, 18*(1), 71–84.

Bhattacharjee, S., Nanda, A. K., & Misra, S. K. (2010). Aging intensity function in reliability analysis. In S. Chakraverty (Ed.), *Proceedings of the International Conference on Challenges and*

Applications of Mathematics in Science and Technology (CAMIST) (pp. 601–607). Macmillan Publishers India Limited, Delhi.

Bhattacharjee, S., Nanda, A. K., & Misra, S. K. (2013). Reliability analysis using ageing intensity function. *Statistics and Probability Letters, 83*, 1364–1371.

Bhattacharjee, S., Nanda, A. K., & Misra, S. K. (2013). Inequalities involving expectations to characterize distributions. *Statistics and Probability Letters, 83*, 2113–2118.

Bhattacharya, A., & Sengupta, D. (1996). On the coefficient of variation of the $\mathcal{L}$- and $\bar{\mathcal{L}}$-classes. *Statistics and Probability Letters, 27*(2), 177–180.

Bhatti, F. A., Hamedani, G. G., Sheng, W., & Ahmad, M. (2018). On extended quadratic hazard rate distributions: Development, properties, characterizations and applications. *Stochastics and Quality Control, 33*(1), 45–60.

Bickel, P. J., & Lehmann, E. L. (1975). Descriptive statistics for nonparametric models I: Introduction. *Annals of Statistics, 3*, 1038–1044.

Billingsley, P. (1968). *Convergence of Probability Measures*. New York: Wiley.

Birnbaum, Z. W., & Saunders, S. C. (1969). A new family of life distributions. *Journal of Applied Probability, 6*(2), 319–327.

Block, H. W., & Savits, T. H. (1976). The $IFRA$ closure problem. *Annals of Probability, 4*(6), 1030–1032.

Block, H. W., & Savits, T. H. (1980). Laplace transforms for classes of life distributions. *Annals of Probability, 8*(3), 465–474.

Block, H. W., & Savits, T. H. (1997). Burn-in. *Statistical Science, 12*(1), 1–19.

Block, H. W., Savits, T. H., & Shaked, M. (1982). Some concepts of negative dependence. *Annals of Probability, 10*, 765–772.

Block, H. W., Savits, T. H., & Singh, H. (1998). The reversed hazard rate function. *Probability in the Engineering and Informational Sciences, 12*, 69–90.

Boland, P. J., & El-Neweihi, E. (1995). Component redundancy versus system redundancy in the hazard rate ordering. *IEEE Transactions on Reliability, 44*(4), 614–619.

Boland, P. J., El-Neweihi, E., & Proschan, F. (1994). Applications of the hazard rate ordering in reliability and order statistics. *Journal of Applied Probability, 31*(1), 180–192.

Bracquemond, C., & Gaudoin, O. (2003). A survey on discrete lifetime distributions. *International Journal of Reliability, Quality and Safety Engineering, 10*(1), 69–98.

Brown, M., & Shanthikumar, J. G. (1998). Comparing the variability of random variables and point processes. *Probability in the Engineering and Informational Sciences, 12*, 425–444.

Bruckner, A. M., & Ostrow, E. (1962). Some function classes related to the class of convex functions. *Pacific Journal of Mathematics, 12*(4), 1203–1215.

Bryson, M. C., & Siddiqui, M. M. (1969). Some criteria for ageing. *Journal of the American Statistical Association, 64*(328), 1472–1483.

Buchholz, H. (1969). *The Confluent Hypergeometric Function*. Berlin: Springer.

Bullen, P. S. (1971). A criterion for n-convexity. *Pacific Journal of Mathematics, 36*(1), 81–98.

Buono, F., Longobardi, M., & Szymkowiak, M. (2020). On generalized reversed aging intensity functions. *Ricerche di Matematica, 71*, 85–108.

Burr, I. W. (1942). Cumulative frequency functions. *Annals of Mathematical Statistics, 13*(2), 215–232.

Caiza, P. D. T., & Ummenhofer, T. (2011). General probability weighted moments for the three-parameter Weibull distribution and their application in S-N curves modelling. *International Journal of Fatigue, 33*(12), 1533–1538.

Cameron, A. C., & Trivedi, P. K. (1986). Econometric models based on count data: Comparisons and applications of some estimators and tests. *Journal of Applied Econometrics, 1*, 29–53.

Cao, J., & Wang, Y. (1991). The NBUC and NWUC classes of life distributions. *Journal of Applied Probability, 28*(2), 473–479.

Capéraà, P. (1988). Tail ordering and asymptotic efficiency of rank tests. *Annals of Statistics, 16*, 470–478.

Carver, H. C. (1919). On the graduation of frequency distributions. *Proceedings of the Casualty Actuarial Society of America, 6*, 52–72.

Carver, H. C. (1921). The mathematical representation of frequency distributions. *Quarterly Publications of the Journal of American Statistical Association, 17*(135), 885–892.

Castillo, J. D., & Perez-Casany, M. (1998). Weighted Poisson distributions for over-dispersion and under-dispersion situations. *Annals of the Institute of Statistical Mathematics, 50*(3), 567–585.

Chakraborty, S. (2015). Generating discrete analogues of continuous probability distributions-a survey of methods and constructions. *Journal of Statistical Distributions and Applications, 2*(6), 1–30.

Chakraborty, S., & Ong, S. H. (2017). Mittag-Leffler function distribution-a new generalization of hyper-Poisson distribution. *Journal of Statistical Distributions and Applications, 4*(6), 1–17.

Chan, W., Proschan, F., & Sethuraman, J. (1991). Convex ordering among functions, with applications to reliability and mathematical statistics. In H. W. Block, A. R. Sampson, & T. H. Savits (Eds.), *Topics in Statistical Dependence*. IMS Lecture Notes (Vol. 16, pp. 121–134).

Chandler, K. N. (1952). The distribution and frequency of record values. *Journal of the Royal Statistical Society, Series B, 14*, 220–228.

Chandra, N. K., & Roy, D. (2001). Some results on reversed hazard rate. *Probability in the Engineering and Informational Sciences, 15*, 95–102.

Chatterjee, A. (1993). *PhD Dissertation*. Calcutta, India: University of Calcutta.

Chatterjee, A., & Mukherjee, S. P. (2001). Equilibrium distribution-its role in reliability theory. In N. Balakrishnan, & C. R. Rao (Eds.), *Handbook of Statistics* (Vol. 20, pp. 105–137). Elsevier.

Chaudhary, A. K., & Kumar, V. (2020). The logistic-Rayleigh distribution with properties and applications. *International Journal of Statistics and Applied Mathematics, 5*(6), 12–19.

Chaudhary, A. K., & Kumar, V. (2021). Arctan exponential extension distribution with properties and applications. *International Journal of Applied Research, 7*(1), 432–442.

Chen, Y. (1994). Classes of life distributions and renewal counting process. *Journal of Applied Probability, 31*(4), 1110–1115.

Chesneau, C., Sharma, V. K., & Bakouch, H. S. (2021a). Extended Topp-Leone family of distributions as an alternative to beta and Kumaraswamy type distributions: Application to glycosaminoglycans concentration level in urine. *International Journal of Biomathematics, 14*(2), 2050088, World Scientific.

Chesneau, C., Tomy, L., & Gillariose, J. (2021). A new modified Lindley distribution with properties and applications. *Journal of Statistics and Management Systems, 24*(7), 1383–1403.

Choudhary, A. K. (2019). Bayesian analysis of two-parameter exponentiated log-logistic distribution. *Silver Jubilee Issue of Pravaha, 25*(1), 1–11.

Chukova, S., & Dimitrov, B. (1992). On distributions having the almost-lack-of-memory property. *Journal of Applied Probability, 29*(3), 691–698.

Civanlar, M. R., & Trussell, H. J. (1986). Constructing membership functions using statistical data. *Fuzzy Sets and Systems, 8*, 1–13.

Cohen, A. C. (1973). The reflected Weibull distribution. *Technometrics, 15*(4), 867–893.

Cohen, A. C., & Whitten, B. J. (1988). *Parameter Estimation in Reliability and Lifespan Models*. Marcel Dekker.

Collett, D. (2015). *Modeling Survival Data in Medical Research*. Chapman and Hall.

Consul, P. C. (1995). Some characterizations of the exponential class of distributions. *IEEE Transactions of Reliability, 44*(3), 403–407.

Consul, P. C., & Famoye, F. (1995). On the generalized negative binomial distribution. *Communications in Statistics-Theory and Methods, 24*, 459–472.

Conway, R., & Maxwell, W. (1962). A queueing model with state dependent service rates. *Journal of Industrial Engineering, 12*, 132–136.

Cooray, K. (2006). Generalization of the Weibull distribution: The odd Weibull family. *Statistical Modelling, 6*(3), 265–277.

Cooray, K., & Ananda, M. M. A. (2008). A generalization of the half-normal distribution with applications to lifetime data. *Communications in Statistics-Theory and Methods, 37*(9), 1323–1337.

Cordeiro, G. M., Cristino, C. T., Hashimoto, E. M., & Ortega, E. M. M. (2013). The beta generalized Rayleigh distribution with applications to lifetime data. *Statistical Papers, 54*, 133–161.

Cordeiro, G. M., & de Castro, M. (2011). A new family of generalized distributions. *Journal of Statistical Computation and Simulation, 81*(7), 883–898.

Cordeiro, G. M., Ortega, E. M. M., & da Cunha, C. C. (2013). The exponentiated generalized class of distributions. *Journal of Data Science, 11*(1), 1–27.

Cordeiro, G. M., Ortega, E. M. M., & Popović, B. (2013). The gamma-Lomax distribution. *Journal of Statistical Computation and Simulation, 85*(2), 305–319.

Cordeiro, G. M., Ortega, E. M. M., & Silva, G. O. (2014). The Kumaraswamy modified Weibull distribution: Theory and applications. *Journal of Statistical Computation and Simulation, 84*(7), 1387–1411.

Cragg, J. G. (1971). Some statistical models for limited dependent variables with application to the demand for durable goods. *Econometrica, 39*, 829–844.

Crowder, M. J., Kimber, A. C., Smith, R. L., & Sweeting, T. J. (1991). *Statistical Analysis of Reliability Data*. Chapman and Hall.

Dagum, C. (1977). A new model of personal income distribution: Specification and estimation. *Economie Appliquée, 30*(3), 413–437.

Dallas, A. C. (1973). Characterization of the exponential distribution. *Bulletin of the Greek Mathematical Society, 14*, 172–175.

Dallas, A. C. (1976). Characterizing the Pareto and power distribution. *Annals of the Institute of Statistical Mathematics, 28*, 491–497.

Dallas, A. C. (1981). Record values and the exponential distribution. *Journal of Applied Probability, 18*(4), 949–951.

Das, S., & Nanda, A. K. (2013). Some stochastic orders of dynamic additive mean residual life model. *Journal of Statistical Planning and Inference, 143*, 400–407.

Das, S., & Nanda, A. K. (2017). Some ageing properties of dynamic additive mean residual life model. *Rashi, 2*(1), 26–33.

Das Gupta, S., & Sarkar, S. K. (1984). On TP_2 and log-concavity. In Y. L. Tong (Ed.), *Inequalities in Statistics and Probability, IMS Lectures Notes, Monograph Series* (Vol. 5, pp. 54–58). Hayward, CA: Institute of Mathematical Statistics.

Davis, D. J. (1952). An analysis of some failure data. *Journal of the American Statistical Association, 47*(258), 113–150.

de Andrade, T. A. N., Bourguignon, M., & Cordeiro, G. M. (2016). The exponentiated generalized extended exponential distribution. *Journal of Data Science, 14*, 393–414.

Deeks, J. (1996). Swots corner: What is an odds ratio? *Bandolier, 25*, 53–54.

Deheuvels, P. (1984). The characterization of distributions by order statistics and record values- a unified approach. *Journal of Applied Probability, 21*(2), 326–334; (Correction in 1985, Vol. 22, No. 4, p. 997).

de Gusmão, F. R. S., Ortega, E. M. M., & Cordeiro, G. M. (2011). The generalized inverse Weibull distribution. *Statistical Papers, 52*(3), 591–619.

Denuit, M., Lefèvre, C., & Shaked, M. (1998). The s-convex orders among real random variables, with applications. *Mathematical Inequalities & Applications, 1*(4), 585–613.

Denuit, M., Lefèvre, C., & Shaked, M. (2000). On the theory of high convexity stochastic orders. *Statistics and Probability Letters, 47*, 287–293.

Dersin, P. (2018). The class of life time distributions with a mean residual life linear in time: Applications to prognostic and health management. In S. Haugen, A. Barros, C. Gulijk, T. Kongsvik, & J. E. Vinnem (Eds.), *Safety and Reliability-Safe Societies in a Changing World*, Proceedings of ESREL, June 17–21, Chapter 138, 7 pages, Trondheim, Norway (pp. 1093–1099).

Deshpande, J. V., Kochar, S. C., & Singh, H. (1986). Aspects of positive ageing. *Journal of Applied Probability, 23*, 48–58.

Deshpande, J. V., Singh, H., Bagai, I., & Jain, K. (1990). Some partial orders describing positive ageing. *Communications in Statistics-Stochastic Models, 6*, 471–481.

Deshpande, J. V., & Suresh, R. P. (1990). Non-monotonic Ageing. *Scandinavian Journal of Statistics, 17*(3), 257–262.

Dey, S., Al-Zahrani, B., & Basloom, S. (2017). Dagum distribution: Properties and different methods of estimation. *International Journal of Advanced Statistics and Probability, 6*, 74–92.

Dharmadhikari, S., & Joag-dev, K. (1988). *Unimodality*. Convexity and Applications: Academic Press Inc.

de Gusmão, F. R. S., Ortega, E. M. M., & Cordeiro, G. M. (2011). The generalized inverse Weibull distribution. *Statistical Papers, 52*, 591–619.

Dias, C. R. B., Alizadeh, M., & Cordeiro, G. M. (2018). The beta Nadarajah-Haghighi distribution. *Hacettepe Journal of Mathematics and Statistics, 47*(5), 1302–1320.

Di Crescenzo, A., & Longobardi, M. (2002). Entropy-based measure of uncertainty in past lifetime distributions. *Journal of Applied Probability, 39*, 434–440.

Dimitrov, B., & Khalil, Z. (1990). On a new characterization of the exponential distribution related to a queueing system with an unreliable server. *Journal of Applied Probability, 27*(1), 221–226.

Ding, J., Tarokh, V., & Yang, Y. (2018). Model selection techniques: An overview. *IEEE Signal Processing Magazine, 35*(6), 16–34.

Domma, F. (2011). Bivariate reversed hazard rate, notions, and measures of dependence and their relationships. *Communications in Statistics-Theory and Methods, 40*, 989–999.

Domma, F., & Cordeiro, G. M. (2013). The beta-Dagum distribution: Definition and properties. *Communications in Statistics-Theory and Methods, 42*(22), 4070–4090.

Du, J., Park, Y., Theera-Ampornpunt, N., McCullough, J. S., & Speedie, S. M. (2012). The use of count data models in biomedical informatics evaluation research. *Journal of the American Medical Informatics Association, 19*, 39–44.

Dubey, S. D. (1960). Contributions to Statistical Theory of Life Testing and Reliability, *PhD Dissertation*, Michigan State University, USA.

Dubey, S. D. (1965). Asymptotic properties of several estimators of the Weibull distribution. *Technometrics, 7*, 423–434.

Dubey, S. D. (1967). Some percentile estimators for Weibull parameters. *Technometrics, 9*(1), 119–129.

Dziubdziela, W., & Kopociński, B. (1976). Limiting properties of the k-th record values. *Applicationes Mathematicae, 15*, 187–190.

Ebrahimi, N. (1996). How to measure uncertainty in the residual life time distribution. *Sankhyā. Series A, 58*, 4856.

Ebrahimi, N., & Kirmani, S. N. U. A. (1996). Some results on ordering of survival functions through uncertainty. *Statistics and Probability Letters, 29*, 167–176.

Ebrahimi, N., & Pellerey, F. (1995). New partial ordering of survival functions based on notion of uncertainty. *Journal of Applied Probability, 32*, 202–211.

Ebrahimi, N., & Zahedi, H. (1992). Memory ordering of survival functions. *Statistics, 23*(4), 337–345.

Efron, B. (1986). Double exponential families and their use in generalized linear regression. *Journal of the American Statistical Association, 81*, 709–721.

El-Bassiouny, A. H., Abdo, N. F., & Shahen, H. S. (2015). Exponential Lomax distribution. *International Journal of Computer Applications, 121*(13), 24–29.

El-Bassiouny, A. H., Abdo, N. F., & Shahen, H. S. (2015). Exponential Lomax Distribution. *International Journal of Computer Applications, 121*(13), 24–29.

El-Bassiouny, A. H., & EL-Damcese, M., Mustafa, A., & Eliwa, M. S. (2015). Characterization of the generalized Weibull-Gompertz distribution based on the upper record values. *International Journal of Mathematics and its Applications, 3*(3), 13–22.

Elbatal, I. (2011). Exponentiated modified Weibull distribution. *Economic Quality Control, 26*, 189–200.

References

Elbatal, I., & Aryal, G. (2013). On the transmuted additive Weibull distribution. *Australian Journal of Statistics, 42*(2), 117–132.

Elbatal, I., Zayed, M., Rasekhi, M., Afify, A. Z., & Iqbal, Z. (2019). A new extended Weibull model for lifetime data. *Journal of Applied Probability and Statistics, 14*(1), 57–73.

Elderton, W. P. (1906). *Frequency Curves and Correlation*. London: Charles and Edwin Layton.

Elderton, W. P., & Johnson, N. L. (1969). *Systems of Frequency Curves*. Cambridge University Press.

Eliwa, M. S., Altun, E., Alhussain, Z. A., Ahmed, E. A., Salah, M. M., Ahmed, H. H., & El-Morshedy, M. (2021). A new one-parameter lifetime distribution and its regression model with applications. *PLOS ONE, 16*(2), 1–19.

Enogwe, S. U., Nwankwo, C. H., & Oti, E. U. (2022). Power Hamza distribution with application to lifetime data. *Journal of Applied Mathematics and Physics, 10*, 31–48.

Epstein, B., & Sobel, M. (1954). Some theorems relevant to life testing from an exponential distribution. *Annals of Mathematical Statistics, 25*(2), 373–381.

Esary, J. D., Marshall, A. W., & Proschan, F. (1970). Some reliability applications of the hazard transform. *SIAM Journal on Applied Mathematics, 18*(4), 849–860.

Esary, J. D., & Proschan, F. (1963). Relationship between system failure rate and component failure rates. *Technometrics, 5*(2), 183–189.

Eugene, N., Lee, M., & Famoye, F. (2002). Beta-normal distribution and its applications. *Communications in Statistics-Theory and Methods, 31*(4), 497–512.

Fagiuoli, E., & Pellerey, F. (1993). New partial orderings and applications. *Naval Research Logistics, 40*, 829–842.

Farr, W. (1860). On the construction of life tables: illustrated by a new life table of the healthy districts of England. *The Assurance Magazine, and Journal of the Institute of Actuaries, 9*(3), 121–141.

Feinstein, A. (1958). *Foundations of Information Theory*. New York: McGraw Hill.

Feller, W. (1971). *An Introduction to Probability Theory and its Applications* (Vol. II). New York: Wiley.

Finner, H., & Roters, M. (1993). Distribution functions and log-concavity. *Communications in Statistics-Theory and Methods, 22*, 2381–2396.

Fishburn, P. C. (1980). Stochastic dominance and moments of distributions. *Mathematics of Operations Research, 5*(1), 94–100.

Fisher, R. A. (1936). The use of multiple measurements in taxonomic problems. *Annals of Eugenics, 7*(2), 179–188.

Franco, M., Ruiz, J. M., & Ruiz, M. C. (2001). On closure of the IFR(2) and NBU(2) classes. *Journal of Applied Probability, 38*, 235–241.

Franco, M., Ruiz, M. C., & Ruiz, J. M. (2003). A note on closure of the ILR and DLR classes under formation of coherent systems. *Statistical Papers, 44*, 279–288.

Gaver, D. P. (1963). Random hazard in reliability problems. *Technometrics, 5*(2), 211–226.

Ghitany, M. E., Al-Awadhi, F. A., & Alkhalfan, L. A. (2007). Marshall-Olkin extended Lomax distribution and its applications to censored data. *Communications in Statistics-Theory and Methods, 36*, 1855–1866.

Ghitany, M. E., Al-Hussaini, E. K., & Al-Jarallah, R. A. (2005). Marshall-Olkin extended Weibull distribution and its application to censored data. *Journal of Applied Statistics, 32*, 1025–1034.

Ghitany, M. E., Al-Mutairi, D. K., Balakrishnan, N., & Al-Enezi, L. J. (2013). Power Lindley distribution and associated inference. *Computational Statistics and Data Analysis, 64*, 20–33.

Ghosh, M., & Razmpour, A. (1982). Estimating the location parameter of an exponential distribution with known coefficient of variation. *Calcutta Statistical Association Bulletin, 31*(3–4), 137–150.

Ghosh, J. K., & Samanta, T. (2001). Model selection-an overview. *Current Science, 80*(9), 1135–1144.

Giles, D. E. (2021). On the estimation of the Topp-Leone distribution, Report of the University of Victoria, Canada, http://web.uvic.ca/~dgiles/downloads/working_papers/Topp-Leone.pdf.

Giri, R. J., Nanda, A. K., Dasgupta, M., Misra, S. K., & Bhattacharjee, M. (2023). On ageing intensity function of some Weibull models. *Communications in Statistics-Theory and Methods, 52*(1), 227–262.

Gohout, W., & Kuhnert, I. (1997). NBUFR closure under the formation of coherent systems. *Statistical Papers, 38*, 243–248.

Gómez, Y. M., Bolfarine, H., & Gómez, H. W. (2014). A new extension of the exponential distribution. *Revista Colombiana de Estadistica, 37*(1), 25–34.

Gómez-Déniz, E. (2010). Another generalization of the geometric distribution. *TEST, 19*, 399–415.

Good, I. J. (1953). The population frequencies of species and the estimation of population parameters. *Biometrika, 40*, 237–264.

Goodarzi, F., Amini, M., Borzadaran, M., & Reza, G. (2017). Characterizations of continuous distributions through inequalities involving the expected values of selected functions. *Applications of Mathematics, 62*(5), 493–507.

Gompertz, B. (1825). On the nature of the function expressive of the law of human mortality, and on a new mode of determining the value of life contingencies. Philosophical transactions of the Royal Society of London, *115*, 513–583.

Govindarajulu, Z. (1966). Characterization of exponential and power distributions. *Scandinavian Actuarial Journal, 1966*(3–4), 132–136.

Govindarajulu, Z. (1977). A class of distributions useful in life testing and reliability. *IEEE Transactions on Reliability, R-26*(1), 67–69.

Greenwood, J. A., Landwehr, J. M., Matalas, N. C., & Wallis, J. R. (1979). Probability weighted moments: Definition and relation to parameters of several distributions expressible in inverse form. *Water Resources Research, 15*(5), 1049–1054.

Griffith, A. A. (1921). The phenomenon of rupture and flow in solids. *Philosophical Transactions of the Royal Society of London, Series A, 221*, 163–198.

Guerrieri, G. (1965). Some characteristic properties of the exponential distribution. *Economist, 24*, 427–437.

Gui, W. (2013). Statistical inferences and applications of the half exponential power distribution. *Journal of Quality and Reliability Engineering, Article ID, 219473*, 1–9.

Gui, L., Sun, J., Liu, Y., Si, Y. J., Dong, J. S., & Wang, X. Y. (2013). Combining model checking and testing with an application to reliability prediction and distribution. In *Proceedings of the International Symposium on Software Testing and Analysis* (pp. 101–111).

Gupta, R. C. (1984). Relationships between order statistics and record values and some characterization results. *Journal of Applied Probability, 21*(2), 425–430.

Gupta, R. C. (1987). On the monotonicity properties of the residual variance and their applications in reliability. *Journal of Statistical Planning and Inference, 16*, 329–335.

Gupta, R. C., & Akman, O. (1995). Mean residual life function for certain types of monotonic ageing. *Stochastic Models, 11*(1), 219–225.

Gupta, R. C., Ghitany, M., & Al-Mutairi, D. (2010). Estimation of reliability from Marshall-Olkin extended Lomax distributions. *Journal of Statistical Computation and Simulation, 80*, 937–947.

Gupta, R. C., & Gupta, R. D. (2004). Generalized skew normal model. *Test, 13*, 501–524.

Gupta, R. C., Gupta, P. L., & Gupta, R. D. (1998). Modeling failure time data by Lehman alternatives. *Communications in Statistics-Theory and Methods, 27*(4), 887–904.

Gupta, R. C., & Kirmani, S. N. U. A. (1987). On order relation between reliability measures. *Communications in Statistics-Stochastic Models, 3*, 149–156.

Gupta, R. D., & Kundu, D. (1999). Generalized exponential distributions. *Australian and New Zealand Journal of Statistics, 41*(2), 173–188.

Gupta, R. D., & Kundu, D. (2001). Exponentiated exponential family: An alternative to gamma and Weibull distributions. *Biometrical Journal, 43*(1), 117–130.

Gupta, R. C., & Langford, E. S. (1984). On the determination of a distribution by its median residual life function: A functional equation. *Journal of Applied Probability, 21*(1), 120–128.

Gupta, R. D., & Nanda, A. K. (2001). Some results on reversed hazard rate ordering. *Communications in Statistics-Theory and Methods, 30*(11), 2447–2457.

Gupta, R. D., & Nanda, A. K. (2002). α- and β-entropies and relative entropies of distributions. *Journal of Statistical Theory and Applications, 1*(3), 177–190.

Gurland, J., & Sethuraman, J. (1994). Reversal of increasing failure rate when pooling failure data. *Technometrics, 36,* 416–418.

Gurland, J., & Sethuraman, J. (1995). How pooling failure data may reverse increasing failure rate. *Journal of the American Statistical Association, 90*(432), 1416–1423.

Hadar, J., & Russel, W. R. (1969). Rules for ordering uncertain prospects. *American Economic Review, 59*(1), 25–34.

Haines, A. L., & Singpurwalla, N. D. (1974). Some contributions to the stochastic characterization of wear. In F. Proschan & R. J. Serfling (Eds.), *Reliability and Biometry: Statistical Analysis of Lifelength* (pp. 47–80). Society for Industrial and Applied Mathematics.

Hall, W. J., & Wellner, J. A. (1981). Mean residual life. In M. Csörgö, D. A. Dawson, J. N. K. Rao, & A. KMd. E. Saleh (Eds.), *Statistics and Related Topics* (pp. 169–184). North-Holland Publishing Company.

Hamed, D., & Alzhaghal, A. (2021). A new class of Lindley distributions: Properties and applications. *Journal of Statistical Distributions and Applications, 8*(11), 1–22.

Hamedani, G. G., & Ahsanullah, M. (2011). Characterizations of the Weibull-geometric distribution. *Journal of Statistical Theory and Applications, 10*(4), 581–590.

Hanook, S., Shahbaz, M. Q., Mohsin, M., & Golam Kibria, B. M. (2013). A note on beta inverse Weibull distribution. *Communications in Statistics-Theory and Methods, 42,* 320–335.

Haupt, E., & Schäbe, H. (1992). A new model for a lifetime distribution with bathtub shaped failure rate. *Microelectronics and Reliability, 32,* 633–639.

Haupt, E., & Schäbe, H. (1994). Constructing lifetime distributions with bathtub shaped failure rate from DFR distributions. *Microelectronics and Reliability, 34*(9), 1501–1508.

Haupt, E., & Schäbe, H. (1997). The TTT transformation and a new bathtub distribution model. *Journal of Statistical Planning and Inference, 60*(2), 229–240.

Hausman, J. A., Hall, B., & Griliches, Z. (1984). Econometric methods for count data with an application to the patents-R&D relationship. *Econometrica, 52*(4), 909–938.

Hazra, N. K., Kuiti, M. R., Finkelstein, M., & Nanda, A. K. (2017). On stochastic comparisons of maximum order statistics from the location-scale family of distributions. *Journal of Multivariate Analysis, 160,* 31–41.

Hazra, N. K., Kuiti, M. R., Finkelstein, M., & Nanda, A. K. (2018). On stochastic comparisons of minimum order statistics from the location-scale family of distributions. *Metrika, 81*(2), 105–123.

Hazra, N. K., Kundu, P. K., & Nanda, A. K. (2018). Some reliability properties of transformed-transformer family of distributions. *American Journal of Mathematical and Management Sciences, 38*(1), 44–56.

Hazra, N. K., & Nanda, A. K. (2016). On some generalized orderings: In the spirit of relative ageing. *Communications in Statistics-Theory and Methods, 45*(20), 6165–6181.

Hazra, N. K., Nanda, A. K., & Shaked, M. (2014). Some aging properties of parallel and series systems with a random number of components. *Naval Research Logistics, 61,* 238–243.

Hendi, M. I., Mashhour, A. F., & Montas, M. A. (1993). Closure of the NBUC class under formation of parallel systems. *Journal of Applied Probability, 30,* 975–978.

Hjorth, U. (1980). A reliability distribution with increasing, decreasing, constant and bathtub shaped failure rates. *Technometrics, 22*(1), 99–107.

Hollander, M., Park, D. H., & Proschan, F. (1986). A class of life distributions for aging. *Journal of the American Statistical Association, 81*(393), 91–95.

Hu, T., He, F., & Khaledi, B. E. (2004). Characterizations of some aging notions by means of dispersion-type or dilation-type variability orders. *Chinese Journal of Applied Probability and Statistics, 20*(1), 66–76.

Hu, T., Kundu, A., & Nanda, A. K. (2001). On generalized orderings and ageing properties with their implications. In Y. Hayakawa, T. Irony, & M. Xie (Eds.), *System and Bayesian Reliability*. Singapore: World Scientific.

Hu, T., Ma, M., & Nanda, A. K. (2003). Moment inequalities for discrete ageing families. *Communications in Statistics-Theory and Methods, 32*(1), 61–90.

Hu, T., Ma, M., & Nanda, A. K. (2004). Characterization of generalized ageing classes by the excess lifetime. *Southeast Asian Bulletin of Mathematics, 28*, 279–285.

Hu, T., Nanda, A. K., Xie, H., & Zhu, Z. (2004). Properties of some stochastic orders: A unified study. *Naval Research Logistics, 51*, 193–216.

Hu, T., & Xie, H. (2002). Proofs of the closure property of $NBUC$ and $NBU(2)$ under convolution. *Journal of Applied Probability, 39*(1), 224–227.

Hussain, I., Abbas, Z., & Ahmad, Z. (2018). Transmuted size-biased exponential distribution and its properties. *Pakistan Journal of Statistics, 34*(2), 99–118.

Isaic-Maniu, A. N., & Voda, V. G. (2008). Generalized Burr-Hatke equation as generator of a homographic failure rate. *Journal of Applied Quantitative Methods, 3*(3), 215–222.

Jamal, F., & Nasir, M. A. (2018). Generalized Burr X family of distributions. *International Journal of Mathematics and Statistics, 19*(1), 56–73.

Jayakumar, K., & Sankaran, K. K. (2021). Exponential intervened Poisson distribution. *Communications in Statistics-Theory and Methods, 50*(13), 3063–3093.

Jiang, R., & Murthy, D. N. P. (1997). Parametric study of competing risk model involving two Weibull distributions. *International Journal of Reliability, Quality and Safety Engineering, 4*(1), 17–34.

Jiang, R., & Murthy, D. N. P. (1999). The exponentiated Weibull family?: A graphic approach. *IEEE Transactions on Reliability, 48*, 68–72.

Jiang, R., Murthy, D. N. P., & Ji, P. (2001). Models involving two inverse Weibull distributions. *Reliability Engineering and System Safety, 73*, 73–81.

Jiang, R., Ji, P., & Xiao, X. (2003). Aging property of univariate failure rate models. *Reliability Engineering and System Safety, 79*, 113–116.

Joag-Dev, K., Kochar, S., & Proschan, F. (1995). A general composition theorem and its applications to certain partial orderings of distributions. *Statistics and Probability Letters, 22*, 111–119.

Joe, H. (1985). Characterizations of life distributions from percentile residual life. *Annals of the Institute of Statistical Mathematics, 37*, 165–172.

Joe, H., & Proschan, F. (1984). Percentile residual life functions. *Operations Research, 32*(3), 668–678.

Johnson, N. L. (1949). Systems of frequency curves generated by methods of translation. *Biometrika, 36*, 149–176.

Johnson, N. L., Kotz, S., & Balakrishnan, N. (1994). *Continuous Univariate Distributions* (Vol. 1). New York: Wiley.

Jones, M. C. (2008). Kumaraswamy's distribution: A beta-type distribution with some tractability advantages. *Statistical Methodology, 6*(1), 70–81.

Jones, M. C. (2020). On univariate slash distributions, continuous and discrete. *Annals of the Institute of Statistical Mathematics, 72*, 645–657.

Jose, K. K., & Abraham, B. (2011). A count data model based on Mittag-Leffler interarrival times. *Statistica, 71*(4), 501–514.

Jose, K. K., & Naik, S. R. (2009). On the q-Weibull distribution and its applications. *Communications in Statistics-Theory and Methods, 38*, 912–926.

Kaas, R., Van Heerwaarden, A. E., & Goovaerts, M. J. (1994). *Ordering of Actuarial Risks, Caire Education Series 1*. Brussels

Kafadar, K. (1988). *Slash distributions. In Encyclopedia of Statistical Sciences*. New York: Wiley.

Kagan, A. M., Linnik, Y. V., & Rao, C. R. (1973). *Characterization Problems in Mathematical Statistics*. New York: Wiley.

Kandil, A. E. M., Kayid, M., & Mahdy, M. M. R. (2010). Median inactivity time function and its reliability properties. *Stochastics and Quality Control, 25*(2), 253–268.

Kao, J. H. K. (1958). Computer methods for estimating Weibull parameters in reliability studies. *IRE Transactions on Reliability and Quality Control, PGRQC-13*, 15–22.

References

Kaplansky, I. (1945). A common error concerning kurtosis. *Journal of American Statistical Association, 40*(230), 259–259.

Karlin, S. (1968). *Total Positivity*. California: Stanford University Press.

Karlin, S. (1982). Some results in optimal partitioning of variance and monotonicity, with truncation level. In G. Kallianpur, P. R. Krishnaiah, & J. K. Ghosh (Eds.), *Statistics and Probability: Essays in Honor of C.R. Rao* (pp. 375–392). North-Holland Publishing Company.

Karlin, S., & Proschan, F. (1960). Pólya type distributions of convolutions. *Annals of Mathematical Statistics, 31*, 721–736.

Kayid, M., & Ahmad, I. A. (2004). On the mean inactivity time ordering with reliability applications. *Probability in the Engineering and Informational Sciences, 18*, 395–409.

Kayid, M., & Izadkhah, M. (2014). Mean inactivity time function, associated ordering and classes of life distributions. *IEEE Transactions on Reliability, 63*(2), 593–602.

Kayid, M., & Shrahili, M. (2023). On the uncertainty properties of the conditional distribution of the past lifetime. Entropy, 25(6). *Article No., 895*, 1–13.

Kebir, Y. (1994). Laplace transform characterization of probabilistic orderings. *Probability in the Engineering and Informational Sciences, 8*(1), 69–77.

Keilson, J., & Sumita, U. (1982). Uniform stochastic ordering and related inequalities. *Canadian Journal of Statistics, 10*, 181–198.

Keller, A. Z., & Kamath, A. R. R. (1982). Alternative reliability models for mechanical systems. In *Proceedings of the 3rd International Conference on Reliability and Maintainability* (pp. 411–415). Scientific Research Publishing.

Kemp, A. W. (1997). Characterizations of a discrete normal distribution. *Journal of Statistical Planning and Inference, 63*(2), 223–229.

Kemp, A. W. (2011). Univariate discrete distributions: An overview. In M. Lovric (Ed.), *International Encyclopedia of Statistical Science* (pp. 1630–1634). Berlin: Springer.

Kendall, M. G., & Buckland, W. R. (1971). *A Dictionary of Statistical Terms*. New York: Hafner Publishing Company.

Khan, M. S. A., Khalique, A., & Abouammoh, A. M. (1989). On estimating parameters in discrete Weibull distribution. *IEEE Transaction on Reliability, 38*(3), 348–350.

Khan, R. A., Bhattacharyya, D., & Mitra, M. (2021). On some properties of the mean inactivity time function. *Statistics and Probability Letters, 170*, 1–9.

Khan, M. S., & King, R. (2013). Transmuted modified Weibull distribution: A generalization of the modified Weibull probability distribution. *European Journal of Pure and Applied Mathematics, 6*(1), 66–68.

Khinchin, A. I. (1957). *Mathematical Foundation of Information Theory*. New York: Dover.

Khodabin, M., & Ahmadabadi, A. R. (2010). Some properties of generalized gamma distribution. *Mathematical Sciences, 4*(1), 9–28.

Kies, J. A. (1958). The strength of glass, Naval Research Laboratory, USA.

Kilany, N. M. (2016). Weighted Lomax distribution. *SpringerPlus, 5*, Article number 1862. https://doi.org/10.1186/s40064-016-3489-2.

Kirmani, S. N. U. A., & Gupta, R. C. (2001). On the proportional odds model in survival analysis. *Annals of the Institute of Statistical Mathematics, 53*(2), 203–216.

Klar, B. (2002). A note on the $\mathscr{L}$-class of life distributions. *Journal of Applied Probability, 39*, 11–19.

Klar, B., & Müller, A. (2003). Characterizations of classes of lifetime distributions generalizing the *NBUE* class. *Journal of Applied Probability, 40*, 20–32.

Klefsjö, B. (1980). Some properties of the HNBUE and HNWUE classes of life distributions. *Statistical Research Report*, Department of Mathematical Statistics, University of Umeå.

Klefsö, B. (1981). HNBUE survival under some shock models. *Scandinavian Journal of Statistics, 8*, 39–47.

Klefsjö, B. (1982). On aging properties and total time on test transforms. *Scandinavian Journal of Statistics, 9*(1), 37–41.

Klefsjö, B. (1983). A useful ageing property based on the Laplace transform. *Journal of Applied Probability, 20*, 615–626.

Klefsjö, B. (1986). TTT transform-a useful tool when analyzing different reliability problems. *Reliability Engineering, 1593*, 231–241.

Kleiber, C., & Kotz, S. (2003). *Statistical Size Distributions in Economics and Actuarial Sciences*. New Jersey: Wiley.

Kochar, S. C. (1979). Distribution-free comparison of two probability distributions with reference to their hazard rates. *Biometrika, 66*, 437–441.

Korkmaz, M. C., & Yousof, H. M. (2017). One parameter odd Lindley exponential model: Mathematical properties and applications. *Stochastics and Quality Control, 32*(1), 25–35.

Kotz, S., & Shanbhag, D. N. (1980). Some new approaches to probability distributions. *Advances in Applied Probability, 12*(4), 903–921.

Kulasekara, K. B., & Tokyn, D. W. (1992). A new discrete distribution with application to survival, dispersal and dispersion. *Communications in Statistics-Simulation and Computation, 21*, 499–518.

Kumar, C. S., & Nair, B. U. (2014). A three parameter hyper-Poisson distribution and some of its properties. *Statistica, 74*(2), 183–198.

Kumaraswamy, P. (1980). Generalized probability density function for double bounded random processes. *Journal of Hydrology, 46*(1–2), 79–88.

Kundu, C., & Ghosh, A. (2017). Inequalities involving expectations of selected functions in reliability theory to characterize distributions. *Communications in Statistics-Theory and Methods, 46*, 8468–8478.

Kundu, C., & Nanda, A. K. (2010). On generalized mean residual life of record values. *Statistics and Probability Letters, 80*, 797–806.

Kundu, C., & Nanda, A. K. (2010). Some reliability properties of the inactivity time. *Communications in Statistics-Theory and Methods, 39*(5), 899–911.

Kundu, C., Nanda, A. K., & Hu, T. (2009). A note on reversed hazard rate of order statistics and record values. *Journal of Statistical Planning and Inference, 139*, 1257–1265.

Kundu, C., Nanda, A. K., & Maiti, S. S. (2010). Some distributional results through past entropy. *Journal of Statistical Planning & Inference, 140*(5), 1280–1291.

Kundu, D., & Nekoukhou, V. (2018). Univariate and bivariate geometric discrete generalized exponential distributions. *Journal of Statistical Theory and Practice, 12*(3), 595–614.

Kupka, J., & Loo, S. (1989). The hazard and vitality measures of ageing. *Journal of Applied Probability, 26*(3), 532–542.

Lai, C. D. (2013). *Generalized Weibull Distributions*. Berlin: Springer.

Lai, C. D. (2013). Issues concerning construction of discrete lifetime models. *Quality Technology and Quantitative Management, 10*(2), 251–262.

Lai, C. D., & Moore, T. (1998). The Beta-integrated failure model. In M. Xie, & D. N. P. Murthy (Eds.), *Proceedings of the International Workshop on Reliability Modeling and Analysis-from Theory to Practice* (pp. 153–159).

Lai, C. D., & Xie, M. (2006). *Stochastic Ageing and Dependence for Reliability*. Springer.

Lai, C. D., Xie, M., & Murthy, D. N. P. (2001). Bathtub-shaped failure rate life distributions. In N. Balakrishnan, & C. R. Rao (Eds.), *Handbook of Statistics* (Vol. 20, pp. 69–104).

Lai, C. D., Xie, M., & Murthy, D. N. P. (2003). Modified Weibull model. *IEEE Transactions on Reliability, 52*, 33–37.

Lai, C. D., & Wang, D. Q. (1995). A finite range discrete life distribution. *International Journal of Reliability, Quality and Safety Engineering, 2*(2), 147–160.

Lariviere, M. A. (2006). A note on probability distributions with increasing generalized failure rate. *Operations Research, 54*(3), 602–604.

Lariviere, M. A., & Porteus, E. L. (2001). Selling to a newsvendor: An analysis of price-only contracts. *Manufacturing and Service Operations Management, 3*(4), 293–305.

Lehmann, E. L. (1955). Ordered families of probability distributions. *Annals of Mathematical Statistics, 26*, 399–419.

Leiva, V. (2015). *The Birnbaum-Saunders Distribution*. New York: Academic Press.

Leiva, V., Ruggeri, F., Saulo, H., & Vivanco, J. F. (2016). A methodology based on the Birnbaum-Saunders distribution for reliability analysis applied to nano-materials. *Reliability Engineering and System Safety, 157*, 192–201.

Lemonte, A. J. (2014). The beta log-logistic distribution. *Brazilian Journal of Probability and Statistics, 28*(3), 313–332.

Lemonte, A. J., Barreto-Souza, W., & Cordeiro, G. M. (2013). The exponentiated Kumaraswamy distribution and its log-transform. *Brazilian Journal of Probability and Statistics, 27*, 31–53.

Lemonte, A., & Cordeiro, G. (2013). An extended Lomax distribution. *Statistics, 47*(4), 800–816.

Lewis, T., & Thompson, J. W. (1981). Dispersive Distributions, and the Connection between Dispersivity and Strong Unimodality. *Journal of Applied Probability, 18*, 76–90.

Li, X., & Kochar, S. C. (2001). Some new results involving the $NBU(2)$ class of life distributions. *Journal of Applied Probability, 38*, 242–247.

Li, X., Li, Z., & Jing, B. (2000). Some results about the $NBUC$ class of life distributions. *Statistics and Probability Letters, 46*, 229–237.

Li, H., & Tian, W. (2020). Slashed Lomax distribution and regression model. *Symmetry, 12*, 1877. https://doi.org/10.3390/sym12111877.

Li, X., & Zuo, M. J. (2004). Stochastic comparison of residual life and inactivity time at a random time. *Stochastic Models, 20*(2), 229–235.

Lillo, R. E. (2005). On the median residual lifetime and its ageing properties: A characterization theorem and its applications. *Naval Research Logistics, 52*, 370–380.

Lillo, R. E., Nanda, A. K., & Shaked, M. (2000). Some shifted stochastic orders. In N. Limnios & M. Nikulin (Eds.), *Recent Advances in Reliability Theory: Methodology, Practice and Inference* (pp. 55–103). Boston, USA: Birkhäuser.

Lillo, R. E., Nanda, A. K., & Shaked, M. (2001). Preservation of some likelihood ratio stochastic orders by order statistics. *Statistics and Probability Letters, 51*, 111–119.

Lindley, D. V. (1958). Fiducial distributions and Bayes theorem. *Journal of the Royal Statistical Society, Series A, 20*, 102–107.

Lindqvist, B. H., & Samaniego, F. J. (2019). Some new results on the preservation of the NBUE and NWUE aging classes under the formation of coherent systems. *Naval Research Logistics, 66*(5), 430–438.

Loh, W. (1984). A new generalization of the class of NBU distributions. *IEEE Transactions on Reliability, 33*(5), 419–422.

Lukacs, E. (1958). Some extensions of a theorem of Marcinkiewicz. *Pacific Journal of Mathematics, 8*(3), 487–501.

Lynch, J. (1999). On conditions for mixtures of increasing failure rate distributions to have increasing rate. *Probability in the Engineering and Informational Sciences, 13*(1), 33–36.

Ma, C. (1999). Uniform stochastic ordering on a system of components with dependent lifetimes induced by a common environment. *Sankhyā, Series A, 61*, 218–228.

Mahdy, M. (2013). Probabilistic properties of discrete mean and variance reversed residual lifetime functions. *Applied Mathematical Sciences, 7*(124), 6167–6179.

Maiti, S. S., & Pramanik, S. (2016). Odds generalized exponentiated Pareto distribution-properties and applications. *Pakistan Journal of Statistics and Operation Research, 12*(2), 257–279.

Maiti, S. S., & Pramanik, S. (2019). Odds X-gamma-G family of distributions. *IAPQR Transactions, 43*(2), 135–163.

Makino, T. (1984). Mean hazard rate and its applications to the normal approximation of the Weibull distribution. *Naval Research Logistics Quarterly, 31*, 1–8.

Marsaglia, G., & Tubilla, A. (1975). A note on the "lack of memoryâŁž property of the exponential distribution. *Annals of Probability, 3*(2), 353–354.

Marshall, A. W., & Olkin, I. (1997). A new method of adding a parameter to a family of distributions with applications to the exponential and Weibull families. *Biometrika, 84*, 641–652.

Marshall, A. W., & Olkin, I. (2007). *Life Distributions*. New York: Springer.

Marshall, A. W., & Proschan, F. (1972). Classes of life distributions applicable in replacement with renewal theory implications. *Proceedings of the Sixth Berkeley Symposium on Mathematical Statistics and Probability, 1*, 395–415.

Marshall, K. T. (1968). Bounds for some generalizations of the $GI/G/1$ queue. *Operations Research, 16*(4), 841–848.

Mazumder, M., & Gaver, D. P. (1984). On the computation of power-generating system reliability indexes. *Technometrics, 26*(2), 173–185.

McCullagh, P. (1980). Regression models for ordinal data with discussion. *Journal of the Royal Statistical Society, Series B, 42*, 109–142.

Mcdonald, J. B. (1984). Some generalized functions for the size distribution of income. *Econometrica, 52*, 647–663.

Mead, M. E., Cordeiro, G. M., Afify, A. Z., & Al-Mofleh, H. (2019). The alpha power transformation family: Properties and applications. *Pakistan Journal of Statistics and Operations Research, 15*(3), 525–545.

Menon, M. V. (1963). Estimation of the shape and scale parameters of the Weibull distribution. *Technometrics, 5*, 175–182.

Misra, S. K., & Bhattacharjee, S. (2018). A case study of ageing intensity function on censored data. *Alexendria Engineering Journal, 57*, 3931–3952.

Mitra, M., & Basu, S. K. (1995). On some properties of the class of distributions that are Laplace dominated by the exponential law and its ramifications. In A. P. Basu, S. K. Basu, & S. P. Mukhopadhyay (Eds.), *Frontiers in Reliability*. World Scientific.

Miziula, P. (2012). Stochastic orders and ageing classes. *Mathematica Applicanda, 40*(1), 105–125.

Moore, T., & Lai, C. D. (1994). The beta failure rate distribution. *Proceedings of the 30th Annual Conference of Operational Research Society of New Zealand* (pp. 339–344).

Mudholkar, G. S., & Kollia, G. D. (1994). Generalized Weibull family: A structural analysis. *Communications in Statistics-Theory and Methods, 23*(4), 1149–1171.

Mudholkar, G. S., & Srivastava, D. K. (1993). Exponentiated Weibull family for analyzing bathtub failure-rate data. *IEEE Transactions on Reliability, 42*(2), 299–302.

Mudholkar, G. S., Srivastava, D. K., & Freimer, M. (1995). The exponentiated Weibull family: A reanalysis of the bus-motor-failure data. *Technometrics, 37*(4), 436–445.

Muhammad, M. (2017). A new lifetime model with a bounded support. *Asian Research Journal of Mathematics, 7*(3), 1–11.

Mukherjee, S. P. (2017). On some finite range life distributions. *Calcutta Statistical Association Bulletin, 69*(1), 103–109.

Mukherjee, S. P., & Chatterjee, A. (1992). Stochastic dominance of higher orders and its implications. *Communications in Statistics-Theory and Methods, 21*(7), 1977–1986.

Mukherjee, S. P., & Islam, A. (1983). A finite range distribution of failure times. *Naval Research Logistics Quarterly, 30*, 487–491.

Mukherjee, S. P., & Maiti, S. S. (1996). A percentile estimator of the inverse Rayleigh parameter. *IAPQR Transactions, 21*, 63–66.

Mukherjee, S. P., & Maiti, S. S. (1996). Reliability from damaged stress and strength data. *Calcutta Statistical Association Bulletin, 46*(1–2), 135–142.

Mukherjee, S. P., & Pal, M. (2002). Distribution of quality adjusted life. *Proceedings of Vth International Symposium on Optimization and Statistics*, Aligarh Muslim University, India.

Mukherjee, S. P., & Pal, M. (2007). Distribution of quality-adjusted life of a product which can exist in three states. *IAPQR Transactions, 32*(2).

Mukherjee, S. P., & Roy, D. (1986). Some characterisations of the exponential and related Life distributions. *Calcutta Statistical Association Bulletin, 35*(3–4), 189–197.

Mukherjee, S. P., & Roy, D. (1987). Failure rate transform of the Weibull variate and its properties. *Communications in Statistics-Theory and Methods, 16*(1), 281–291.

Mukherjee, S. P., & Roy, D. (1989). Properties of classes of probability distributions based on the concept of reciprocal co-ordinate subtangent. *Calcutta Statistical Association Bulletin, 38*, 169–180.

References

Mukherjee, S. P., & Roy, D. (1993). A versatile failure time distribution. *Microelectronics Reliability, 33*(7), 1053–1056.

Mukherjee, S. P., & Sasmal, B. C. (1984). Estimation of Weibull parameters using fractional moments. *Calcutta Statistical Association Bulletin, 33*(3–4), 179–186.

Mukherjee, S. P., & Sinha, S. K. (1978). *Weibull parameter estimation through log-linear regression. Unpublished Technical Report.* Canada: University of Manitoba.

Mullahy, J. (1986). Specification and testing of some modified count data models. *Journal of Econometrics, 33*, 341–365.

Müller, A. (1997). Stochastic orders generated by integrals: A unified approach. *Advances in Applied Probability, 29*(2), 414–428.

Murthy, D. N. P., Xie, M., & Jiang, R. (2004). *Weibull Models.* Wiley Interscience.

Murthy, V. K. (1968). A new method of estimating the Weibull shape parameter, *Technical Report ARL 68-0076*, April, Aerospace Research Laboratories United States Air Force, Wright Patterson Air Force Base, Ohio.

Muse, A. H., Mwalili, S. M., & Ngesa, O. (2021). On the log-logistic distribution and its generalizations: A survey. *International Journal of Statistics and Probability, 10*(3), 93–125.

Muth, E. J. (1977). Reliability models with positive memory derived from the mean residual life. In C. P. Tsokos & I. Shimi (Eds.), *Theory and Applications of Reliability* (Vol. II, pp. 401–436). USA: Academic Press.

Muth, E. J. (1980). Memory as a property of probability distributions. *IEEE Transactions on Reliability, R, 29*(2), 713–716.

Nadarajah, S., Bakouch, H. S., & Tahmasbi, R. (2011). A generalized Lindley distribution. *Sankhyā, Series B, 73*, 331–359.

Nadarajah, S., & Haghighi, F. (2011). An extension of the exponential distribution. *Statistics, 45*(6), 543–558.

Nadarajah, S., & Kotz, S. (2003). Moments of some J-shaped distributions. *Journal of Applied Statistics, 30*(3), 311–317.

Nadarajah, S., & Kotz, S. (2005). On some recent modifications of Weibull distribution. *IEEE Transactions on Reliability, 54*(4), 561–562.

Nadarajah, S., & Kotz, S. (2006). The beta exponential distribution. *Reliability Engineering and System Safety, 91*(6), 689–697.

Nagaraja, H. N. (1977). On a characterization based on record values. *Australian Journal of Statistics, 19*(1), 70–73.

Nagarjuna, V. B. V., Vardhan, R. V., & Chesneau, C. (2021). Kumaraswamy generalized power Lomax distribution and its applications. *Stats, 4*, 28–45. https://doi.org/10.3390/stats4010003.

Nair, N. U. (1983). A measure of memory for some discrete distributions. *Journal of Indian Statistical Association, 24*, 141–147.

Nair, K. R. M., & Rajesh, G. (2000). Geometric vitality function and its application to reliability. *IAPQR Transactions, 25*(1), 1–8.

Nair, N. U., Sankaran, P. G., & Balakrishnan, N. (2013). *Quantile-Based Reliability Analysis.* Birkhäuser.

Nair, N. U., Sreedharan, S., & Sandhya, E. (2022). Memory of distributions: A renewal theoretic approach. *Journal of the Indian Society for Probability and Statistics, 23*(4), 173–185.

Nair, N. U., Sunoj, S. M., & Rajesh, G. (2021). Some aspects of reversed hazard rate and past entropy. *Communications in Statistics-Theory and Methods, 50*(9), 2106–2116.

Nakagawa, T., & Osaki, S. (1975). The discrete Weibull distribution. *IEEE Transactions on Reliability, R-24*(5), 300–301.

Nanda, A. K. (1995). Stochastic orders in terms of Laplace transforms. *Calcutta Statistical Association Bulletin, 45*(179–180), 195–201.

Nanda, A. K. (1998). On some stochastic order relations and their applications, *PhD Dissertation*, Panjab University, Chandigarh, India.

Nanda, A. K. (2000). Generalized ageing classes in terms of Laplace transforms. *Sankhyā, Series A, 62*(2), 258–266.

Nanda, A. K. (2006). Properties of generalized residual entropy, *Statistical Methods*, Special Issue on Proceedings of the National Seminar on Modelling and Analysis of Life Time Data (pp. 23–26).

Nanda, A. K. (2010). Characterization of distributions through failure rate and mean residual life functions. *Statistics and Probability Letters, 80*, 752–755.

Nanda, A. K., Bhattacharjee, S., & Alam, S. S. (2004). Study of reliability properties of statistical models in terms of aging intensity function. *IEEE India Annual Conference, INDICON* (pp. 229–231).

Nanda, A. K., Bhattacharjee, S., & Alam, S. S. (2005), Proportional mean residual life model in reliability analysis. In N. Ravichandran (Ed.), *Proceedings of the 37th Operations Research Society of India (ORSI) Annual Convention*-Vision 2020: The Strategic Role of Operational Research (pp. 682–694). New Delhi: Allied Publishers.

Nanda, A. K., Bhattacharjee, S., & Alam, S. S. (2006). On up shifted reversed mean residual life order. *Communications in Statistics-Theory and Methods, 35*(8), 1513–1523.

Nanda, A. K., Bhattacharjee, S., & Alam, S. S. (2006). Properties of proportional mean residual life model. *Statistics and Probability Letters, 76*(9), 880–890.

Nanda, A. K., Bhattacharjee, S., & Alam, S. S. (2007). Properties of aging intensity function. *Statistics and Probability Letters, 77*, 365–373.

Nanda, A. K., Bhattacharjee, S., & Balakrishnan, N. (2010). Mean residual life function, associated orderings and properties. *IEEE Transactions on Reliability, 59*(1), 55–65.

Nanda, A. K., & Chowdhury, S. (2019). Shannon's entropy and its generalizations towards statistics, reliability and information science during 1948–2018. arXiv:1901.09779v1.

Nanda, A. K., Chowdhury, S., Gayen, S., & Bhattacharjee, S. (2024). A new bivariate distribution with uniform marginals. *Communications in Statistics-Theory and Methods, 53*(19), 6918–6943.

Nanda, A. K., & Chowdhury, S. (2021). Shannon's entropy and its generalizations towards statistical inference in last seven decades. *International Statistical Review, 89*(1), 167–185.

Nanda, A. K., & Das, S. (2011). Dynamic proportional hazard rate and reversed hazard rate models. *Journal of Statistical Planning and Inference, 141*, 2108–2119.

Nanda, A. K., & Das, S. (2012). Stochastic orders of the Marshall-Olkin extended distribution. *Statistics and Probability Letters, 82*, 295–302.

Nanda, A. K., & Das, S. (2013). Some ageing properties of Marshall-Olkin extended distribution. *International Journal of Mathematics and Statistics, 13*(1), 93–107.

Nanda, A. K., Das, S., & Balakrishnan, N. (2013). On dynamic proportional mean residual life model. *Probability in the Engineering and Informational Sciences, 27*(4), 553–588.

Nanda, A. K., & Kundu, A. (2009). On generalized stochastic orders of dispersion-type. *Calcutta Statistical Association Bulletin, 61*, 155–182.

Nanda, A. K., & Gupta, R. D. (2001), Some properties of reversed hazard rate function. *Statistical Methods, 3*(2), 108–124. [Errata: *Statistical Methods, 6*(1), 90–91 (2004)].

Nanda, A. K., Hazra, N. K., Al-Mutairi, D. K., & Ghitany, M. E. (2017). On some generalized ageing orderings. *Communications in Statistics-Theory and Methods, 46*(11), 5273–5291.

Nanda, A. K., & Jain, K. (1996). A note on two new ageing properties. *IAPQR Transactions, 21*(2), 109–117.

Nanda, A. K., & Jain, K. (1999). Some weighted distribution results on univariate and bivariate cases. *Journal of Statistical Planning and Inference, 77*, 169–180.

Nanda, A. K., Jain, K., & Singh, H. (1996). On closure of some partial orderings under mixtures. *Journal of Applied Probability, 33*, 698–706.

Nanda, A. K., Jain, K., & Singh, H. (1996). Properties of moments for s-order equilibrium distributions. *Journal of Applied Probability, 33*, 1108–1111.

Nanda, A. K., Jain, K., & Singh, H. (1998). Preservation of some partial orderings under the formation of coherent systems. *Statistics and Probability Letters, 39*, 123–131.

Nanda, A. K., & Kundu, A. (2009). On generalized stochastic orders of dispersion-type. *Calcutta Statistical Association Bulletin, 61*(241–244), 155–182.

References

Nanda, A. K., & Kundu, A. (2011). On improvement and deterioration of a repairable system under generalized stochastic orders. *Calcutta Statistical Association Bulletin, 63*, 259–272.

Nanda, A. K., & Maiti, S. S. (2006). Generalized residual information, loglikelihood and an intrinsic residual life distribution measure. *Statistical Methods*, Special Issue on Proceedings of the National Seminar on Modelling and Analysis of Life Time Data (pp. 77–86).

Nanda, A. K., & Maiti, S. S. (2007). Rényi information measure for a used item. *Information Sciences, 177*, 4161–4175.

Nanda, A. K., Maiti, S. S., Kundu, C., & Kundu, A. (2019). Parameter estimates of general failure rate model: A Bayesian approach. *Journal of Computational and Applied Mathematics, 351*, 317–330.

Nanda, A. K., Misra, N., Paul, P., & Singh, H. (2005). Some properties of order statistics when sample size is random. *Communications in Statistics-Theory and Methods, 34*(11), 2105–2113.

Nanda, A. K., & Paul, P. (2003). Test for reversed hazard rate function. *Calcutta Statistical Association Bulletin, 54*(215–216), 181–193.

Nanda, A. K., & Paul, P. (2004a). An observation from DMRL life distribution. *Statistical Methods, 6*(1), 57–65. [Errata: *Statistical Methods, 6*(2), 235 (2004).]

Nanda, A. K., & Paul, P. (2004b). A comprehensive study on past entropy. *Proceedings of the IEEE Annual Conference*, INDICON, (pp. 113–116).

Nanda, A. K., & Paul, P. (2006). Some properties of past entropy and their applications. *Metrika, 64*(1), 47–61.

Nanda, A. K., & Sengupta, D. (2005). Discrete life distributions with decreasing reversed hazard. *Sankhyā, 67*(1), 106–125.

Nanda, A. K., & Shaked, M. (2001). The hazard rate and reversed hazard rate orders, with applications to order statistics. *Annals of the Institute of Statistical Mathematics, 53*(4), 853–864.

Nanda, A. K., & Shaked, M. (2008). Partial orderings and aging properties of order statistics when the sample size is random: A brief review. *Communications in Statistics-Theory and Methods, 37*(11), 1710–1720.

Nanda, A. K., Singh, H., Misra, N., & Paul, P. (2003). Reliability properties of reversed residual lifetime. *Communication in Statistics-Theory and Methods, 32*(10), 2031–2042.

Navarroa, J., del Águila, Y., Sordo, M. A., & Suárez-Llorens, A. (2014). Preservation of reliability classes under the formation of coherent systems. *Applied Stochastic Models in Business and Industry, 30*, 444–454.

Navarro, J., & Shaked, M. (2010). Some properties of the minimum and the maximum of random variables with joint logconcave distributions. *Metrika, 71*, 313–317.

Nayak, S. S. (1981). Characterization based on record values. *Journal of Indian Statistical Association, 19*, 123–127.

Nekoukhou, V., Alamatsaz, M. H., & Bidram, H. (2012). A discrete analog of the generalized exponential distribution. *Communications in Statistics-Theory and Methods, 41*, 2000–2013.

Nekoukhou, V., & Bidram, H. (2015). The exponentiated discrete Weibull distribution. *SORT, 39*(1), 127–146.

Nikulin, M., & Haghighi, F. (2006). A chi-squared test for the generalized power Weibull family for the head-and-neck cancer censored data. *Journal of Mathematical Sciences, 133*(3), 1333–1341.

O'Brien, G. L. (1984). Stochastic dominance and moment inequalities. *Mathematics of Operations Research, 9*(3), 475–477.

Ogunde, A. A., Chukwu, A. U., & Oseghale, I. O. (2023). The Kumaraswamy generalized inverse Lomax distribution and applications to reliability and survival data. *Scientific African, 19*, Article No. e01483.

Oguntunde, P. E., Khaleel, M. A., Ahmed, M. T., Adejumo, A. O., & Odetunmibi, O. A. (2017). A new generalization of the Lomax distribution with increasing, decreasing, and constant failure rate. *Modelling and Simulation in Engineering, 2017*(1), Article ID 6043169, 6 pages.

Owoloko, E. A., Oguntunde, P. E., & Adejumo, A. O. (2015). Performance rating of the transmuted exponential distribution: An analytical approach. *SpringerPlus, 4*, Article No. 818.

Padgett, W. J., & Spurrier, J. D. (1985). On discrete failure models. *IEEE Transactions on Reliability, R-, 34*(3), 253–256.

Paluszny, A., & Nicholls, P. P. (1978). Predicting time dependent reliability of ceramic rotors. In J. J. Burke, E. N. Lenoe, & R. N. Katz (Eds.), *Ceramics for High Performance Applications*. Massachusets: Brook Hill Publications.

Pappas, V., Adamidis, K., & Loukas, S. (2012). A family of lifetime distributions. *International Journal of Quality, Statistics, and Reliability, Article ID 760687*, 6 pages.

Pararai, M., Warahena-Liyanage, G., & Oluyede, B. O. (2014). A new class of generalized inverse Weibull distribution with applications. *Journal of Applied Mathematics & Bioinformatics, 4*(2), 17–35.

Pararai, M., Warahena-Liyanage, G., & Oluyede, B. O. (2015). A new class of generalized power Lindley distribution with applications to lifetime data. *Theoretical Mathematics & Applications, 5*(1), 53–96.

Pardo, L., Salicrú, M., Menéndez, M. L., & Morales, D. (1995). Divergence measures based on entropy functions and statistical inference. *Sankhyā, Series B, 57*(3), 315–337.

Park, D. H. (2003). Class of NBU-t_0 life distribution. In H. Pham (Ed.), *Handbook of Reliability Engineering* (pp. 181–197). London: Springer.

Park, J., & Boring, R. L. (2023). *Investigation of human reliability analysis methods for analyzing pre-initiators* (pp. 1–7). U.S: Department of Energy, Idaho National Laboratory.

Patil, G. P., & Seshadri, V. (1964). Characterization theorems for some univariate probability distributions. *Journal of the Royal Statistical Society, Series B, 26*(2), 286–292.

Pečarić, J. E., Proschan, F., & Tong, Y. L. (1992). *Convex Functions, Partial Orderings, and Statistical Applications*. New York: Academic Press.

Pellerey, F., Shaked, M., & Zinn, J. (2000). Nonhomogeneous Poisson process and logconcavity. *Probability in the Engineering and Informational Sciences, 14*, 353–373.

Pfeifer, D. (1982). Characterizations of exponential distributions by independent non-stationary record increments. *Journal of Applied Probability, 19*(1), 137–135.

Pham, H. (2002). A Vtub-shaped hazard rate function with applications to system safety. *International Journal of Reliability and Applications, 3*, 1–16.

Pham, H., & Lai, C. D. (2007). On recent generalizations of the Weibull distribution. *IEEE Transactions on Reliability, 56*, 454–458.

Phani, K. K. (1987). A new modified Weibull distribution function. *Communications of American Ceramic Society, 70*, 182–184.

Pillai, R. N. (1990). On Mittag-Leffler Functions and related Distributions. *Annals of the Institute of Statistical Mathematics, 42*(1), 157–161.

Pillai, R. N., & Jayakumar, K. (1995). Discrete Mittag-Leffler distributions. *Statistics and Probability Letters, 23*(3), 271–274.

Pokhrel, K., Aryal, G. R., Kafle, R. C., Tharu, B., & Khanal, N. (2022). Mcdonald-G Poisson family of distributions. *Statistica, 82*(2), 119–144.

Rady, E. A., Hassanein, W. A., & Elhaddad, T. A. (2016). The power Lomax distribution with an application to bladder cancer data. *SpringerPlus, 5*, 1838.

Rahman, M. M., Al-Zahrani, B., Shahbaz, S. H., & Shahbaz, M. Q. (2020). Transmuted probability distributions: A review. *Pakistan Journal of Statistics and Operation Research, 16*(1), 83–94.

Raja Rao, B., Alhumoud, J. M., & Damaraju, C. V. (2006). Setting the clock back to zero property of a family of bivariate life distributions. *International Mathematical Forum, 1*(34), 1691–1707.

Raja Rao, B., & Talwalker, S. (1990). "Setting the clock back to zeroâŁž property of a family of life distributions. *Journal of Statistical Planning and Inference, 24*(3), 347–352.

Rajarshi, M. B., & Rajarshi, S. M. (1988). Bathtub distributions: A review. *Communications in Statistics-Theory and Methods, 17*, 2597–2622.

Rajesh, G., & Nair, K. R. M. (1998). Residual entropy function in discrete time. *Far East Journal of Theoretical Statistics, 2*(1), 43–57.

References

Ramos, M. W. A., Cordeiro, G. M., Marinho, P. R. D., Dias, C. R. B., & Hamedani, G. G. (2013). The Zografos-Balakrishnan log-logistic distribution: Properties and applications. *Journal of Statistical Theory and Applications, 12*(3), 225–244.

Rao, M., Chen, Y., & Vemuri, B. C. (2004). Cumulative residual entropy?: A new measure of information. *IEEE Transactions on Information Theory, 50*(6), 1220–1228.

Rasekhi, M., Alizadeh, M., Altun, E., Afify, A. Z., & Ahmad, M. (2017). The modified exponential distribution with applications. *Pakistan Journal of Statistics, 33*(5), 383–398.

Reinhardt, H. E. (1968). Characterizing the exponential distribution. *Biometrics, 24*(2), 437–439.

Rényi, A. (1961). On measures of entropy and information. *Proceedings of the 4th Berkeley Symposium on Mathematical Statistics and Probability* (Vol. 1, pp. 547–561).

Righter, R., Shaked, M., & Shanthikumar, J. G. (2009). Intrinsic aging and classes of nonparametric distributions. *Probability in the Engineering and Informational Sciences, 23*, 563–582.

Ristic, M. M., & Balakrishnan, N. (2012). The gamma-exponentiated exponential distribution. *Journal of Statistical Computation and Simulation, 82*(8), 1191–1206.

Roknabadi, A. H. R. (2000). Some discrete life models, *PhD Dissertation*, Ferdowsi University of Mashhad, Iran.

Roknabadi, A. H. R. (2006). Telescopic families of discrete life distribution. *Conference on Ordered Statistical Data (ORSD)*, Ferdowsi University of Mashhad, Iran.

Roknabadi, A. H. R., Borzadaran, G. R. M., & Khorashadizadeh, M. (2009). Some aspects of discrete hazard rate function in telescopic families. *Economic Quality Control, 24*(1), 35–42.

Rolski, T. (1975). Mean residual life. *Bulletin of the International Statistical Institute, 46*, 266–270.

Rosaiah, K., Kantam, R. R. L., & Kumar, S. (2006). Reliability test plans for exponentiated log-logistic distribution. *Economic Quality Control, 21*(2), 279–289.

Rossini, A. J., & Tsiatis, A. A. (1996). A semiparametric proportional odds regression model for the analysis of current status data. *Journal of the American Statistical Association, 91*(434), 713–721.

Roy, D. (1984). A characterization of the generalized gamma distribution. *Calcutta Statistical Association Bulletin, 33*, 137–141.

Roy, D. (1988). Life distributions–characterization and classifications, *PhD dissertation*, University of Calcutta, Kolkata.

Roy, D. (1989). Characterisation of Gumbel's bivariate expoenential and Lindley and Singpurwalla's bivariate Lomax distributions. *Journal of Applied Probability, 27*, 886–889.

Roy, D. (2004). Discrete Rayleigh distribution. *IEEE Transactions on Reliability, 53*(2), 255–260.

Roy, D., & Dasgupta, T. (2001). A discretizing approach for evaluating reliability of complex systems under stress-strength model. *IEEE Transactions on Reliability, 50*(2), 145–150.

Roy, D., & Ghosh, T. (2009). A new discretization approach with application in reliability estimation. *IEEE Transactions on Reliability, 58*(3), 456–461.

Roy, D., & Gupta, P. L. (1992). Classification of discrete lives. *Microelectronics and Reliability, 32*(10), 1459–1473.

Roy, D., & Mukherjee, S. P. (1986). A note on characterizations of the Weibull distribution. *Sankhyā, Series A, 48*(2), 250–253.

Ruiz, J. M., & Navarro, J. (1994). Characterization of distributions by relationships between failure rate and the mean residual life. *IEEE Transactions of Reliability, 43*(4), 640–644.

Samaniego, F. J. (1985). On closure of the IFR class under formation of coherent systems. *IEEE Transactions on Reliability, 34*(1), 69–72.

Samaniego, F. J. (2007). *System Signatures and their Applications in Engineering Reliability*. USA: Springer.

Samanta, M. (1985). On estimating the location parameter of an exponential distribution with known coefficient of variation. *Calcutta Statistical Association Bulletin, 34*(1–2), 43–49.

Sangsanit, Y., & Bodhisuwan, W. (2016). The Topp-Leone generator of distributions: Properties and inferences. *Songklanakarin Journal of Science and Technology, 38*(5), 537–548.

Sankaran, P. G., Nair, N. U., & Hitha, N. (1996). Some characterizations of the geometric law. *Aligarh Journal of Statistics, 15*, 49–54.

Sarhan, A. M. (2009). Generalized quadratic hazard rate distribution. *International Journal of Applied Mathematics and Statistics, 14*(509), 94–109.

Sarhan, A. M., & Kundu, D. (2009). Generalized linear failure rate distributions. *Communications in Statistics-Theory and Methods, 38*(5), 642–660.

Sarhan, A. M., & Zaindin, M. (2009). Modified Weibull distribution. *Applied Sciences, 11*, 123–136.

Sato, H., Ikota, M., Sugimoto, A., & Masuda, H. (1999). A new defect distribution metrology with a consistent discrete exponential formula and its applications. *IEEE Transactions on Semiconductor Manufacturing, 12*(4), 409–418.

Saunders, I. W., & Moran, P. A. P. (1978). On the quantiles of the gamma and F distributions. *Journal of Applied Probability, 15*(2), 426–432.

Schmittlein, D. C., & Morrison, D. G. (1981). The median residual lifetime?: A characterization theorem and an application. *Operations Research, 29*(2), 392–399.

Schöniger, A., Wöhling, T., Samaniego, L., & Nowak, W. (2014). Model selection on solid ground?: Rigorous comparison of nine ways to evaluate Bayesian model evidence. *Water Resources Research, 50*(12), 9484–9513.

Sclove, S. L. (1994). Small sample and large sample statistical model selection criteria. *Springer Lecture Notes in Statistics, 89*, 31–39.

Seki, T., & Yokoyama, S. (1996). Robust parameter-estimation using the bootstrap method for the 2-parameter Weibull distribution. *IEEE Transactions on Reliability, 45*(1), 34–41.

Sen, A. (2005). Linear failure rate distribution. In S. Kotz, N. Balakrishnan, C. Read, & B. Vidakovic (Eds.), *Encyclopedia of Statistical Sciences* (2nd ed., Vol. 6, pp. 4212–4217). New Jersey: Wiley.

Sen, S., Maiti, S. S., & Chandra, N. (2016). The X-gamma distribution: Statistical properties and applications. *Journal of Modern Applied Statistical Methods, 15*(1), 774–788.

Sengupta, D. (2013). Testing for and against ageing. *Journal of Indian Statistical Association, 51*(1), 231–253.

Sengupta, D., & Nanda, A. K. (1997). Log-concave and concave distributions in reliability, Technical Report No. ASD/97/6, Applied Statistics Unit, Indian Statistical Institute, Calcutta.

Sengupta, D., & Nanda, A. K. (1999). Log-concave and concave distributions in reliability. *Naval Research Logistics, 46*, 419–433.

Sengupta, D., & Nanda, A. K. (2010). The proportional reversed hazards regression model. *Journal of Applied Statistical Sciences, 18*(4), 461–476.

Sengupta, D., Singh, H., & Nanda, A. K. (1999). *The proportional reversed hazards model*. Applied Statistics Division, Indian Statistical Institute, Calcutta: Technical Report.

Seshadri, V., & Patil, G. P. (1964). A characterization of a bivariate distribution by the marginal and the conditional distributions of the same component. *Annals of the Institute of Statistical Mathematics, 15*, 215–221.

Shaked, M. (1982). Dispersive ordering of distributions. *Journal of Applied Probability, 19*, 310–320.

Shaked, M., & Shanthikumar, J. G. (1991). Dynamic multivariate mean residual life functions. *Journal of Applied Probability, 28*(3), 613–629.

Shaked, M., & Shanthikumar, J. G. (1994). *Stochastic Orders and their Applications*. Boston: Academic Press.

Shaked, M., & Shanthikumar, J. G. (2007). *Stochastic Orders*. New York: Springer.

Shaked, M., Shanthikumar, J. G., & Valdez-Torres, J. B. (1995). Discrete hazard rate functions. *Computers and Operations Research, 22*, 391–402.

Shaked, M., & Wong, T. (1995). Preservation of stochastic orderings under random mapping by point processes. *Probability in the Engineering and Informational Sciences, 9*, 563–580.

Shams, T. M. (2013). The Kumaraswamy-generalized Lomax distribution. *Middle-East Journal of Scientific Research, 17*(5), 641–646.

Shanker, R. (2015). Akash distribution and its applications. *International Journal of Probability and Statistics, 4*(3), 65–75.

Shanker, R., & Fesshaye, H. (2016). On modeling of lifetime data using Akash, Shanker, Lindley and exponential distributions. *Biometrics and Biostatistics International Journal, 3*(6), 214–224.

Shanmugam, R. (1985). An intervened Poisson distribution and its medical application. *Biometrics, 41*(4), 1025–1029.

Shanmugam, R., Bartolucci, A. A., & Singh, K. P. (2002). The analysis of neurologic studies using an extended exponential model. *Mathematics and Computers in Simulation, 59*(1–3), 81–85.

Shannon, C. E. (1948). A mathematical theory of communications. *Bell System Technical Journal, 27*, 379–423.

Shaw, W. (2007). The alchemy of probability distributions: Beyond Gram-Charlier and Cornish-Fisher expansions, and skew normal or kurtotic normal distributions, Corpus ID: 17633144.

Shaw, W., & Buckley, I. R. C. (2009). The alchemy of probability distributions: Beyond Gram-Charlier expansions, and a skew-kurtotic-normal distribution from a rank transmutation map. https://arxiv.org/pdf/0901.0434.

Shimizu, R., & Davies, L. (1981). General characterization theorems for the Weibull and the stable distributions. *Sankhyā, Series A, 43*(3), 282–310.

Siddiqui, S. A., Balkrishan, Gupta, & S., & Subharwal, M. (1992). A finite range failure model. *Microelectronics Reliability, 32*(10), 1453–1457.

Siddiqui, M. M., & Cağlar, M. (1994). Residual lifetime distribution and its applications. *Microelectronics Reliability, 34*(2), 211–227.

Siddiqui, S. A., Dwivedi, S., Dwivedi, P., & Alam, M. (2016). Beta exponentiated Mukherjii-Islam distribution: Mathematical study of different properties. *Global Journal of Pure and Applied Mathematics, 12*(1), 951–964.

Siddiqui, S. A., Jain, S., Siddiqui, I., Khan, K., & Alam, M. (2016). Characterization and development of a new failure model. *Journal of Theoretical and Applied Information Technology, 86*(1), 87–95.

Siddiqui, S. A., Subharwal, M., Gupta, S., & Balakrishan, N. (1994). Finite range survival model. *Microelectronics Reliability, 34*(8), 1377–1380.

Silcock, H. (1954). The phenomenon of labour turnover. *Journal of the Royal Statistical Society, Series A, 117*(4), 429–440.

Sindhu, T. N. (2002). An extended Pearson system useful in reliability analysis, *PhD Dissertation*, Cochin University of Science and Technology, India.

Sindhu, T. N., & Atangana, A. (2021). A reliability analysis incorporating exponentiated inverse Weibull and inverse power law. *Quality and Reliability Engineering International, 37*(6), 2399–2422.

Sindhu, T. N., Thomas, G., & Gillariose, J. (2020). A generalization of the exponentiated rectangular distribution. *IAPQR Transactions, 44*(2), 174–200.

Singh, H. (1989). On partial orderings of life distributions. *Naval Research Logistics, 36*, 103–110.

Singh, S. K., & Maddala, G. S. (1976). A function for size distribution of incomes. *Econometrica, 44*(5), 963–970.

Singh, V. P., & Zhang, L. (2022). Generalized beta Lomax distribution. In *Generalized Frequency Distributions for Environmental and Water Engineering* (pp. 187–207). Cambridge University Press. To be published.

Singpurwalla, N. D. (1995). Survival in dynamic environments. *Statistical Science, 10*(1), 86–103.

Song, K. S. (2001). Rényi information, loglikelihood and an intrinsic distribution measure. *Journal of Statistical Planning and Inference, Information Sciences, 177*, 4161–4175.

Spiegelhalter, D. J., Best, N. G., Carlin, B. P., & Van Der Linde, A. (2002). Bayesian measures of model complexity and fit. *Journal of Royal Statistical Society, Series B, 64*, 583–639.

Srivastava, M. S. (1967). A characterization of the exponential distribution. *American Mathematical Monthly, 74*, 414–416.

Stacy, E. W. (1962). A generalization of the gamma distribution. *Annals of Mathematical Statistics, 33*(3), 1187–1192.

Staff, P. J. (1967). The displaced Poisson distribution-region B. *Journal of the American Statistical Association, 62*(318), 643–654.

Stein, W. E., & Dattero, R. (1984). A new discrete Weibull distribution. *IEEE Transactions on Reliability, 33*(2), 196–197.

Stoyan, D. (1983). *Comparison Methods for Queues and Other Stochastic Models*. New York: Wiley.
Stoyanov, J., & Al-Sadi, M. H. M. (2004). Properties of classes of life distributions based on the conditional variance. *Journal of Applied Probability, 41*(4), 953–960.
Stuart, A., & Ord, J. K. (1994). *Kendall's Advanced Theory of Statistics (Vol-I)*. Wiley.
Sunoj, S. M., Nair, N. U., Nanda, A. K., & Rasin, R. S. (2020). Ageing intensity function for conditionally specified models. *American Journal of Mathematical and Management Sciences, 39*(4), 329–344.
Sunoj, S. M., & Rasin, R. S. (2017). A quantile-based study on ageing intensity function. *Communications in Statistics-Theory and Methods, 47*(22), 5474–5484.
Surles, J. G., & Padgett, W. J. (2001). Inference for reliability and stress-strength for a scaled Burr Type X distribution. *Lifetime Data Analysis, 7*(2), 187–200.
Szymkowiak, M. (2018). Characterizations of distributions through aging intensity. *IEEE Transactions on Reliability, 67*, 446–458.
Szymkowiak, M. (2019). Measures of ageing tendency. *Journal of Applied Probability, 56*(2), 358–383.
Tahir, M., Cordeiroz, G., Mansoorx, M., & Zubair, M. (2015). The Weibull-Lomax distribution: Properties and applications. *Hacettepe Journal of Mathematics and Statistics, 44*(2), 461–480.
Talwalker, S. (1977). On the lack of memory property of the exponential distribution. *Bulletin of the International Statistical Institute, 47*, 581–584.
Tata, M. N. (1969). On outstanding values in a sequence of random variables. *Zeitschrift für Wahrscheinlichkeitstheorie und verwandte Gebiete, 12*, 9–20.
Teicher, H. (1961). Identifiability of mixtures. *Annals of Mathematical Statistics, 32*, 244–248.
Teicher, H. (1963). Identifiability of finite mixtures. *Annals of Mathematical Statistics, 34*, 1256–1269.
Titterington, D., Smith, A., & Markov, U. (1985). *Statistical Analysis of Finite Mixtures*. New York: Wiley.
Tobin, J. (1958). Liquidity preference as behavior towards risk. *The Review of Economic Studies, 25*(2), 65–86.
Topp, C. W., & Leone, F. C. (1955). A family of J-shaped frequency functions. *Journal of the American Statistical Association, 50*(269), 209–219.
Vasicek, O. (1976). A test of normality based on sample entropy. *Journal of Royal Statistical Society, Series B, 38*(1), 54–59.
Vaupel, J. W., & Yashin, A. I. (1985). Heterogeneity's ruses: Some surprising effects of selection on population dynamics. *American Statistician, 39*(3), 176–185.
Viertl, R. (1988). *Statistical Methods in Accelerated Life Testing*. Gottingen: Vandenhoeck and Ruprecht.
Viertl, R. (2009). On reliability estimation using fuzzy lifetime data. *Journal of Statistical Planning and Inference, 139*(5), 1750–1755.
Walker, K. (1970). The effect of stress ratio during crack propagation and fatigue for 2024–T3 and 7075–T6 aluminium. In M. S. Rosenfeld (Ed.), *Effects of Environment and Complex Load History on Fatigue Life* (pp. 1–14). ASTM International.
Wang, W. Y. (1996). Life distribution classes and two unit standby redundant system, *PhD Dissertation*, Chinese Academy of Science, Beijing.
Wang, W. Y. (1998). Life distribution classes and two-unit standby redundant system, *PhD Dissertation*, Chinese Academy of Sciences, Beijing.
Wang, A., Müller, H., & Capra, W. B. (1998). Analysis of oldest-old mortality: Lifetables revisited. *Annals of Statistics, 26*(1), 126–163.
Wang, Y. H., & Srivastava, R. C. (1980). A characterization of the exponential and related distributions by linear regression. *Annals of Statistics, 8*(1), 217–220.
Weibull, W. (1951). A statistical distribution function of wide applicability. *Journal of Applied Mechanics, 18*, 293–297.

Westergren, A., Karlsson, S., Anderson, P., Ohlsson, O., & Hallberg, I. R. (2001). Eating difficulties, need for assisted eating, nutritional status and pressure ulcers in patients admitted for stroke rehabilitation. *Journal of Clinical Nursing, 10*, 257–269.

Wheeler, R. E. (1980). Quantile estimators of Johnson curve parameters. *Biometrika, 67*, 725–728.

Whitmore, G. A. (1970). Third-degree stochastic dominance. *American Economic Review, 60*(3), 457–459.

Wiener, N. (1948). *Cybernetics*. New York: The MIT Press and Wiley.

Wolff, R. W. (1989). *Stochastic Modelling and the Theory of Queues*. Englewood Cliffs, NJ: Prentice Hall.

Xekalaki, E. (1983). Hazard functions and life distributions in discrete time. *Communications in Statistics-Theory and Methods, 12*(21), 2003–2009.

Xekalaki, E., & Dimaki, C. (2005). Identifying the Pareto and Yule distributions by properties of their reliability measures. *Journal of Statistical Planning and Inference, 131*, 231–252.

Xie, M., & Lai, C. D. (1995). Reliability analysis using an additive Weibull model with bathtub shaped failure rate function. *Reliability Engineering and System Safety, 52*, 87–93.

Xie, M., Tang, Y., & Goh, T. N. (2002). A modified Weibull extension with bathtub shaped failure rate function. *Reliability Engineering and System Safety, 76*(3), 279–285.

Yan, L., Kang, D., & Wang, H. (2021). Further results of the TTT transform ordering of order n. *Symmetry, 13*, 1960–1991.

Yousof, H. M., Altun, E., Ramires, T. G., Alizadeh, M., & Rasekhi, M. (2018). A new family of distributions with properties, regression models and applications. *Journal of Statistics and Management Systems, 21*(1), 163–188.

Yousof, H. M., Korkmaz, M. C., & Sen, S. (2021). A new two-parameter lifetime model. *Annals of Data Science, 8*(1), 91–106.

Yue, D., & Cao, J. (2000). Some results on the residual Life at random time. *Acta Mathematicae Applicatae Sinica, 16*, 435–443.

Zahedi, H. (1991). Proportional mean remaining life models. *Journal of Statistical Planning and Inference, 29*, 221–228.

Zahui, L., & Xiaohu, L. (1998). $IFR*t_0$ and $NBU*t_0$ classes of life distributions. *Journal of Statistical Planning and Inference, 70*, 191–200.

Zamani, Z., Borzadaran, G. R. M., & Amini, M. (2017). On a new positive dependence concept based on the conditional mean inactivity time order. *Communications in Statistics-Theory and Methods, 46*(4), 1779–1787.

Zhang, S., & Cheng, H. (2010). Testing for increasing mean inactivity time. *Statistics, 44*(5), 467–476.

Zografos, K., & Balakrishnan, N. (2009). On families of beta- and generalized gamma-generated distributions and associated inference. *Statistical Methodology, 6*, 344–362.

Index

A
Akaike information criterion, 310
Akash distribution, 66
Almost-lack-of-memory property, 58
Amoroso distribution, 86
Arc tan exponential distribution , 82

B
Basic composition formula, 195
Bathtub-shaped, 23
Bayes information criterion, 311
Beta-binomial distribution, 103
Bridge criterion, 311
Bull's eye experiment, 55

C
Carleman's criterion, 18
Coefficient of excess, 19
Coherent system, 6
Cold standby, 8, 11
Confluent hypergeometric function, 82, 106

Constant Failure Rate (CFR), 22, 23
Convex transform order, 124
Critical parameter, 2, 298

D
Decreasing Failure Rate (DFR), 27, 31, 134

Decreasing Generalized Mean Residual Life (DGMRL), 27, 28
Decreasing in Failure Rate Average (DFRA), 23
Decreasing Likelihood Ratio (DLR), 24
Decreasing Reversed Hazard Rate (DRHR), 23–25
Design, 2
Deviance information criterion, 311
Disability adjusted life expectancy, 298
Disability Adjusted Years of Life, 298
Dispersive order, 217
Displaced Poisson distribution, 106

E
Equilibrium distribution, 164
Erlang distribution, 30
Expected stopped time, 127
Exponential Lomax distribution, 70
Exponentiated generalized Weibull distribution, 291
Exponentiated Kumaraswamy Weibull distribution, 294
Exponentiated Lemonte distribution, 294
Exponentiated rectangular distribution, 136
Exponentiated truncated log-Weibull distribution, 291
Extended exponential class, 114
Extended Weibull distribution, 292

Index

F
Final prediction error criterion, 310
Fisk distribution, 71
Flexible Weibull extension distribution, 293

Force of mortality, 44, 45
Frailty parameters, 54
Fuzzy number, 299

G
Galton distribution, 55
Generalized loss-of-memory property, 275
Generalized mean residual life, 27
Generalized power Weibull distribution, 293

GoLomax distribution, 70

H
Hamza distribution, 89
Higher order stochastic dominance, 172
HMRL order, 176
Hot standby, 7

I
Inactivity time, 118
Increasing Failure Rate (IFR), 22, 25, 27, 30, 44, 134
Increasing Generalized Failure Rate (IGFR), 28
Increasing Generalized Mean Residual Life (IGMRL), 27
Increasing in Failure Rate Average (IFRA), 23
Increasing Likelihood Ratio (ILR), 24, 162, 163
Increasing Reversed Hazard Rate (IRHR), 24
Intervened exponential distribution, 80
Intervention parameter, 80
Inverse Weibull distribution, 290, 291
Irrelevant component, 6

K
Kumaraswamy Inverse Weibull distribution, 290
Kumaraswamy Modified Weibull distribution, 294

L
Lack-of-memory property, 58
Lehmann Type I distribution, 53
Lehmann Type II distribution, 53
Leptokurtic, 20
Linear exponential distribution, 63
Linear Failure Rate (LFR), 22, 44
Linear failure rate distribution, 63
L-moments, 17
Log-concave, 23, 24, 43, 44
Log-convex, 24
Loglog distribution, 294
Lomax distribution, 28, 69
Lorenz curve, 123
Loss function, 309
Loss-of-memory property, 29, 32, 57, 58, 263, 276

M
Matching spares, 160
Mean time to failure, 22
Membership function, 299
Memoryless property, 32, 33
Mesokurtic, 20
Mielke beta-kappa distribution, 99
Mission plan, 2
Mortality order, 156

N
New Better than Used (NBU), 134
New Worse than Used (NWU), 134

O
Odd Weibull distribution, 293

P
Pareto Type I distribution, 67
Pareto Type II, 69
Percentile residual life, 29
Perfect memory, 33
Performance parameter, 1
Periodic failure rate property, 58
Platykurtic, 20
Polya frequency function, 44
Power Hamza distribution, 90
Power law distribution, 136
Power Lindley distribution, 292
Probability weighted moments, 16

Product reliability, 3
Proportional hazards model, 53
Proportional odds model, 162
Proportional reversed hazards model, 53
Putting the clock back to zero property, 29

Q
Quality-adjusted life, 298
Quality of Conformance, 2
Quality of design, 2
Quality of performance, 3, 8, 298
Quality parameter, 2
Quota-share treaty, 227

R
Rate of mortality, 44
Rayleigh distribution, 22
Reciprocal coordinate sub-tangent, 253
Reciprocal Weibull distribution, 290
Reflected Kumaraswamy-generated distributions, 55
Relevant component, 6
Residual entropy, 37
Resilience parameters, 54
Reversed Weibull distribution, 290
Riemann zeta distribution, 106

S
Scoring function, 309
Shelf-life, 57, 99
Slashed distribution, 82
Stationary renewal excess distribution, 164
Structure function, 6
Sub-Poisson distribution, 106
Sup-entropy, 36
Super-Poisson distribution, 106

T
Theory of errors, 55
Tilt parameter, 52
TP_2 function, 181
Transformed-Transformer family, 87

W
Warm standby, 8
Weak likelihood ratio order, 176
Weighted exponential distribution, 80

Z
Zipf's law distribution, 106

GPSR Compliance
The European Union's (EU) General Product Safety Regulation (GPSR) is a set of rules that requires consumer products to be safe and our obligations to ensure this.

If you have any concerns about our products, you can contact us on

ProductSafety@springernature.com

In case Publisher is established outside the EU, the EU authorized representative is:

Springer Nature Customer Service Center GmbH
Europaplatz 3
69115 Heidelberg, Germany

www.ingramcontent.com/pod-product-compliance
Lightning Source LLC
Chambersburg PA
CBHW050859090625
27894CB00013BA/19